Energy Materials for Ingestible and Implantable Healthcare Devices

For a list of titles in this series see: rsc.li/detection-science

How to obtain future titles on publication:
A standing order plan is available for this series. A standing order will bring delivery of each new volume immediately on publication.

For further information please contact:
Book Sales Department, Royal Society of Chemistry, Thomas Graham House, Science Park, Milton Road, Cambridge, CB4 0WF, UK
Telephone: +44 (0)1223 420066, Fax: +44 (0)1223 420247
Email: booksales@rsc.org
Visit our website at books.rsc.org

Energy Materials for Ingestible and Implantable Healthcare Devices

Edited by

Santanu Patra
Drug Delivery and Sensing (IDUN), Department of Health Technology,
Technical University of Denmark, Ørsted Plads, Kgs. Lyngby 2800, Denmark
Email: sanpa@dtu.dk

and

Anja Boisen
Drug Delivery and Sensing (IDUN), Department of Health Technology,
Technical University of Denmark, Ørsted Plads, Kgs. Lyngby 2800, Denmark
Email: aboi@dtu.dk

ROYAL SOCIETY OF CHEMISTRY

Detection Science Series No. 33

Hardback ISBN: 978-1-83767-525-8
PDF ISBN: 978-1-83767-985-0
EPUB ISBN: 978-1-83767-986-7
Print ISSN: 2052-3068
Electronic ISSN: 2052-3076

A catalogue record for this book is available from the British Library

The Royal Society of Chemistry is a charity, registered in England and Wales, Number 207890, and a company incorporated in England by Royal Charter (Registered No. RC000524), registered office: Burlington House, Piccadilly, London W1J 0BA, UK, Telephone: +44 (0)20 7437 8656.

For further information see our website at www.rsc.org

For general enquiries, please contact books@rsc.org

For EU product safety enquiries, please email books@rsc.org or contact Royal Society of Chemistry Worldwide (Germany) GmbH, Römischer Hof, Unter den Linden 10, 10117 Berlin.

Printed in the United Kingdom by CPI Group (UK) Ltd, Croydon, CR0 4YY, UK

Preface

Implantable and ingestible medical devices offer a practical and dependable method for the real-time diagnosis, treatment, and monitoring of various diseases. Implantable technologies, including pacemakers, glucose monitoring sensors, and neurostimulators, are designed to be implanted within the body for continuous therapeutic monitoring and necessary interventions. On the other hand, ingestible devices, mostly equipped with cameras, sensors, and actuators, navigate the gastrointestinal (GI) tract, facilitating non-invasive diagnostics and targeted drug delivery within the GI tract. Both technologies can monitor vital signs, identify irregularities, implement therapeutic interventions, and offer significant insights into patients' health, facilitating personalized treatment programs. The power source is the most important factor influencing the efficacy and operational lifespan of implantable or ingestible devices. When selecting a power source, it is essential to consider its output power, capacity, size, and safety/toxicity to the human body.

This book, **Energy Materials for Ingestible and Implantable Healthcare Devices**, discusses the essential requirements by providing a thorough review of various energy harvesting and storage technologies, along with power management systems, specifically designed for biomedical applications. The book aims to address the gaps in present research by offering an updated review of the status, challenges, and prospective solutions for energy materials in healthcare. All the chapters in this book explore the integration of materials science, microelectronics, and biomedical engineering, offering an in-depth study of current energy requirements, cutting-edge technologies, challenges, and future perspectives. This book is an essential resource for researchers, engineers, clinicians, and individuals interested in bioelectronics, medical devices, and energy demands, providing a guide to an autonomous future where medical devices integrate effortlessly with the human body.

Detection Science Series No. 33
Energy Materials for Ingestible and Implantable Healthcare Devices
Edited by Santanu Patra and Anja Boisen

Published by the Royal Society of Chemistry, www.rsc.org

The chapters are organized for better representation, with Chapter 1 providing the underlying framework by explaining the crucial interaction of energy materials and healthcare technology, particularly highlighting the increasing energy requirements of implantable and ingestible medical devices. Chapter 2 defines the requirements for implantable and ingestible bioelectronic devices, analyzes recent developments, and presents a summary of current research while emphasizing their various applications. Chapter 3 connects ingestible and implantable devices, outlining their power requirements and possible options. In Chapter 4, the fundamental characteristics of materials used in energy-transfer systems are presented, followed by an exploration of device characterization to highlight the significance of design and material selection. Chapters 5 and 6 discuss potential energy-harvesting methods, including mechanical and biochemical energy harvesting, while Chapters 7 and 8 address energy storage technologies, such as batteries and supercapacitors, for biomedical devices. Chapters 9 and 10 focus on power management and wireless transfer. The trends and challenges related to integrating energy materials with biomedical devices are covered in Chapter 11. Chapter 12, the last chapter, summarizes the discussion of the book and outlines the challenges along with a prospective insight for future research. We believe this book will serve as a significant resource, updating readers on the current scenario while also inspiring more research and contributions to the advancement of energy materials in healthcare.

We are very thankful to all the authors of the chapters for their very informative and passionate contributions to this book. Special thanks are expressed to Janet Freshwater, Liangqi Li, and the entire team at the Royal Society of Chemistry for their dedicated support and cooperation throughout the process of compiling and publishing this book.

Santanu Patra
Anja Boisen
Lyngby, Denmark

Acknowledgements

We acknowledge funding from the Novo Nordisk Foundation Challenge Programme in Energy materials with biological applications (EMGUT): grant ref. No. NNF22OC0072961 and European Research Council in Foldable, REconfigurable & Jagged devices for enhanced drug Absorption/seeding (FREJA), grant ref. No. 101054945.

Detection Science Series No. 33
Energy Materials for Ingestible and Implantable Healthcare Devices
Edited by Santanu Patra and Anja Boisen

Published by the Royal Society of Chemistry, www.rsc.org

Contents

Detection Science Series No. 33
Energy Materials for Ingestible and Implantable Healthcare Devices
Edited by Santanu Patra and Anja Boisen

Published by the Royal Society of Chemistry, www.rsc.org

CHAPTER 1

Overview and Need for Energy Materials in Healthcare

NGHIA DINH HUYNH,*[a] NICHOLAS WALLER,[a] JACK PERRIS,[b] DANIEL M. MULVIHILL[b] AND NIKOLAJ GADEGAARD*[a]

[a] James Watt School of Engineering, University of Glasgow, Glasgow, G12 8QQ, UK; [b] Materials and Manufacturing Research Group, James Watt School of Engineering, University of Glasgow, Glasgow G12 8QQ, UK
*Emails: Nghia.Huynh@glasgow.ac.uk; Nikolaj.Gadegaard@glasgow.ac.uk

1.1 Overview and Energy Needs for Ingestible/Implantable Medical Devices (IMDs)

This chapter presents the intersection between energy materials and healthcare technologies, with an emphasis on ingestible and implantable biomedical devices (IMDs). The exploration of energy needs and the technological landscape in healthcare, specifically focusing on IMDs, highlights the increasing role of energy-efficient technologies and innovations that drive the performance, durability, and functionality of these devices. Implantable medical devices are critical in managing chronic conditions, monitoring vital signs, delivering therapies, and restoring functionality to impaired biological systems. To understand these, a schematic is presented in Figure 1.1, detailing the types of IMDs, energy needs, current and emerging technologies in IMDs, and energy management challenges.

Additionally, the current market and emerging research of IMDs classified by the dependent/nondependent power supply are presented (see Figure 1.2). The IMDs are categorised as battery-dependent and battery-free to drive

Detection Science Series No. 33
Energy Materials for Ingestible and Implantable Healthcare Devices
Edited by Santanu Patra and Anja Boisen

Published by the Royal Society of Chemistry, www.rsc.org

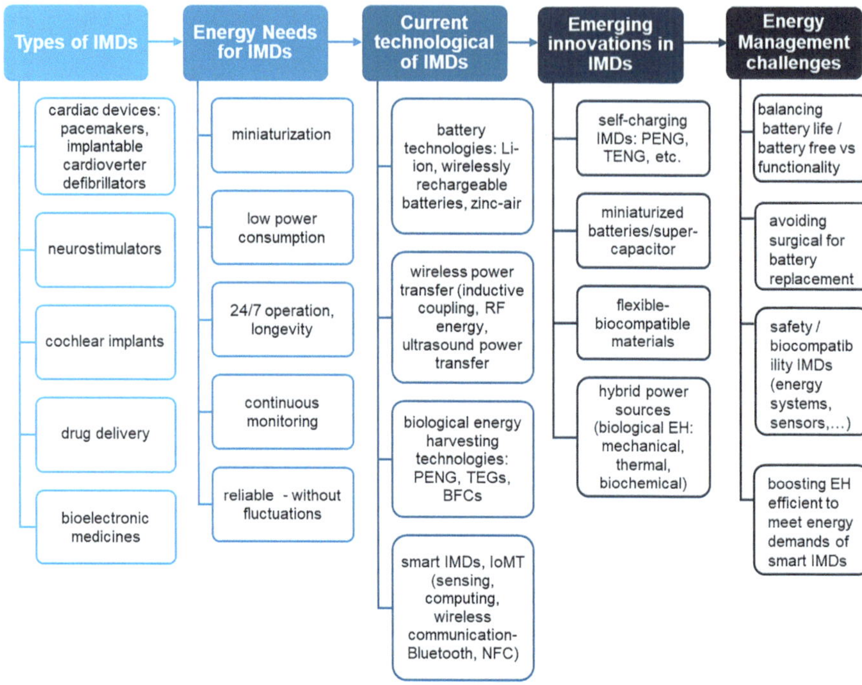

Figure 1.1 The exploration of the technological landscape and energy needs in implantable medical devices (IMDs), from current to emerging trends and challenges in energy management.

medical devices. In this illustration, the IMDs applied in neuro-stimulation, endoscopic capsules, and pacemaker operating based on the power source might need to be designed with a biocompatible battery and a tiny super-capacitor such as a bio-supercapacitor (bio-SC) localised in a blood vessel. Furthermore, recent research exhibits the battery-free, self-powered IMDs in the human body for brain stimulation, muscle stimulation, blood pressure sensing, pacemakers, and bladder pressure sensing using new energy harvesting methods such as triboelectric nanogenerators (TENGs) and piezoelectric nanogenerators (PENGs).

1.2 Energy Materials and Power Management Systems in Healthcare

1.2.1 Energy Materials – An Introduction

Various energy harvesting methods, such as thermal energy from body heat, movement-based (mechanical), and chemical energy from body fluids, are built from different types of energy materials for harvesting and further energy storage. Based on the methods and design of the energy

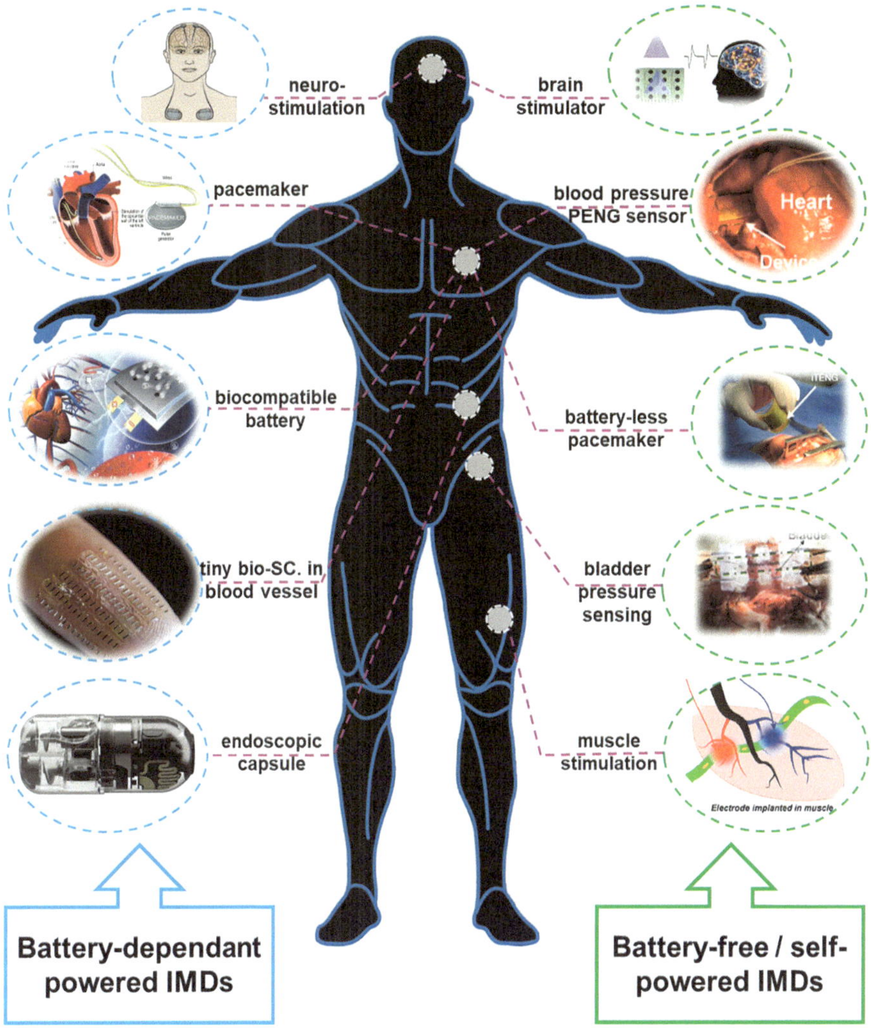

Figure 1.2 The current market and emerging research of IMDs classified by the need for a power supply (battery/supercapacitor and self-powered IMDs). "Neuro-stimulation" reproduced from ref. 1 with permission from Springer Nature, Copyright 2021. "Pacemaker" ref. 2 "Biocompatible battery" reproduced from ref. 3 with permission from Elsevier, Copyright 2024. "Tiny bio-SC. in blood vessel" reproduced from ref. 4, https://doi.org/10.1038/s41467-021-24863-6, under the terms of the CC BY 4.0 license, http://creativecommons.org/licenses/by/4.0/. "Endoscopic capsule PillBot" reproduced from ref. 5 with permission from Endiatx. "Brain stimulator" reproduced from ref. 6 with permission from Elsevier, Copyright 2020. "Blood pressure PENG sensor" reproduced from ref. 7 with permission from Elsevier, Copyright 2016. "Battery-less pacemaker" reproduced from ref. 8, https://doi.org/10.1038/s41467-019-09851-1, under the terms of the CC BY 4.0 license, http://creativecommons.org/licenses/by/4.0/. "Bladder pressure sensing" reproduced from ref. 9 with permission from American Chemical Society, Copyright 2018. "Muscle stimulation" reproduced from ref. 10 with permission from American Chemical Society, Copyright 2019.

harvester device, the energy materials selection should be considered carefully. For example, the triboelectric series (see Figure 1.3) is presented here for a wide range of materials selection to determine a high-output TENG device.

1.2.1.1 Energy Harvesting and Storage Materials

Biomedical electronic devices intended for ingestion and implantation are required to be biocompatible, miniaturised, mechanically compliant, and capable of autonomous operation over extended periods. These systems critically depend on reliable power sources that are compact and durable, capable of both scavenging and storing energy within the human body. This section highlights key material classes utilised in energy harvesting and storage, with an emphasis on their applicability to biomedical environments.

To achieve sustained functionality, IMDs are frequently integrated with microscale energy storage components such as micro-supercapacitors and micro-batteries. These components, when designed with biocompatible materials, play an essential role in enabling continuous and regulated energy supply for next-generation biomedical electronics. The following discussion outlines major classes of energy harvesting materials – including triboelectric, piezoelectric, and biofuel cell materials – alongside their corresponding storage counterparts.

The efficiency of TENGs in healthcare applications is fundamentally governed by the properties of the triboelectric materials used, particularly triboelectric charge density and their positions in the triboelectric series.[11] The surface charge density of the materials directly impacts the energy harvesting performance – a critical factor in TENG design where surface interactions determine operational efficiency.[12] Materials such as polytetrafluoroethylene (PTFE) and polydimethylsiloxane (PDMS) are widely employed due to their strong charge affinity and proven compatibility with biomedical applications.[13] These polymers also support micro- and nanoscale surface modifications, which can significantly enhance surface charge density.[14] Current research focuses on optimising biocompatible polymers that maintain high output and stability within biological environments.

Material selection for TENGs is guided by the triboelectric series (see Figure 1.3), which ranks materials according to their electron-donating or electron-accepting tendencies during contact electrification. Contact between two materials results in charge transfer, where one becomes positively charged (electron donor) and the other negatively charged (electron acceptor). The magnitude of this effect correlates with the relative positions of the materials in the series. For instance, positively ranked materials such as human skin, glass, and nylon tend to donate electrons, while materials such as silicone, PTFE, and rubber at the negative end act as electron acceptors. The potential difference generated during contact electrification increases

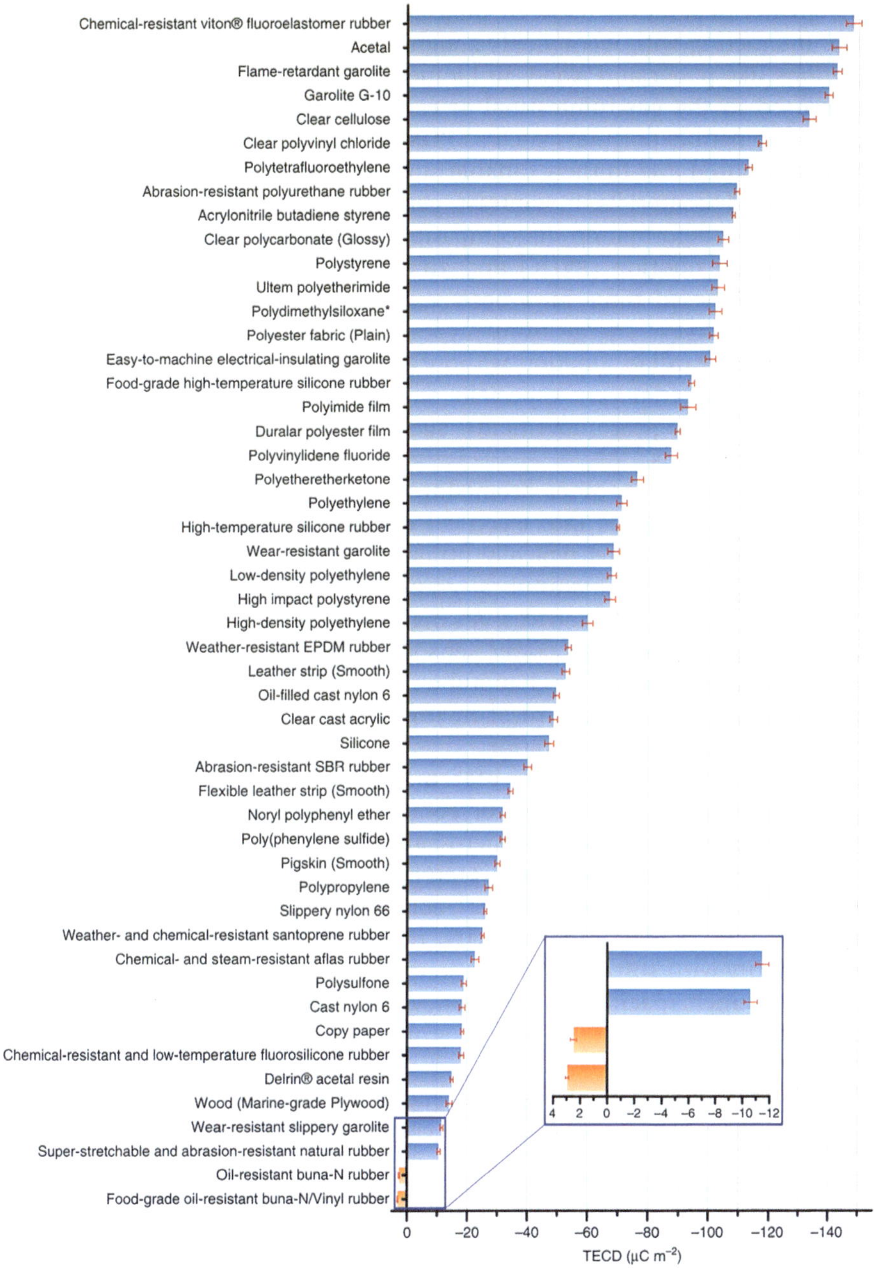

Figure 1.3 Quantified triboelectric series for over 50 polymers, with error bars representing the range within one standard deviation.[11] Reproduced from ref. 11, https://doi.org/10.1038/s41467-019-09461-x, under the terms of the CC BY 4.0 license, http://creativecommons.org/licenses/by/4.0/.

with greater separation between materials on the series. As an example, PTFE interacting with wool yields a significantly higher voltage than combinations like paper and wood. In biomedical electronic devices, especially those used in the gastrointestinal or cardiovascular environment, flexible and biocompatible polymers such as PDMS, alginate, chitosan, and PEDOT:PSS are favoured due to their tunable surfaces and mechanical softness.[15]

Piezoelectric materials, in contrast, generate electric charges in response to mechanical deformation. Biocompatible piezoelectric candidates include polyvinylidene fluoride (PVDF) and zinc oxide (ZnO) nanowires. For instance, PVDF-based films have been successfully implemented in cardiac implants to harvest energy from myocardial motion, enabling long-term monitoring of cardiac function without the need for external power sources.[16] These materials are especially suitable for harvesting energy from rhythmic physiological movements such as respiration, heartbeat, and muscle contractions.[17]

Biofuel cells offer another promising approach by converting chemical energy from body fluids into electrical energy *via* enzymatic or microbial reactions. These systems typically utilise materials such as carbon-based electrodes (*e.g.*, carbon nanotubes, graphene), noble metal catalysts, and enzyme immobilization matrices. An ingestible glucose biofuel cell, for example, has been shown to power sensors by oxidising glucose present in intestinal fluids using carbon nanotube-modified electrodes.[18] This strategy leverages the body's own chemical environment, enabling fully autonomous and internally sustained operation.

Energy harvested through the aforementioned technologies must be effectively stored for intermittent or continuous use. Miniaturised storage systems, including solid-state micro-batteries and micro-supercapacitors are indispensable in this regard. Materials such as hydrogels, graphene, activated carbon, and Zn^{2+}- or Li^{+}-compatible polymers are frequently employed due to their high capacity, rapid charge/discharge cycles, and compatibility with soft electronic systems.[19] Notably, a Zn-alginate hydrogel-based supercapacitor has been developed for transient implantable applications, offering stable energy storage and safe biodegradation following functional use.[20]

In all cases, the materials used for energy harvesting and storage in biomedical devices must meet additional critical criteria:

- Biocompatibility with surrounding tissues and bodily fluids
- Resistance to sterilization-induced degradation
- Hypoallergenic properties to avoid immunogenic responses
- Antimicrobial resistance to reduce the risk of infection and device colonization

These considerations ensure that energy systems not only perform efficiently but also integrate seamlessly within the dynamic physiological environment of the human body.

1.2.1.2 Enhancing Surface Charge Density

Surface charge density enhancement in TENGs follows three primary strategies: material selection and properties, surface engineering and functionalisation, and operational condition optimisation.

- Material Selection and Properties

  Beyond basic triboelectric series considerations, composite materials are frequently engineered to maximise surface charge density. These composites strategically combine materials with complementary triboelectric properties to optimise charge separation. For example, PTFE-PDMS or PTFE-PVDF (polyvinylidene fluoride) combinations can significantly enhance charge transfer efficiency while maintaining the biocompatibility necessary for healthcare applications.[21] These composite systems often demonstrate superior resilience to environmental conditions encountered in biological systems.
- Surface Engineering and Functionalisation

  Surface engineering techniques, such as nano-structuring and micropatterning, can increase the effective contact area between triboelectric surfaces, thereby enhancing the output charge. By creating microscopic surface textures, engineers can improve the contact efficiency and retention of surface charges. For example, electrospun nanofibres or plasma-etched patterns improve triboelectric performance by providing larger and more flexible contact points.[21]

  Surface functionalisation through chemical modification further enhances charge trapping capabilities.[22] The strategic introduction of functional groups through chemical treatments can significantly improve charge retention duration, which is particularly important for implantable medical devices requiring stable energy output over extended periods.
- Operational Condition Optimisation

  Operational optimisation techniques, particularly charge pumping, can substantially enhance TENG performance.[23] This process involves applying controlled voltage bias to pre-charge material surfaces, effectively increasing the baseline charge density. Such pre-charging strategies are especially valuable in healthcare applications where consistent output is required despite irregular mechanical inputs.

1.2.1.3 Hybrid Systems and *In Vivo* Challenges

The integration of hybrid energy harvesting and storage systems in IMDs is essential to overcome the limitations of standalone energy harvesters or storage units. Some energy harvesters, such as TENGs, PENGs, and biofuel cells, can provide sustainable and reliable power. However, their intermittent and low output performance cannot ensure the continuous operation of biomedical devices. In contrast, micro-supercapacitors and micro-batteries alone cannot sustain in long-term generation without external charging. A hybrid system effectively addresses this limitation by enabling energy scavenging during periods of physiological activity while simultaneously storing the harvested energy for later controlled use. By combining harvesting and storage, the system provides a more stable and sustained power supply for critical IMDs in healthcare applications.

Although holding great potential, there are still several challenges obstructing the realization of high-performance energy scavenging *in vivo* applications. These are the limitations of power density, output degradation in moist and dynamic biological environments inside the human body, mechanical matching between human soft tissues and IMD materials, and immune responses. For example, triboelectric materials exhibit the advantages of lightweight design and high output performance under mechanical stimuli but tend to be degraded significantly and show poor stability in aqueous or wet environments. In the case of piezoelectric materials, some piezo ceramics provide good energy conversion, nevertheless, they may create cytotoxicity risks, leading to the limitation of their applicability in long-term operation IMDs.[24] In addition, the low-frequency and irregular mechanical stimuli of physiological movements are a big obstacle to maintaining or boosting efficient energy conversion.

To overcome these problems, biodegradable encapsulations, soft, elastic materials with high charge trapping, and tailoring the surface to micro-nanostructures can enhance the output performance of energy harvester devices. The surface functionalisation and microstructural design are also able to improve the biocompatibility, maintain the stability of device performance, and engage well with soft tissues. Besides, the development of advanced power management systems also plays a crucial role in regulating and optimising the harvested electrical energy, bringing hybrid self-powered systems for long-term biomedical use.

1.2.2 Power Management Systems

1.2.2.1 Power Management System Requirements

The translation of mechanical energy captured by TENGs into stable, usable power requires sophisticated power management systems (PMSs). These systems must address three fundamental challenges: rectification of the generated current, voltage regulation, and energy storage (see Figure 1.4a). The rectification process converts the inherently alternating current output of TENGs into direct current, while voltage regulation ensures the output matches specific device requirements. Advanced power converter topologies – including buck, boost, and flyback configurations – are implemented to ensure efficient energy storage and delivery with minimal losses. The development of these systems enables TENGs to provide consistent, regulated power even under variable operating conditions, a crucial requirement for healthcare applications where reliability is paramount.

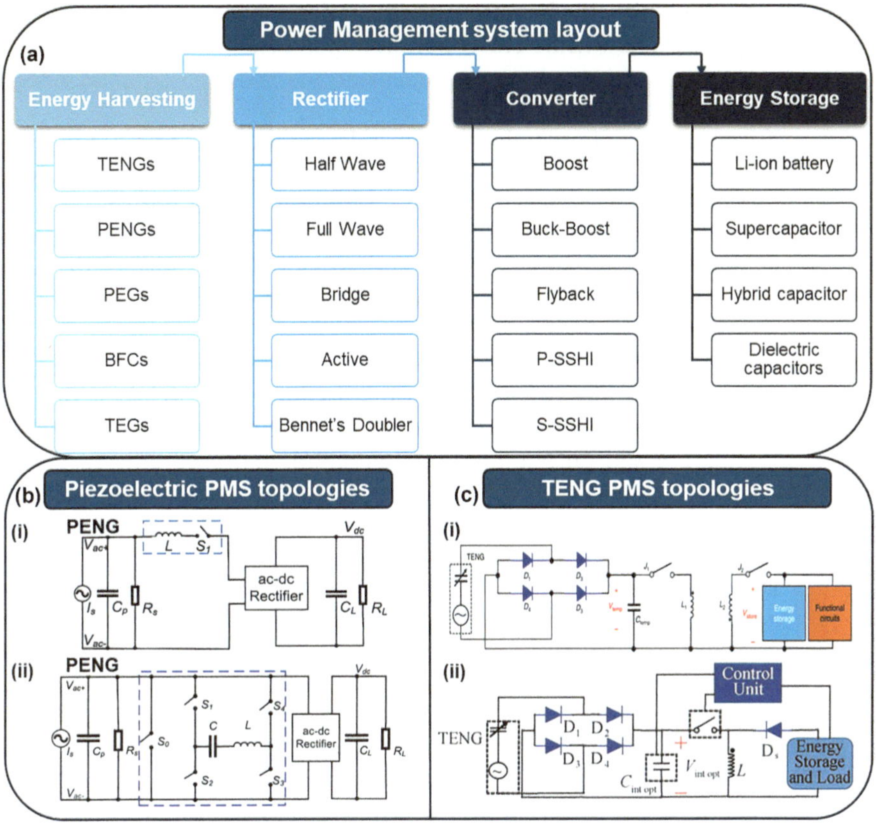

Figure 1.4 (a) Overview of the energy harvesting process including triboelectric nanogenerators (TENGs), piezoelectric nanogenerators (PENGs), pyroelectric nanogenerators (PEGs), biofuel cells (BFCs), and thermoelectric generators (TEGs), moving through rectification and conversion to energy storage. Common energy sources like TENGs and PENGs are rectified using half-wave, full-wave, or active circuits, then converted *via* buck, boost, or flyback converters and stored in systems such as lithium-ion batteries, supercapacitors, or dielectric capacitors. (b) Examples of piezoelectric PMS topologies: (i) parallel synchronised switch harvesting on inductor (P-SSHI) and (ii) Synchronized switch harvesting on capacitor inductor (SSHCI).[25] Reproduced from ref. 25, https://doi.org/10.1016/j.nanoen.2024.110196, under the terms of the CC BY 4.0 license, https://creativecommons.org/licenses/by/4.0/. (c) Examples of TENG PMS topologies: (i) flyback converter design.[26] Reproduced from ref. 26, https://doi.org/10.1038/ncomms9975, under the terms of the CC BY 4.0 license, https://creativecommons.org/licenses/by/4.0/. (ii) Buck-boost converter design.[27] Reproduced from ref. 27 with permission from IOP Publishing, Copyright 2017.

1.2.2.2 Rectification Circuits for TENGs

To convert the TENG's alternating current (AC) into a direct current (DC) that can be stored or used directly, rectification circuits are essential. Each type of rectification offers specific benefits and limitations that make it suitable for different applications within TENG-powered systems.

- Half-wave rectification
  - Advantages: a simple circuit that only requires a single diode. It is effective for minimal power requirements and is efficient in low-demand applications where a smooth output is not critical.
  - Disadvantages: it is less efficient than other forms of rectification as it only utilises one half-cycle of the AC signal. This results in lower power output and a pulsating DC that may be insufficient for devices requiring stable power.
- Full-wave rectification
  - Advantages: full-wave rectification uses both half-cycles of the AC signal, doubling the frequency of the output pulses and providing a higher average output power. This setup improves efficiency and reduces the need for large filtering components. Full-wave rectification is, therefore, suitable for TENG applications requiring a steadier and higher DC output.
  - Disadvantages: the circuit design is more complex, typically involving additional diodes or a bridge rectifier configuration, increasing both cost and size. Full-wave rectification may not be as compact as other options, making it less ideal for extremely space-constrained devices.
- Active rectification
  - Advantages: active rectifiers use transistors or other active components to minimise voltage drops across diodes, providing higher efficiency and a more stable output voltage. This type of rectification is especially useful in TENG applications where maximising energy output is critical, such as in devices requiring a consistent energy supply.
  - Disadvantages: these circuits are more complex and costly to implement. Active rectification also requires a small power source to operate the control circuitry, which is counterproductive to having a self-powered medical device. The added complexity makes it less suitable for TENG applications where minimal power requirements are prioritised.

1.2.2.3 Converter Types

Power converter selection significantly impacts TENG/PENG system performance in healthcare applications. Four primary topologies have demonstrated particular utility, as follows.

- Parallel Synchronised Switch Harvesting on Inductor (P-SSHI) (see Figure 1.4b(i))
  - Advantages: P-SSHI is a synchronised switch-based topology that effectively reduces energy losses in PENGs by switching between states in sync with the input signal. This converter boosts the harvested power and is especially beneficial for low-power applications

needing enhanced energy efficiency. P-SSHI can significantly increase PENGs' efficiency by reducing the reactive power loss.
   - Disadvantages: the P-SSHI topology is complex to implement due to its synchronised switching mechanism and requires precise control circuitry. This added complexity can increase both the cost and size of the system, potentially limiting its application to more basic PENG setups.
- Synchronized Switch Harvesting on Capacitor Inductor (SSHCI) (see Figure 1.4b(ii))
   - Advantages: SSHCI is optimised to increase energy storage efficiency by synchronising the switch operation with the PENG signal cycle. This topology allows for improved power transfer efficiency to the storage unit and is particularly effective in applications with intermittent energy demands, such as biosensors.
   - Disadvantages: SSHCI circuits are complex and require additional components to maintain the synchronisation, which can increase energy consumption. This complexity makes SSHCI more suited for applications where high efficiency is prioritised over simplicity.
- Flyback Converter (see Figure 1.4c(i))
   - Advantages: flyback converters provide electrical isolation between the TENGs and the device, a crucial feature for medical applications. This converter type supports both voltage step-up and step-down transformations, making it highly flexible and suitable for implantable medical devices or biosensors where isolation and patient safety are priorities.
   - Disadvantages: flyback converters use transformers, making them larger and sometimes noisier, which can limit their applicability in very compact devices. Additional filtering may be required to produce a stable output, and the circuit's complexity can increase costs and development time.
- Buck-Boost Converter (see Figure 1.4c(ii))
   - Advantages: the buck-boost converter offers a versatile solution by providing both step-up and step-down voltage regulation. This topology is ideal for TENG applications where the output voltage varies and may either exceed or fall below device requirements, such as in wearable devices where energy generation depends on movement.
   - Disadvantages: buck-boost converters can introduce switching noise and often require filtering to produce a stable output. They may also be less efficient at higher power levels, making them more suitable for low-power applications rather than consistently high-demand settings.

The triboelectric properties of materials, enhanced through strategic surface modification and functionalisation, form the foundation for efficient energy harvesting. Meanwhile, sophisticated power management systems, incorporating advanced rectification and conversion techniques, ensure this harvested energy can be effectively utilised. The synergy between materials science and electronic engineering in this field continues to advance, pushing the boundaries of what's possible in self-powered medical devices.

As research progresses, particular emphasis remains on optimising charge density through novel material combinations while developing increasingly efficient power management architectures. These parallel developments are crucial for realising the full potential of TENGs in healthcare applications, where reliability and efficiency are paramount.

1.2.3 Summary of Energy Harvesting, Power Management Systems, and Storage

A comprehensive summary of energy harvesting, power management, and storage technologies is presented in Table 1.1. Technologies are described in terms of their working principles, biomedical suitability, medical applications, advantages, limitations, and supporting references. Triboelectric nanogenerators (TENGs), piezoelectric nanogenerators (PENGs), and pyroelectric nanogenerators (PEGs) represent key mechanical and thermal energy harvesting platforms. In addition, biofuel cells (BFCs) provide biochemical energy conversion using enzymatic or microbial oxidation of internal body fluids. Power management systems include passive rectifiers (half-wave and full-wave diode circuits), active rectifiers (MOSFET- or transistor-based), direct current to direct current (DC–DC) converters such as buck-boost and flyback topologies, and synchronised switch harvesting circuits, including parallel synchronised switch harvesting on inductor (P-SSHI) and synchronized switch harvesting on capacitor inductor (SSHCI). Storage technologies include sodium-ion micro-batteries, edible zinc-carbon-based micro-supercapacitors, solid-state lithium micro-batteries for long-term implantable applications, and transient biodegradable batteries designed for short-duration use in gastrointestinal (GI) environments. Applications such as vagus nerve stimulation (VNS), deep brain stimulation (DBS), biosensing, and temporary diagnostic platforms are also highlighted. This integrated summary supports the optimisation of design across energy input, regulation, and storage in physiologically constrained medical contexts.

1.2.4 Advanced Energy Harvesting Materials for IMDs

1.2.4.1 Implantable Medical Devices

1.2.4.1.1 Triboelectric Nanogenerators (TENGs). Triboelectric nanogenerators, or TENGs, are devices that can convert mechanical energy into electrical energy based on the coupling effect of triboelectric and electrostatic induction. Contact electrification is the process of charge transfer between two materials in contact. The materials must have two distinct electrical affinities to allow one material to donate electrons while the other will receive electrons. This creates a charge imbalance between the surfaces in contact. The electrical induction allows for the potential difference (voltage) to develop across the two electrodes placed on either side of the contacting surfaces. This can drive electrons to an external circuit,

Table 1.1 Comparative summary of energy harvesting, power management, and energy storage technologies for use in implantable and ingestible medical devices.

Technology	Descriptor	Suitability	Applications	Advantages	Limitations
TENGs[28,29]	Biomechanical harvesters *via* contact electrification (PTFE, PDMS); thin, flexible	Implantable, ingestible	VNS, pacemakers, drug delivery	Lightweight, low-cost, biodegradable	Moisture-sensitive, needs rectification
PENGs[30,31]	Strain-induced harvesters using ZnO, PVDF	Implantable, ingestible	Pulse sensors, vital monitoring	High sensitivity, biocompatible	Lower output from safe materials, toxicity in others
PEGs[32,33]	Harvest energy from thermal fluctuations (PVDF, $BaTiO_3$)	Implantable	Thermal/inflammatory monitoring	Passive, flexible form	Requires thermal gradient, low output
BFCs[34,35]	Enzymatic/microbial energy from fluids (sweat, GI)	Ingestible	GI sensors, drug delivery	Self-powered, fluid-activated	Enzyme degradation, low voltage
Passive rectifiers	Diode-based half/full-wave rectification	Implantable, ingestible	Basic AC to DC conversion	Simple, low-cost	High ripple, low efficiency
Active rectifier[36,37]	MOSFET/transistor-based high-efficiency rectifier	Implantable	Energy conditioning in IMDs	High efficiency, low dropout	Needs control power, complex
DC–DC converter[36,38,39]	Flyback, buck/boost for voltage regulation	Implantable, ingestible	Voltage control for storage/load	Flexible voltage scaling	Larger size, noise filtering needed
P-SSHI/SSHCI[25]	Synchronous switch harvesting for PENGs	Implantable, ingestible	Enhanced piezo power extraction	Boosts transfer efficiency	Circuit complexity, timing needed
Micro-batteries[40,41]	Na-ion micro-batteries for long-term storage	Implantable	Pacemakers, biosensors	High energy density, reliable	Non-rechargeable, surgical removal
Supercapacitors[42,43]	Zn/C devices for short bursts	Ingestible, implantable	GI sensors, temporary systems	Fast response, biodegradable	Low energy density, short life
Transient batteries[44,45]	Biodegradable Mg/Zn–silk systems	Ingestible	Single-use GI sensors	Resorbable, safe, no surgery	Short life, low energy

generating a current.[51] Repeated actions of surfaces within a TENG device, such as contact and separation, sliding, and vibrating, will generate an electrical output that can be used to power devices such as renewable energy, wearable electronics, self-powered sensors, and healthcare devices.[52]

Ordinary healthcare devices are typically powered by lithium-ion batteries. These batteries have the inherent problems of being difficult to recycle, large

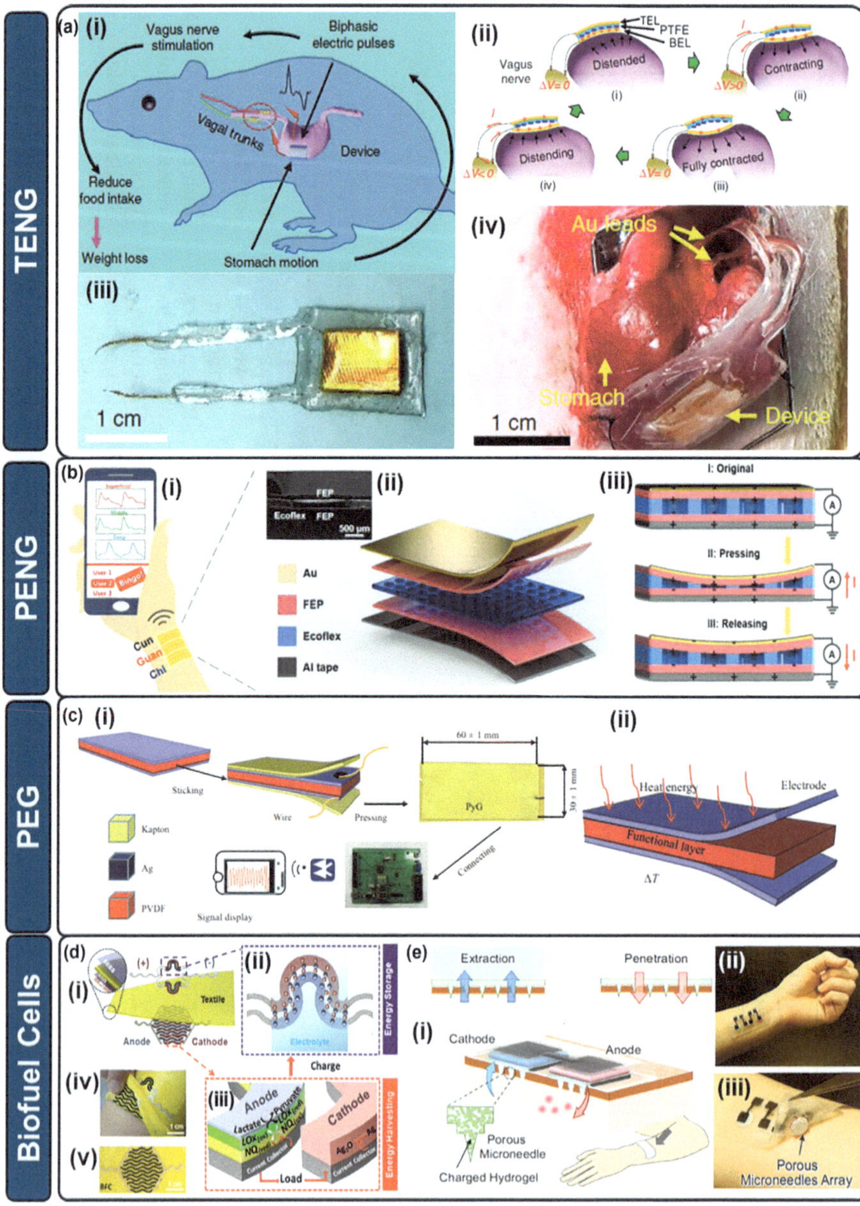

compared to the devices they power, and eventually deplete and need replacing. Devices powered by an ordinary battery/power source require continual replacement, which adds considerable physical and financial strain to the patient. TENG devices can alleviate the problems of conventional power sources and power healthcare devices such as pacemakers, vital monitoring systems, and drug delivery systems[8,53,54] by harnessing the physical movements of the organs and other parts of the human body, allowing these healthcare devices to become self-powered by biomechanical energy. Recent advances in materials mean that the TENG devices used to power healthcare devices can be biocompatible and biodegradable if necessary. TENG devices are also lightweight and flexible by nature due to being made of thin sheets of material; this makes them ideal for implantation into the human body as they will cause less physical stress at the site of insertion. They are also typically inexpensive as they can be manufactured from simple materials.[55]

An implanted vagus nerve stimulation (VNS) device (see Figure 1.5a) based on flexible TENGs in human body organs for weight control was introduced by Yao *et al.*[46] The VNS device is composed of a biocompatible and flexible

Figure 1.5 (a) The correlated vagus nerve stimulation (VNS) system and its biocompatibility. (i) Operation principle of the correlated VNS system schematically showing the pathway for biphasic electric signal generation and VNS. (ii) Schematics of the working principle of a VNS device under different stomach motion stages. (iii) A packaged VNS device. (iv) An implanted VNS device with Au leads being connected to anterior and posterior vagal trunks.[46] Reproduced from ref. 46, https://doi.org/10.1038/s41467-018-07764-z , under the terms of the CC BY 4.0 license, http://creativecommons.org/licenses/by/4.0/. (b) Piezoelectric device for detecting pulse waves. (i) Concept of a health monitoring system to detect pulse waves. (ii) Schematic diagram of the device – this contains an FEP/Ecoflex/FEP structure piezoelectric film. (iii) Basic operation of the piezoelectric pulse wave sensor.[47] Reproduced from ref. 47 with permission from John Wiley & Sons, Copyright 2018 Wiley-VCH Verlag GmbH & Co. KGaA, Weinheim. (c) (i) Pyroelectric device design and associated circuit design. Used for physiological monitoring. (ii) Capacitor style device structure and thermal radiation theoretical principle.[48] Reproduced from ref. 48 with permission from Springer Nature, Copyright 2021. (d) Schematic illustrations and images of a wearable biofuel cell (BFC) – supercapacitor (SC) harvesting–storage hybrid device. (i) Schematic device design and basic electrical components. (ii–iii) Basic schematic of energy generation by a BFC from the lactate in sweat to charge the SC. (iv) Photograph of a wearable wristband showing both the BFC and SC. (v) Photograph of a biofuel cell.[49] Reproduced from ref. 49 with permission from the Royal Society of Chemistry. (e) Transdermal electroosmosis using a porous microneedle array patch. (i) Conceptual illustration of the device used for electroosmotic penetration and extraction for efficient drug delivery and analysis of interstitial fluid. The photograph shows the device structure – (ii) four biobattery cells in series and (iii) porous microneedle.[50] Reproduced from ref. 50, https://doi.org/10.1038/s41467-021-20948-4, under the terms of the CC BY 4.0 license, http://creativecommons.org/licenses/by/4.0/.

TENG, which is attached to the surface of the stomach. The electric signals provided by this nanogenerator aim to stimulate the vagal afferent fibres, leading to a reduction of food intake and control of the body weight of a rat. The photograph of the device (see Figure 1.5a(iii), its localisation (see Figure 1.5a(i)) in a rat's stomach, and the working principle of VNS (see Figure 1.5a(ii)) illustrate the structure and the connection of two Au leads to the vagus nerve in the rat's stomach, which can generate different output voltages at varying input frequencies, with amplitudes ranging from 0.05 to 0.12 V. This work demonstrates outstanding weight control results with a device that is battery-free and less invasive as compared to the gastric bypass, sleeve gastrectomy, *etc.*.

1.2.4.1.2 Piezoelectric Nanogenerators (PENGs). A piezoelectric nanogenerator (PENG) is a similar type of device to a TENG as it converts mechanical energy into electrical energy. However, a PENG makes use of the piezoelectric effect. PENG devices will always have a piezoelectric material to allow the device to generate electrical energy from mechanical stress such as compression, stretching, and bending. A PENG device can harness the active movements within a human body to create mechanical stress on the active piezoelectric material, which in turn creates electricity. The piezoelectric effect that generates the electrical energy will typically only occur in crystalline materials. Only certain crystalline structures allow the piezoelectric effect to occur. The structures that facilitate this unique effect have charged particles with the ability to shift in the crystal lattice when mechanical stress is applied. This movement of particles creates a separation of charges, generating an electrical field.[56] This electrical field can then be used to drive a current, creating electrical energy that can be harvested to power many devices, such as self-powered healthcare devices including sensors for vitals monitoring and pacemaker systems.[57,58]

PENG devices generate electrical energy through the ambient periodic motions of the body. They have the benefits of allowing healthcare devices to power themselves and remove the bulky lithium-ion batteries from the respective systems. PENGs have the advantage of being highly sensitive to small mechanical inputs, making them ideal devices to generate an effective level of electrical energy to power medical devices from the ambient movements of organs. Also, like TENG devices, PENGs can be manufactured from relatively low-cost materials that are biocompatible. However, there can be a trade-off with the piezoelectrical materials used within healthcare devices. For instance, zinc oxide (ZnO) has good biocompatibility but a lower piezoelectric coefficient, so it will give reduced electrical output compared with the active materials in traditional piezoelectric devices. Some of the more effective piezoelectric materials are toxic and can pollute the internal environment of the body, so care must be taken when selecting materials for implantable PENG devices.[59]

The PENG device (see Figure 1.5b) that is used for active health monitoring is positioned outside of the body (see Figure 1.5(i)). The device architecture (see Figure 1.5(ii)) consists of fluorinated ethylene propylene (FEP) layers that sandwich an EcoFlex thin film with an Au electrode and an Al tape electrode. The device construction and materials allow for mechanical flexibility and high responsivity to the subtle movements of the arterial pulse on a human wrist. This PENG device can provide high precision and power stability, making the device capable of being integrated into medical assessment technologies used to detect common health issues such as irregular heartbeat rhythms.[47] As well as detecting health problems, the health monitoring system can differentiate pulse signatures between individuals and measure blood pressure using the pulse sensing system. The system is completely self-powered so there is no need for external power supplies.

1.2.4.1.3 Pyroelectric Nanogenerators (PEGs). Pyroelectric nanogenerators (PEGs) have drawn interest from the healthcare device community due to their potential to convert waste heat into electrical energy.[60] The pyroelectric effect is the driving force in PEG devices. The pyroelectric effect allows certain materials to generate an electric charge in response to temperature changes. This phenomenon is driven by the spontaneous polarisation within pyroelectric materials. Thermal fluctuations result in a temporary voltage across the material. The generated charge can be harvested using electrodes, converting thermal energy into electrical energy.

Pyroelectric materials are categorised into four types, each with distinct characteristics and applications:

Single Crystals: materials such as lithium niobate ($LiNbO_3$) and triglycine sulfate (TGS). Applications include thermal imaging and infrared detection.

Inorganic Thin Films: thin films, such as barium strontium titanate (BST), are valued for their low thermal capacity and flexibility, ideal for integration into wearable electronics and sensors.

Ceramics: pyroelectric ceramics, including lead zirconate titanate (PZT) and barium titanate ($BaTiO_3$), are essential materials for developing high-performance self-powered sensors.

Organic Polymers and Composites: polymers like polyvinylidene fluoride (PVDF) offer flexibility and ease of fabrication, though they typically have lower pyroelectric coefficients compared to inorganic materials.[61]

Applications of pyroelectric nanogenerators include energy harvesting and self-powered sensors, which are both critical areas for healthcare devices. Energy harvesting PEGs can convert waste heat from industrial processes, automotive engines, and electronic devices into electrical energy. This conversion not only enhances energy efficiency but also reduces the environmental impact of waste heat.[62] PEGs can power sensors for environmental monitoring, structural health assessment, and biomedical applications. These types of sensors operate without the need for external power sources, enabling more versatile and sustainable sensor networks.[63] The PEG sensor (see Figure 1.5c) is used for healthcare vitals monitoring and is powered by thermal loading produced from the external environment and heat from the human body.[48]

Pyroelectric nanogenerators represent a significant advancement in energy harvesting technologies, offering a sustainable solution for converting waste heat into usable electrical energy. The ongoing research and development in pyroelectric materials are paving the way for more efficient and versatile applications, from industrial energy recovery to self-powered sensors and wearable electronics.[60] With continued innovation and interdisciplinary collaboration, PENGs hold the potential to make a substantial impact on the future of sustainable energy and smart technology systems.

1.2.4.1.4 Biofuel Cells (BFCs). Biofuel cells (BFCs) are innovative devices that convert chemical energy from biofuels into electricity through bio-electrochemical reactions using biocatalysts such as enzymes, organelles, or microorganisms. Wearable biofuel cells (W-BFCs) are designed to be flexible and stretchable, enabling integration into wearable devices while maintaining functionality under deformation. These cells can harness energy from body fluids like sweat, tears, and interstitial fluids, making them ideal for continuous and sustainable power supply in wearable applications. W-BFCs are advantageous due to their biocompatibility, environmental sustainability, and potential for miniaturisation. Applications of W-BFCs extend beyond energy harvesting to include self-powered biosensing, controlled drug delivery, and smart wound dressings, highlighting their versatility in health monitoring and medical treatments. Recent advances in materials and fabrication techniques have significantly enhanced the performance and durability of W-BFCs, addressing key challenges and paving the way for their broader use in various wearable technologies.[64]

An example of an activated BFC[49] (see Figure 1.5d) is described as a sweat-based, wearable hybrid energy device combining a supercapacitor (SC) and a BFC. This device is screen-printed on both sides of a stretchable textile. It can harvest biochemical energy from sweat and store it for use in wearable electronics. The BFC component utilises lactate from sweat, enzymatically oxidising it to generate electricity, which is then stored in the SC. The SC employs MnO_2/carbon nanotube composites for high capacitance and stability. The device is designed to maintain stable performance despite mechanical deformations like bending, twisting, and stretching, making it suitable for integration into smart textiles and wearable electronics. The successful on-body testing highlights its potential for practical applications in healthcare, fitness, security, and environmental monitoring.

Drug delivery systems are also possible using BFC-related devices. The device (see Figure 1.5e) demonstrates a minimally invasive transdermal drug delivery and interstitial fluid extraction method using a porous microneedle array patch.[50] The patch incorporates ion-conductive porous microneedles (PMNs) modified with a charged hydrogel. This design offers three key advantages: reduced transdermal resistance due to low-invasive skin puncture, enhanced transport of larger molecules *via* interconnected micropores, and generation of electroosmotic flow (EOF). The generated EOF significantly improves transdermal molecular penetration and extraction, similar to the effects of external pressure. Experiments using pig skin demonstrated enhanced delivery of a model drug (dextran) and efficient glucose extraction. Importantly, the researchers successfully powered the transdermal EOF system using a fructose/oxygen enzymatic biobattery, paving the way for a completely organic, self-powered iontophoresis patch. The study shows that the EOF can be tuned by adjusting the density of charged hydrogel within the microneedles, offering the potential for optimised delivery of various molecules depending on size and charge.

1.2.4.2 Ingestible Medical Devices

Ingestible medical devices are aimed at the area of gastrointestinal (GI) diagnostics, monitoring, and therapy. These devices are compact, biocompatible capsules embedded with sensors, microcontrollers, and communication systems that are designed to navigate the GI tract. Ingestible medical devices collect and transmit data or deliver targeted interventions. Their appeal lies in their non-invasive nature, potential for real-time data acquisition, and ability to access otherwise hard-to-reach regions of the digestive system.

The last two decades have seen significant momentum in this field. Traditional capsule endoscopy, exemplified by the FDA-approved PillCam,[65] offered the first glimpse into wireless GI imaging. However, as microelectronics and materials science advanced, ingestible devices evolved beyond passive imaging. Recent systems now include biochemical sensors that measure pH, gases, and pressure; actuators capable of drug delivery or electrical stimulation; and onboard AI chips that can process diagnostic images internally.[66] Table 1.2 presents a list of example ingestible medical devices from both commercial and research projects.

The design of ingestible medical devices can be split into seven key areas: sensors/actuators, integrated circuits, communication systems, power, packaging, localisation, and locomotion (see Figure 1.6a). Each module involves trade-offs between size, functionality, and energy consumption. For instance, optical sensors (see Figure 1.6b) offer precise imaging but require significant power, whereas electrochemical sensors can be more compact and energy efficient.[67] Recent innovations include devices that harvest energy from GI fluids or wireless power transfer, reducing reliance on bulky batteries. Self-powered devices utilising TENG and PENG technologies are also of critical interest to the field.[68]

Technical feasibility alone does not guarantee clinical adoption. There is an emphasis in the research community that despite the growing variety of ingestible devices, few have been integrated into routine clinical workflows.[66] Challenges include unclear cost-effectiveness, regulatory hurdles,

Table 1.2 Examples of ingestible medical devices from research and commercial development.

Device name/description	Developer/ institution	Primary use/function
PillCam (M2A capsule)[65]	Medtronic	Video capsule endoscopy for imaging the GI tract; various models target oesophagus (ESO), small intestine (SB3), and colon (COLON2)
CapsoCam/CapsoCam Plus[70]	CapsoVision	Video capsule that captures panoramic images of the GI tract; CapsoCam Plus stores data onboard for post-excretion retrieval
OMOM capsule[71]	Jinshan Science & Technology	Video endoscopy capsule with duplex data transmission for real-time control and adjustments
EndoCapsule[72]	Olympus	Video capsule endoscopy with real-time imaging capability
MiroCam[73]	IntroMedic	Wireless endoscopic capsule using human body communication (HBC) instead of RF for data transmission
Bravo™ pH Capsule[74]	Medtronic	Clips onto oesophageal mucosa to monitor pH for GERD diagnostics over 48–96 hours
SmartPill™[75]	Medtronic	Measures pH, pressure, and temperature to assess GI transit time and motility
HemoPill[76]	Ovesco Endoscopy	Detects GI bleeding using optical absorption differences in red and violet light
Atmo Gas Capsule[77]	Atmo Biosciences	Senses O_2, H_2, and CO_2 gases in the GI tract for functional and microbial assessment
3D Transit system[78]	Motilis Medica	Electromagnetic tracking system that measures GI motility and regional transit time
SonoPill[79]	University of Glasgow	Ultrasound-based capsule for transmural imaging and drug delivery
SOMA Capsule[80]	MIT	Self-orienting geometry for optimal alignment with target tissue without active steering
Tethered Confocal Endomicroscopy Capsule[81]	Research (*e.g.*, UC San Diego)	High-resolution microscopic imaging of mucosal and submucosal structures
Magnetically Controlled Capsule Endoscopy (MCE)[82]	Ankon Technologies (China)	Uses magnetic steering and a string for controlled navigation, especially in the stomach

and difficulty interpreting novel data streams. The success of capsule endoscopy provides a strong foundation, but broader applications such as targeted drug delivery or microbiota sampling still require further validation through trials and clinician training.

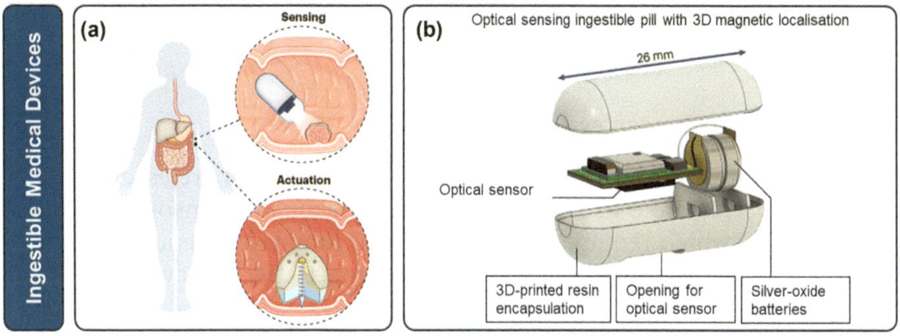

Figure 1.6 (a) Smart ingestible capsules equipped with sensing and actuation functions, (b) ingestible pill featuring optical sensing and 3D magnetic localisation. Reproduced from ref. 67 with permission from Springer Nature, Copyright 2024.

McRae *et al.*[69] proposed a useful framework for understanding ingestible devices based on their intended residency time in the GI tract: short term (<24 hours), medium term (days to weeks), and long term (months or more). Each timescale presents distinct engineering and safety challenges. Short-term devices, like video endoscopes, need only to survive for hours and can afford higher power consumption. In contrast, medium- and long-term devices must resist corrosion, maintain mechanical integrity, and ensure biocompatibility over extended periods. Long-term devices especially demand new materials with ultra-low moisture permeability and fatigue-resistant designs to avoid breakdown during repetitive GI motility cycles.

Despite these challenges, the future of ingestible devices is promising. Potential applications extend beyond diagnosis to include real-time gut microbiome monitoring, personalised drug regimens, and closed-loop therapeutic systems that react to physiological cues. Advances in materials science, such as biodegradable electronics, could also lead to transient devices that safely dissolve after use, eliminating the need for retrieval.[66]

In summary, ingestible medical devices are transitioning from single-purpose capsules to sophisticated diagnostic therapeutic platforms. Engineers continue to solve the challenges of power, miniaturisation, and biocompatible materials, and as clinicians grow more comfortable integrating these tools into practice, these devices are poised to become central to next-generation GI care.

1.2.5 Advanced Energy Storage Materials for IMDs

To power the implantable electronic device inside the human body, energy storage is essential for the device. Among the energy storage sources, micro-batteries and micro-supercapacitors are popular, and it is miniaturised energy sources used in implantable medical devices like pacemakers and biosensors. Micro-batteries store energy for continuous, long-term power, while micro-supercapacitors deliver quick bursts of energy when needed.

Together, they help these devices function reliably, safely, and efficiently within the body.

1.2.5.1 Micro-batteries for IMDs

In powering implantable medical devices, micro-batteries are essential and have an important role in IMDs such as pacemakers, biosensors, and neurostimulators, which require tiny, reliable, and long-lasting energy sources. For these devices, micro-batteries must meet the following critical requirements:

- Biocompatibility: the materials used in the battery must not cause adverse reactions in the human body. This ensures safe, long-term implantation without harming surrounding tissues.
- Energy Density: due to limited space in IMDs, micro-batteries must store as much energy as possible in a very small volume. This maximises device lifespan, reduces the need for replacement surgeries, and ultimately lowers costs for patients.
- Reliability and Stability: it is essential to maintain consistent performance over long periods, as implantable devices typically operate continuously or intermittently for several years.
- Safety is Paramount: when it comes to implantation, ensuring safety is crucial. The battery must be engineered to mitigate risks such as leakage, short-circuiting, or overheating, all of which could harm human tissue.

To achieve the above requirements, researchers continually work to develop micro-battery materials and structures that meet these standards. An example of such innovation is the symmetric Na-ion micro-battery with sphere-in-network heteronanomat electrodes. In this work, the authors demonstrate biocompatibility, high reliability, and high energy density, tailored for use in implantable medical devices.[83] They develop electrodes made from a novel sphere-in-network heteronanomat structure, which provides a robust framework for ion storage and transport. This micro-battery design ensures long-term stability, efficient operation, and safe implantation, addressing key challenges in developing sustainable power sources for bioelectronics. This innovation is particularly significant for its potential to enhance the reliability and functionality of IMDs.

A visual overview of a micro-battery's structure, materials, and integration for intended applications (see Figure 1.7a) is composed of a schematic of the spere-in-network heteronanomat electrodes and their integration into the Na-ion micro-battery. This structure illustrates the unique properties of the heteronanomat materials, which support efficient ion transport. In a further application, it illustrates the battery in a potential biomedical device with an emphasis on size, energy density, and biocompatibility.

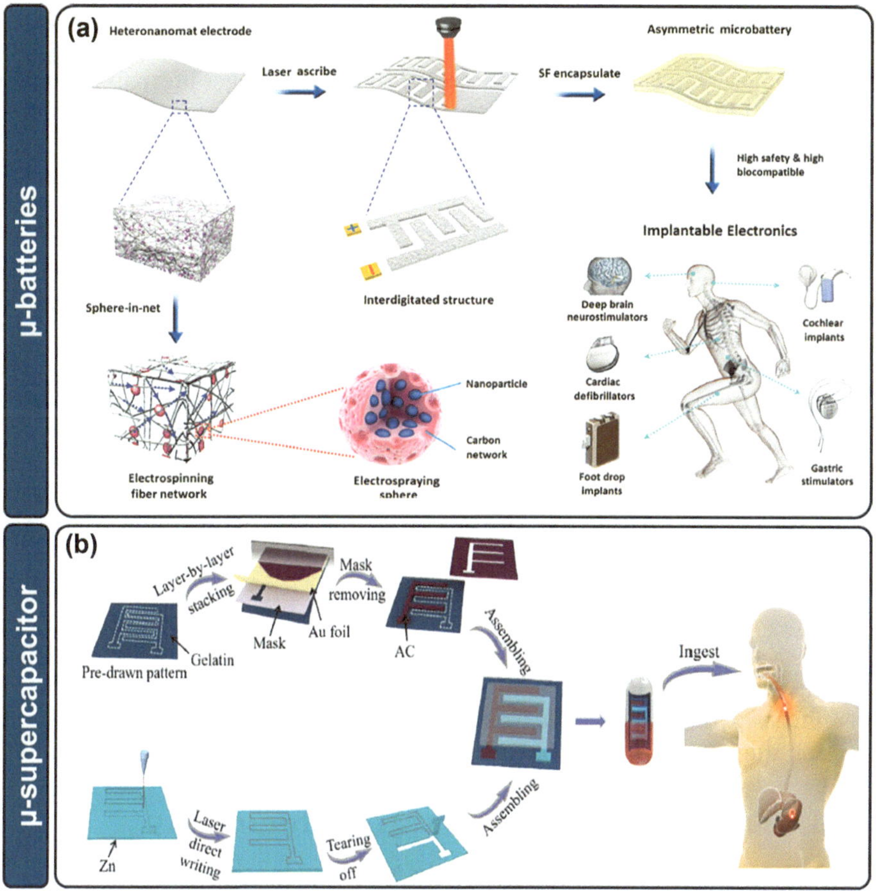

Figure 1.7 (a) Process diagram for the production of a flexible symmetric sodium-ion micro-battery and its applications in implantable electronics. The battery is based on heterogeneous nanoelectrodes and biocompatible electrolytes.[83] Reproduced from ref. 83 with permission from American Chemical Society, Copyright 2018. (b) Design and fabrication schematic of an edible Zn-ion-based micro-supercapacitor.[84] Reproduced from ref. 84 with permission from American Chemical Society, Copyright 2022.

1.2.5.2 Micro-supercapacitors for IMDs

Besides the micro-battery mentioned above, micro-supercapacitors are advanced energy storage devices that provide quick bursts of power and are ideal for use in IMDs. A few of their key features, characteristics, and requirements are:

- Quick Charge and Discharge: micro-supercapacitors can rapidly charge and discharge energy, suitable for IMDs that require short, high bursts of power.

- Power Density: a high energy in a small volume, *i.e.*, a big power density, which is key for compact medical applications.
- Longevity in Use: micro-supercapacitors have a longer lifespan compared to traditional batteries, allowing for sustained performance over time.
- Biocompatibility: materials used in micro-supercapacitors must be designed to be biocompatible to ensure safety when implanted in the human body.

The above features make micro-supercapacitors a promising option for powering various implantable medical devices, such as sensors and stimulators, providing greater safety and efficient operation. There are many works focused on super-capacitors nowadays, in this section, a novel approach to energy storage using an edible zinc-ion micro-supercapacitor designed for potential use in the human stomach is introduced.[84] The micro-supercapacitor demonstrates an exceptionally high energy density and is made from safe, biodegradable materials. It can be consumed and utilised within the body, providing a sustainable power source for biomedical applications while minimising the environmental impact.

The zinc-ion micro-supercapacitor design and structure (see Figure 1.7b) includes a small size that is suitable for *in vivo* applications. In this design, the edible active carbon micro-cathode and the zinc micro-anode demonstrate how these elements are integrated into a compact and functional device. The figure emphasises the micro-supercapacitor's potential for safe ingestion and uses within the human body, showcasing its suitability for biomedical applications.

1.3 Future of IMDs – Toward Autonomous

While significant advancements have been made, there are several challenges that researchers and manufacturers must address in the energy landscape of IMDs (see Figure 1.8):

- Battery Life *vs.* Device Functionality: as IMDs become more sophisticated (*e.g.*, incorporating wireless connectivity, artificial intelligence (AI), or real-time diagnostics), they consume more energy. Balancing functionality with a long battery life remains a significant challenge.
- Surgical Interventions for Battery Replacement: one of the key limitations in today's IMDs is the need for surgical replacement of batteries after several years. This increases the risk of complications for patients and the healthcare costs.
- Safety and Biocompatibility: any power source used in IMDs must be biocompatible to avoid adverse reactions within the body. The battery or capacitor should not be leaked, degraded, or cause harmful reactions.
- Efficient Energy Harvesting: while energy harvesting technologies (piezoelectric, thermoelectric, biochemical) hold promise, they currently generate small amounts of power. Scaling these technologies to meet the energy demands of modern IMDs is an ongoing challenge.

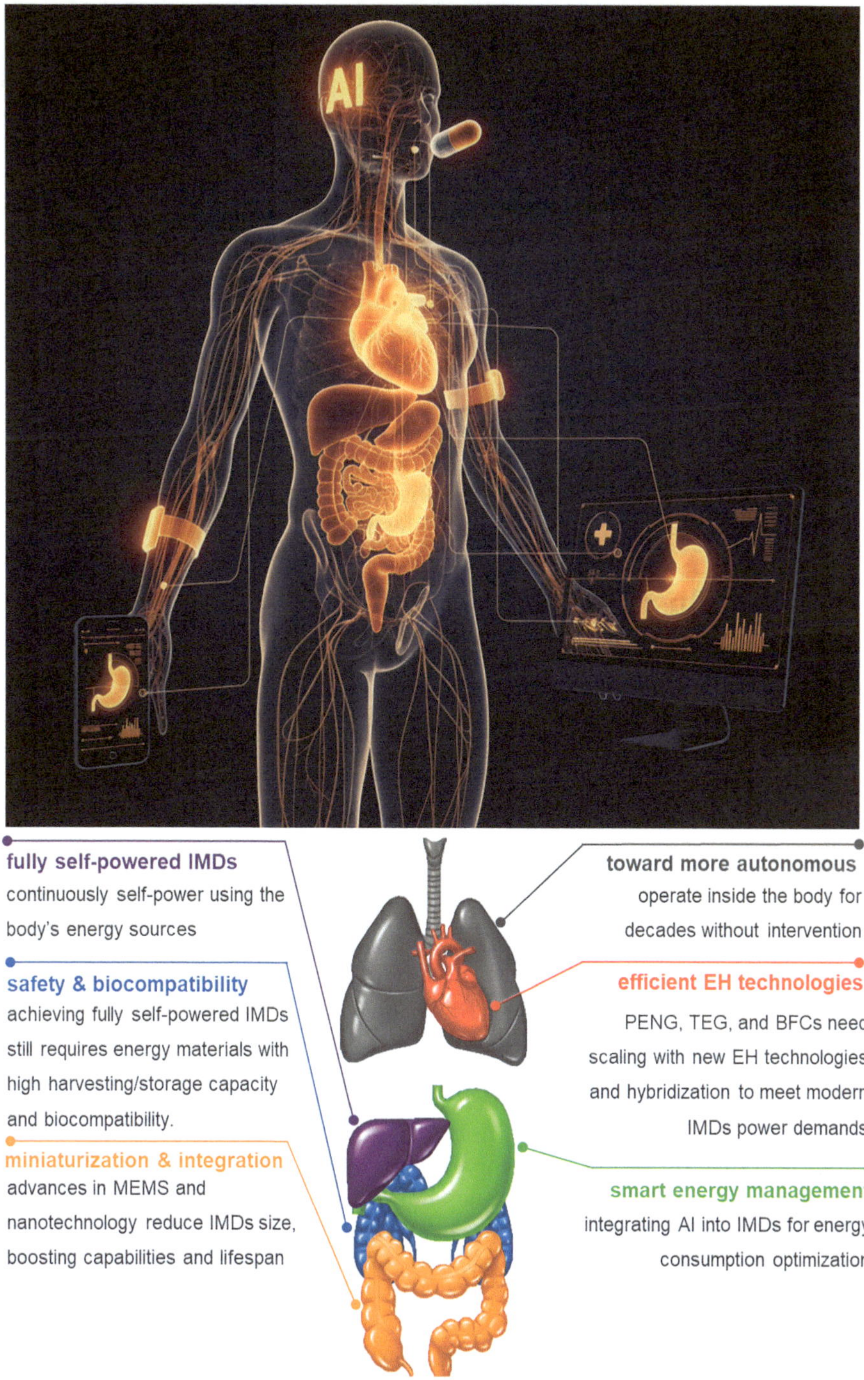

Figure 1.8 Future directions in IMD energy solutions. (The upper panel was generated using Google Gemini [*via* Google Cloud], while the lower panel incorporates visual elements adapted from a template designed by PoweredTemplate.com.)

From the challenges in energy management for IMDs, the future directions of IMDs are set. It should move toward more autonomous, efficient, and safe power solutions that can operate inside the body for decades without intervention, including:

- Fully Self-powered IMDs: the goal is to develop IMDs that can continuously power themselves through the body's energy sources, such as movement, heat, or biochemical reactions, eliminating the need for batteries.
- AI and Smart Energy Management: integrating AI into IMDs for dynamic energy management could allow devices to optimise their energy consumption by adjusting their functionality based on the patient's condition or energy availability.
- Miniaturisation and Integration: further advances in microelectromechanical systems (MEMS) and nanotechnology will continue to reduce the size of IMDs while enhancing their capabilities, leading to less invasive devices with longer lifespans.

1.4 Conclusion

In this chapter, an overview of the current development and emerging innovation of energy materials in implantable and ingestible medical devices has been presented and outlined. It has highlighted the increasing energy demand in modern IMDs, and also discussed the limitations of the commercial, traditional power sources, as a result outlining the motivation behind the integration of energy materials between harvesting and storage technologies. Some of the key materials, such as triboelectric, piezoelectric, and biofuel cell systems, were reviewed in view of their suitability, compatibility, and safety for biomedical applications, along with the compatibility of storage materials like micro-supercapacitors and micro-batteries.

In addition, the need for a hybrid system for integration between energy harvesting and storage materials is emphasised, especially addressing the low-output and intermittent, unstable operation of standalone harvesters. Advances in these hybrid solutions, allowing continuous harvesting with controlled, regulated storage, offer an autonomous, stable, and robust operation. Despite significant advancements, critical challenges remain, including issues related to biocompatibility, degradation of material, limited power density, and mechanical mismatch with biological tissues. To address these limitations, numerous strategies, such as material optimisation, 2D/3D surface structural engineering, and the integration of advanced power management systems, offer viable directions for future research and application.

This chapter concluded by highlighting key future directions in IMD design, including self-powered operation, AI-enhanced energy control, and refined miniaturisation; that together promise to shape a new era of intelligent and sustainable healthcare technologies.

Abbreviations

AC	Alternating current
AI	Artificial intelligence
BFCs	Biofuel cells
bio-SC	Bio-supercapacitor
BST	Barium strontium titanate
DC	Direct current
EOF	Electroosmotic flow
FEP	Fluorinated ethylene propylene
IMDs	Implantable medical devices
MEMS	Microelectromechanical systems
PDMS	Polydimethylsiloxane
PEGs	Pyroelectric nanogenerators
PENGs	Piezoelectric nanogenerators
PMNs	Porous microneedles
PMS	Power management systems
P-SSHI	Parallel synchronised switch harvesting on inductor
PTFE	Polytetrafluoroethylene
PVDF	Polyvinylidene fluoride
PZT	Lead zirconate titanate
SC	Supercapacitor
SSHCI	Synchronized switch harvesting on capacitor inductor
TENGs	Triboelectric nanogenerators
TGS	Triglycine sulfate
VNS	Vagus nerve stimulation
W-BFCs	Wearable biofuel cells

Acknowledgements

This work was supported by the Novo Nordisk Foundation Challenge Programme in Energy Materials with Biological Applications (EMGUT): grant ref. No. NNF22OC0072961. The work was also supported by the Engineering and Physical Sciences Research Council (EPSRC) in the UK through the "Next Generation Energy Autonomous Textile Fabrics based on Triboelectric Nanogenerators" grant (EP/V003380/1).

References

1. J. K. Krauss, N. Lipsman, T. Aziz, A. Boutet, P. Brown and J. W. Chang, *et al.*, Technology of deep brain stimulation: current status and future directions, *Nat. Rev. Neurol.*, 2020, **17**, 75–87.
2. Cardiac Physiological Pacing for Pacemaker (CPP) | WWMG Blog. https://www.wwmedgroup.com/blog/cardiac-physiological-pacing-for-pacemaker-cpp/ (accessed 23 May2025).

3. Y. Lv, X. Liu, J. Liu, S. Wu, S. Sun and P. Wu, *et al.*, Implantable and biocompatible Na-O2 battery, *Chem*, 2024, **10**, 1885–1896.
4. Y. Lee, V. K. Bandari, Z. Li, M. Medina-Sánchez, M. F. Maitz and D. Karnaushenko, *et al.*, Nano-biosupercapacitors enable autarkic sensor operation in blood, *Nat. Commun.*, 2021, **12**, 1–10.
5. PillBotTM by Endiatx: Remote-Controlled, Disposable Endoscopy Pill. https://endiatx.com/ (accessed 24 May 2025).
6. H. Guan, D. Lv, T. Zhong, Y. Dai, L. Xing and X. Xue, *et al.*, Self-powered, wireless-control, neural-stimulating electronic skin for *in vivo* characterization of synaptic plasticity, *Nano Energy*, 2020, **67**, 104182.
7. X. Cheng, X. Xue, Y. Ma, M. Han, W. Zhang and Z. Xu, *et al.*, Implantable and self-powered blood pressure monitoring based on a piezoelectric thinfilm: Simulated, *in vitro* and *in vivo* studies, *Nano Energy*, 2016, **22**, 453–460.
8. H. Ouyang, Z. Liu, N. Li, B. Shi, Y. Zou and F. Xie, *et al.*, Symbiotic cardiac pacemaker, *Nat. Commun.*, 2019, **10**, 1–10.
9. F. Arab Hassani, R. P. Mogan, G. G. L. Gammad, H. Wang, S. C. Yen and N. V. Thakor, *et al.*, Toward Self-Control Systems for Neurogenic Underactive Bladder: A Triboelectric Nanogenerator Sensor Integrated with a Bistable Micro-Actuator, *ACS Nano*, 2018, **12**, 3487–3501.
10. J. Wang, H. Wang, N. V. Thakor and C. Lee, Self-Powered Direct Muscle Stimulation Using a Triboelectric Nanogenerator (TENG) Integrated with a Flexible Multiple-Channel Intramuscular Electrode, *ACS Nano*, 2019, **13**, 3589–3599.
11. H. Zou, Y. Zhang, L. Guo, P. Wang, X. He and G. Dai, *et al.*, Quantifying the triboelectric series, *Nat. Commun.*, 2019, **10**, 1–9.
12. Z. L. Wang, On Maxwell's displacement current for energy and sensors: the origin of nanogenerators, *Mater. Today*, 2017, **20**, 74–82.
13. H. Wang, J. Cheng, Z. Wang, L. Ji and Z. L. Wang, Triboelectric nanogenerators for human-health care, *Sci. Bull.*, 2021, **66**, 490–511.
14. I. Aazem, R. Walden, A. Babu and S. C. Pillai, Surface patterning strategies for performance enhancement in triboelectric nanogenerators, *Results Eng.*, 2022, **16**, 100756.
15. C. Wang, C. Wang, Z. Huang and S. Xu, Materials and Structures toward Soft Electronics, *Adv. Mater.*, 2018, **30**, 1801368.
16. C. Hu, K. Behdinan and R. Moradi-Dastjerdi, PVDF Energy Harvester for Prolonging the Battery Life of Cardiac Pacemakers, *Actuators*, 2022, **11**, 187.
17. G. T. Hwang, M. Byun, C. K. Jeong and K. J. Lee, Flexible piezoelectric Thin-Film energy harvesters and nanosensors for biomedical applications, *Adv. Healthcare Mater.*, 2015, **4**, 646–658.
18. A. Zebda, S. Cosnier, J. P. Alcaraz, M. Holzinger, A. Le Goff and C. Gondran, *et al.*, Single Glucose Biofuel Cells Implanted in Rats Power Electronic Devices, *Sci. Rep.*, 2013, **3**, 1–5.
19. W. Fan, F. Wang, X. Xiong, B. Song, T. Wang and X. Cheng, *et al.*, Recent advances in functional materials and devices for Zn-Ion hybrid supercapacitors, *NPG Asia Mater.*, 2024, **16**, 1–20.

20. H. Sheng, L. Jiang, Q. Wang, Z. Zhang, Y. Lv and H. Ma, *et al.*, A soft implantable energy supply system that integrates wireless charging and biodegradable Zn-ion hybrid supercapacitors, *Sci. Adv.*, 2023, **9**, DOI: 10.1126/sciadv.adh8083.
21. C. Cao, Z. Li, F. Shen, Q. Zhang, Y. Gong and H. Guo, *et al.*, Progress in techniques for improving the output performance of triboelectric nanogenerators, *Energy Environ. Sci.*, 2024, **17**, 885–924.
22. Y. Liu, J. Mo, Q. Fu, Y. Lu, N. Zhang and S. Wang, *et al.*, Enhancement of Triboelectric Charge Density by Chemical Functionalization, *Adv. Funct. Mater.*, 2020, **30**, 2004714.
23. Y. Bai, L. Xu, S. Lin, J. Luo, H. Qin and K. Han, *et al.*, Charge Pumping Strategy for Rotation and Sliding Type Triboelectric Nanogenerators, *Adv. Energy Mater.*, 2020, **10**, 2000605.
24. K. K. Poon, M. C. Wurm, D. M. Evans, M. A. Einarsrud, R. Lutz and J. Glaum, Biocompatibility of (Ba,Ca)(Zr,Ti)O_3 piezoelectric ceramics for bone replacement materials, *J. Biomed. Mater. Res., Part B*, 2020, **108**, 1295–1303.
25. N. Almarri, J. Chang, W. Song, D. Jiang and A. Demosthenous, Piezoelectric energy harvesting and ultra-low-power management circuits for medical devices, *Nano Energy*, 2024, **131**, 110196.
26. S. Niu, X. Wang, F. Yi, Y. S. Zhou and Z. L. Wang, A universal self-charging system driven by random biomechanical energy for sustainable operation of mobile electronics, *Nat. Commun.*, 2015, **6**, 1–8.
27. D. Bao, L. Luo, Z. Zhang and T. Ren, A power management circuit with 50% efficiency and large load capacity for triboelectric nanogenerator, *J. Semicond.*, 2017, **38**, 095001.
28. S. Wang, L. Lin and Z. L. Wang, Triboelectric nanogenerators as self-powered active sensors, *Nano Energy*, 2015, **11**, 436–462.
29. Y. Yang, H. Zhang, R. Liu, X. Wen, T. C. Hou and Z. L. Wang, Fully enclosed triboelectric nanogenerators for applications in water and harsh environments, *Adv. Energy Mater.*, 2013, **3**, 1563–1568.
30. Q. Xu, J. Wen and Y. Qin, Development and outlook of high output piezoelectric nanogenerators, *Nano Energy*, 2021, **86**, 106080.
31. Q. Shi, T. Wang and C. Lee, MEMS Based Broadband Piezoelectric Ultrasonic Energy Harvester (PUEH) for Enabling Self-Powered Implantable Biomedical Devices, *Sci. Rep.*, 2016, **6**, 1–10.
32. D. Zhang, H. Wu, C. R. Bowen and Y. Yang, Recent Advances in Pyroelectric Materials and Applications, *Small*, 2021, **17**, 2103960.
33. H. Ryu and S. W. Kim, Emerging Pyroelectric Nanogenerators to Convert Thermal Energy into Electrical Energy, *Small*, 2021, **17**, 1903469.
34. J. Wang, K. Ren, Y. Zhu, J. Huang, S. A. Liu and E. R. Rene, *et al.*, A Review of Recent Advances in Microbial Fuel Cells: Preparation, Operation, and Application, *BioTech*, 2022, **11**, 44.
35. V. Mathur and R. Mukherji, Advances in biofuel cell research and future prospects, in *Biofuel Cells and Energy Generation*, 2025, pp. 27–65.

36. H. Zhou, G. Liu, J. Zeng, Y. Dai, W. Zhou and C. Xiao, *et al.*, Recent Progress of Switching Power Management for Triboelectric Nanogenerators, *Sensors*, 2022, **22**, 1668.
37. T. Hu, H. Wang, D. Bamgboje and W. Harmon, Power Management Systems for Triboelectric Nanogenerators, in *Handbook of Triboelectric Nanogenerators*, 2023, pp. 387–420.
38. C. Fang, T. Tong, T. Bu, Y. Cao, S. Xu and Y. Qi, *et al.*, Overview of Power Management for Triboelectric Nanogenerators, *Adv. Intell. Syst.*, 2020, **2**, 1900129.
39. M. Prudente, L. L. Pfitscher, G. Emmendoerfer, E. F. Romaneli and R. Gules, Voltage multiplier cells applied to non-isolated DC-DC converters, *IEEE Trans. Power Electron.*, 2008, **23**, 871–887.
40. H. Yoo, M. Mahato, W. Oh, J. Ha, H. Han and C. W. Ahn, *et al.*, Exploring role of microbatteries in enhancing sustainability and functionality of implantable biosensors and bioelectronics, *Biosens. Bioelectron.*, 2024, **260**, 116419.
41. P. Li, M. Liao, J. Li, L. Ye, X. Cheng and B. Wang, *et al.*, Rechargeable Micro-Batteries for Wearable and Implantable Applications, *Small Struct.*, 2022, **3**, 2200058.
42. C. V. V. Muralee Gopi, S. Alzahmi, V. Narayanaswamy, K. V. G. Raghavendra, B. Issa and I. M. Obaidat, A review on electrode materials of supercapacitors used in wearable bioelectronics and implantable biomedical applications, *Mater. Horiz.*, 2025, **12**, 4092–4132.
43. N. Chodankar, C. Padwal, H. D. Pham, K. (K. en) Ostrikov, S. Jadhav and K. Mahale, *et al.*, Piezo-supercapacitors: A new paradigm of self-powered wellbeing and biomedical devices, *Nano Energy*, 2021, **90**, 106607.
44. A. Dutta and H. Cheng, Pathway of transient electronics towards connected biomedical applications, *Nanoscale*, 2023, **15**, 4236–4249.
45. N. Mittal, A. Ojanguren, M. Niederberger, E. Lizundia, N. Mittal and A. Ojanguren, *et al.*, Degradation Behavior, Biocompatibility, Electrochemical Performance, and Circularity Potential of Transient Batteries, *Adv. Sci.*, 2021, **8**, 2004814.
46. G. Yao, L. Kang, J. Li, Y. Long, H. Wei and C. A. Ferreira, *et al.*, Effective weight control *via* an implanted self-powered vagus nerve stimulation device, *Nat. Commun.*, 2018, **9**, 1–10.
47. Y. Chu, J. Zhong, H. Liu, Y. Ma, N. Liu and Y. Song, *et al.*, Human Pulse Diagnosis for Medical Assessments Using a Wearable Piezoelectret Sensing System, *Adv. Funct. Mater.*, 2018, **28**, 1803413.
48. J. He, S. Li, X. Hou, Y. Zhou, H. Li and M. Cui, *et al.*, A non-contact flexible pyroelectric sensor for wireless physiological monitoring system, *Sci. China Inf. Sci.*, 2022, **65**, 1–11.
49. J. Lv, I. Jeerapan, F. Tehrani, L. Yin, C. A. Silva-Lopez and J. H. Jang, *et al.*, Sweat-based wearable energy harvesting-storage hybrid textile devices, *Energy Environ. Sci.*, 2018, **11**, 3431–3442.
50. S. Kusama, K. Sato, Y. Matsui, N. Kimura, H. Abe and S. Yoshida, *et al.*, Transdermal electroosmotic flow generated by a porous microneedle array patch, *Nat. Commun.*, 2021, **12**, 1–11.

51. F. R. Fan, Z. Q. Tian and Z. Lin Wang, Flexible triboelectric generator, *Nano Energy*, 2012, **1**, 328–334.
52. C. Zhao, H. Feng, L. Zhang, Z. Li, Y. Zou and P. Tan, *et al.*, Highly Efficient In Vivo Cancer Therapy by an Implantable Magnet Triboelectric Nanogenerator, *Adv. Funct. Mater.*, 2019, **29**, 1808640.
53. F. Yi, Z. Zhang, Z. Kang, Q. Liao and Y. Zhang, Recent Advances in Triboelectric Nanogenerator-Based Health Monitoring, *Adv. Funct. Mater.*, 2019, **29**, 1808849.
54. Z. Liu, J. Nie, B. Miao, J. Li, Y. Cui and S. Wang, *et al.*, Self-Powered Intracellular Drug Delivery by a Biomechanical Energy-Driven Triboelectric Nanogenerator, *Adv. Mater.*, 2019, **31**, 1807795.
55. G. Khandelwal, N. P. Maria Joseph Raj and S. J. Kim, Triboelectric nanogenerator for healthcare and biomedical applications, *Nano Today*, 2020, **33**, 100882.
56. J. Zhang, C. Wang and C. Bowen, Piezoelectric effects and electromechanical theories at the nanoscale, *Nanoscale*, 2014, **6**, 13314–13327.
57. Q. Shi, T. Wang and C. Lee, MEMS Based Broadband Piezoelectric Ultrasonic Energy Harvester (PUEH) for Enabling Self-Powered Implantable Biomedical Devices, *Sci. Rep.*, 2016, **6**, 1–10.
58. R. Yang, Y. Qin, C. Li, G. Zhu and Z. L. Wang, Converting biomechanical energy into electricity by a muscle-movement- driven nanogenerator, *Nano Lett.*, 2009, **9**, 1201–1205.
59. M. A. Parvez Mahmud, N. Huda, S. H. Farjana, M. Asadnia and C. Lang, Recent Advances in Nanogenerator-Driven Self-Powered Implantable Biomedical Devices, *Adv. Energy Mater.*, 2018, **8**, 1701210.
60. R. Mondal, M. A. M. Hasan, J. M. Baik and Y. Yang, Advanced pyroelectric materials for energy harvesting and sensing applications, *Mater. Today*, 2023, **66**, 273–301.
61. C. R. Bowen, J. Taylor, E. Leboulbar, D. Zabek, A. Chauhan and R. Vaish, Pyroelectric materials and devices for energy harvesting applications, *Energy Environ. Sci.*, 2014, **7**, 3836–3856.
62. R. Singh, S. Dogra, S. Dixit, N. I. Vatin, R. Bhardwaj and A. K. Sundramoorthy, *et al.*, Advancements in thermoelectric materials for efficient waste heat recovery and renewable energy generation, *Hybrid Adv.*, 2024, **5**, 100176.
63. Y. Ji, Y. Wang and Y. Yang, Photovoltaic–Pyroelectric–Piezoelectric Coupled Effect Induced Electricity for Self-Powered Coupled Sensing, *Adv. Electron. Mater.*, 2019, **5**, 1900195.
64. H. Wu, Y. Zhang, A. L. Kjøniksen, X. Zhou and X. Zhou, Wearable Biofuel Cells: Advances from Fabrication to Application, *Adv. Funct. Mater.*, 2021, **31**, 2103976.
65. S. N. Adler and Y. C. Metzger, PillCam COLON capsule endoscopy: Recent advances and new insights, *Ther. Adv. Gastroenterol.*, 2011, **4**, 265–268.
66. P. A. Thwaites, C. K. Yao, E. P. Halmos, J. G. Muir, R. E. Burgell and K. J. Berean, *et al.*, Review article: Current status and future directions of

ingestible electronic devices in gastroenterology, *Aliment. Pharmacol. Ther.*, 2024, **59**, 459–474.
67. A. Abdigazy, M. Arfan, G. Lazzi, C. Sideris, A. Abramson and Y. Khan, End-to-end design of ingestible electronics, *Nat. Electron.*, 2024, **7**, 102–118.
68. S. Y. Yang, V. Sencadas, S. S. You, N. Z. X. Jia, S. S. Srinivasan and H. W. Huang, *et al.*, Powering Implantable and Ingestible Electronics, *Adv. Funct. Mater.*, 2021, **31**, 2009289.
69. J. C. McRae, P. Jastrzebska-Perfect and G. Traverso, Challenges and opportunities for ingestible electronics across timescales, *Device*, 2023, **1**, 100055.
70. About CapsoCam Plus - Capsovision. https://capsovision.com/capsocam-plus/ (accessed 23 May 2025).
71. OMOM Capsule Endoscopy Manufacturer, High Definition Endoscope System Camera | Jinshan_JINSHAN. https://www.jinshangroup.com/product/omom-hd-capsule-endoscopy-camera/ (accessed 23 May 2025).
72. N. Hosoe, K. Watanabe, T. Miyazaki, M. Shimatani, T. Wakamatsu and K. Okazaki, *et al.*, Evaluation of performance of the Omni mode for detecting video capsule endoscopy images: A multicenter randomized controlled trial, *Endosc. Int. Open*, 2016, **04**, E878–E882.
73. Product | IntroMedic. https://www.intromedic.com/eng/item/item_010100_view.asp?search_kind=&gotopage=1&no=3 (accessed 23 May 2025).
74. Bravo™ Reflux Testing System | Medtronic (UK). https://www.medtronic.com/covidien/en-gb/products/reflux-testing/bravo-reflux-testing-system.html# (accessed 23 May 2025).
75. SmartPill™ Motility Testing System | Medtronic (UK). https://www.medtronic.com/covidien/en-gb/products/motility-testing/smartpill-motility-testing-system.html# (accessed 23 May 2025).
76. Ovesco Endoscopy AG - HemoPill®. https://ovesco.com/hemopill/ (accessed 23 May 2025).
77. The Capsule – Atmo Biosciences. https://www.atmobiosciences.com/the-capsule/ (accessed 23 May 2025).
78. Nom De L'entreprise. https://www.motilis.com/ (accessed 23 May 2025).
79. Sonopill. https://www.gla.ac.uk/research/az/sonopill/about/ (accessed 23 May 2025).
80. Z. Zhang and L. Shang, Smart ingestible devices: Orally delivering macromolecules and beyond, *Matter*, 2021, **4**, 3379–3381.
81. W. Shreffler, J. S. Sauk, J. Leung, A. Katz, N. S. Nishioka and Q. Yuan, *et al.*, Tethered confocal endomicroscopy capsule for diagnosis and monitoring of eosinophilic esophagitis, *Biomed. Opt. Express*, 2014, **5**, 197–207.
82. X. Jiang, J. Pan, Z. S. Li and Z. Liao, Standardized examination procedure of magnetically controlled capsule endoscopy, *VideoGIE*, 2019, **4**, 239.

83. G. Zhang, F. Geng, T. Zhao, F. Zhou, N. Zhang and S. Zhang, *et al.*, Biocompatible Symmetric Na-Ion Microbatteries with Sphere-in-Network Heteronanomat Electrodes Realizing High Reliability and High Energy Density for Implantable Bioelectronics, *ACS Appl. Mater. Interfaces*, 2018, **10**, 42268–42278.
84. K. Chen, L. Yan, Y. Sheng, Y. Ma, L. Qu and Y. Zhao, An Edible and Nutritive Zinc-Ion Micro-supercapacitor in the Stomach with Ultrahigh Energy Density, *ACS Nano*, 2022, **16**, 15261–15272.

CHAPTER 2

Implantable and Ingestible Technologies

ERDEM CIL, ICARO VELOSO SOARES AND DENYS NIKOLAYEV*

IETR – CNRS, Univ Rennes, Rennes, France
*Email: d@deniq.com

2.1 Introduction

Recent breakthroughs in electronics have driven significant advancements in wireless communications, leading to notable progress in the field over the past decade. Advances in electronics, combined with substantial progress in biosensor technologies, biointerface design, and materials science over the past few years, now enable the development of highly sophisticated long-lasting medical devices that can operate within the human or animal body with reduced surgical intervention frequency and scale.[1–6] These devices are not only miniaturized but also equipped with smart capabilities, enabling them to monitor, diagnose, and treat various conditions as they operate inside the body. Moreover, thanks to long-term biocompatible materials and refined interface and electrode technologies, such as conductive polymers, these innovations ensure that such devices can operate safely and efficiently without causing damage to surrounding tissues.

The integration of wireless miniaturized devices into a network serves as the foundation for body-centric communications through wireless body area networks (WBANs). WBANs represent a specialized form of wireless networking technology that facilitates short-range communication between devices in close proximity to the body.[7] These devices are interconnected through reliable wireless communication links, and they can be positioned

Detection Science Series No. 33
Energy Materials for Ingestible and Implantable Healthcare Devices
Edited by Santanu Patra and Anja Boisen

Published by the Royal Society of Chemistry, www.rsc.org

in various locations depending on the target application: placed within the body (in-body devices), attached to the skin or worn (on-body devices), or placed nearby in the surrounding environment (off-body devices).[8] With the interconnection between the elements, WBANs allow for continuous monitoring and data exchange, which provide real-time insights into the physiological conditions of the user.

WBANs have found applications in various aspects of daily life, including healthcare, entertainment, and fitness.[9] In healthcare, in particular, WBANs stand out as a ground-breaking technology that enables the real-time, remote monitoring and analysis of both patients' and healthy users' physiological data.[10] Devices, whether embedded in-body or worn on-body, can be strategically positioned on different parts of the body to continuously sense or record critical health parameters, such as heart rate, body core temperature, or oxygen levels. The collected information is then wirelessly transmitted to an off-body device, where it can be analysed in real time by health experts or the users themselves.[11] Such continuous monitoring through WBANs plays an essential role in modern healthcare systems, allowing for early detection of health issues and more effective management of chronic conditions.[12] In this way, the use of these technologies also reduces medical costs by minimizing the need for frequent hospital visits and post-surgical clinical interventions.[13] Moreover, they contribute to the quality of life for patients by offering minimally invasive continuous monitoring while imposing minimal restrictions on their daily activities.

Based on the method used to insert the devices into the body, in-body devices utilized in medical applications can be classified into three primary categories: implantable, ingestible, and injectable devices.[14]

(1) **Implantable devices:** implantable devices are typically inserted into the body *via* surgical procedures, namely, they are "implanted" inside the body. They play a crucial role in a range of medical fields, primarily targeting diagnostic, therapeutic, and assistive applications.[15] Depending on the target medical application, these devices can be placed in various parts of the body, including the brain,[16] chest,[17] and joints like the knee,[18] to continuously monitor, measure, and stimulate vital physiological functions. The devices implanted in the brain, for example, are used for neurostimulation or neurorecording, where they track neural activity for treating neurological conditions. Neuroprosthetic technologies, another brain-centered application, are used to restore motor and sensory functions lost due to paralysis or amputation.[19] These advanced devices can decode intended movements by analysing cortical signals, allowing users to control prosthetic limbs directly through 'brain-to-body' interfaces. This enables them to perform crucial tasks such as reaching and manipulating objects, restoring independence and functional mobility.[20] Moreover, devices placed in the chest, such as pacemakers or defibrillators, can regulate cardiac functions such as heart rhythms without the need for

conventional leads. In joints like the knee, implantable devices can monitor biomechanical movements, sense the development of post-surgical infections, or deliver targeted therapies for orthopaedic conditions. Moreover, implantable pressure-monitoring devices provide convenient managing of intracranial or cardiovascular pressure imbalances.[21] Lastly, implantable drug-delivery systems enable controlled, localized release of medication in the body.[22]

(2) **Ingestible devices:** ingestible devices have a compact, capsule-like structure similar to that of conventional pills, allowing them to be easily swallowed. Once swallowed, these devices advance through the gastrointestinal (GI) tract to perform their task. Innovative ingestible devices offer significant advantages across various health applications, particularly in the fields of wireless capsule endoscopy (WCE) and GI sensing and biotelemetry. For instance, in wireless endoscopy, these devices can be used to capture high-quality images of the GI tract, transmit real-time video, and deliver targeted medications if required.[23] In this way, ingestible devices provide non-invasive methods for visualizing the GI tract, allowing healthcare professionals to investigate the tract and locate abnormalities without the need for conventional procedures such as endoscopy and colonoscopy. Furthermore, ingestible devices are crucial in the biotelemetry of physiological parameters in the GI tract, such as pH levels and temperature. This involves continuous monitoring of these critical metrics and the wireless transmission of the collected data to an external device for further analysis.[24] Another recent trend in ingestible bioelectronics focuses on advancing microbiome analysis by targeting the small intestine, an area not fully captured through traditional bacteriological and parasitological analysis. Innovative ingestible devices are being developed to non-invasively sample the gut's microbial environment,[25] which significantly influences the host's general health. These devices incorporate pH-sensitive mechanisms that activate sampling only in specific gut regions, enhancing sample accuracy by preventing contamination from downstream microbiota. By integrating lightweight, biocompatible materials and advanced micro-valves, these capsules are designed to improve safety and functionality.[26] Additionally, ingestible electronics with enhanced sensor, circuit, and communication capabilities are enabling real-time, location-aware data collection and therapeutic delivery.[27] By providing continuous information on these critical parameters, ingestible devices allow healthcare experts to take medical decisions more conveniently regarding diagnosis and treatment of the GI tract.

(3) **Injectable devices:** injectable devices are typically micro-devices that are injected into the body by means of needles. This method presents a promising approach for in-body devices due to its minimally invasive nature, which substantially reduces potential negative effects on the body compared to conventional surgical operations.[28] The diverse

application fields of injectable devices include continuous, real-time monitoring of critical parameters such as heart rate, movement, and core body temperature in animal subjects[29] as well as monitoring animal hibernation,[30] which can provide scientists with valuable data for ecological studies. Moreover, injectable devices can operate as therapeutic neurostimulators,[31] delivering targeted electrical impulses to modulate neural activity and treat various neurological conditions.

Although injectable devices have emerged relatively recently, the versatility of their application fields highlights their significant potential in advancing medical monitoring and treatment methodologies. For instance, injectable nanoparticles, constructed from organic and inorganic nanomaterials, can support applications ranging from cancer treatment to tissue engineering by enabling precise therapeutic and diagnostic interventions at the cellular level.[32] 'Neural dust' technology represents a potential breakthrough in bioelectronics, where ultra-small, ultrasound-powered motes allow for real-time neural activity monitoring and stimulation.[33] These piezoelectric-powered devices can be injected and positioned in the body for targeted neural stimulation, pushing the boundaries of neuromodulation with reduced surgical complexity. The concept 'Neurodots' also offers potential for multisite neuromodulation, advancing brain stimulation approaches by targeting complex, interconnected neural networks.[34] These novel injectable technologies highlight a significant potential to transform medical monitoring and treatment methodologies through precision and minimally invasive design.

In this chapter, we will examine implantable and ingestible devices by firstly focusing on their fundamental structure and the conventional components necessary for their functionality. We will overview various challenges encountered during the design and development of these devices, emphasizing critical factors such as material selection and size constraints, which are essential for ensuring biocompatibility and effective performance within the body. Additionally, we will review the latest designs and application fields of implantable and ingestible devices by highlighting innovative approaches in the literature. Table 2.1 summarizes key features of some of the reviewed designs to provide a clearer overview. Although the primary focus of this chapter is on implantable and ingestible technologies, we will also briefly discuss some injectable device examples to provide a more comprehensive overview of in-body systems. Finally, considering potential applications and emerging technologies, we will discuss future perspectives in this field.

2.2 Components of Implantable and Ingestible Devices

The design and translation of implantable and ingestible devices is an inherently multidisciplinary process. These devices require the integration

Table 2.1 Comparison between novel implementations of implantable and ingestible bioelectronics for different medical applications.

Ref.	Application	Device description	Biocompatible materials	Footprint	Communication link	Power budget	Powering	Comments
71	Wireless and leadless myocardial stimulation	Self-assembled microtubular coil for wireless and battery-free cardiac pacing	Polyimide and styrene–ethylene–butylene–styrene (SEBS) polymer	20 mm, Ø2.5 mm	Magnetic coupled, Bluetooth	~ 8 mW	Wireless-powered, inductive coupling	The microtubular design allows for intravascular implantation and stimulation of a non-beating heart, restoring myocardial contraction and blood circulation.
72	Monitoring hemodynamics and cardiovascular health	An implantable piezoelectric sensor system is used to monitor hemo-dynamics with a "growable" sheath around arterial vessels	Polyvinylidene fluoride and nanofibrous elastic polyurethane	(13 × 24 × 8) mm	Bluetooth	130 mW	Wireless-powered, inductive coupling	The growable sheath avoids growth constriction while providing stable monitoring, enabling long-term implantation.
73	Hemodynamic monitoring in the inferior vena cava (IVC)	Passive crown-shaped antenna for accurate measurement of the cross-sectional area of the IVC	Nickel-titanium alloy (nitinol)	<Ø2.5 cm	Magnetic coupled	—	Wireless-powered, inductive coupling	The sensor provides real-time information on IVC size and is more sensitive to volume loading than right atrial pressure.
74	Diagnosis of cardiovascular diseases	3D-printed bioelectrode array for continuous electro-physiological mapping of dynamically beating rat hearts	Hydrogels made of conducting polymers, polystyrene sulfonate	—	—	—	—	Multi-material direct ink writing 3D printing allows for the fabrication of implantable bioelectrodes with improved features over traditional methods.

75	Real-time control and monitoring of cardiac function	A thin-film array with multiple micro-scale inorganic LEDs for optical stimulation and electrodes for electrical recording	Pyralux AP8535R encapsulated with parylene-C	40 mm	Infrared	32 mW	Wireless-powered, inductive coupling	The design of the thin-film array can be customized to fit the shape of different animal hearts. This personalized approach ensures optimal contact with the heart tissue and improves the effectiveness of stimulation and recording.
76	Detection of biomarkers in bodily fluids	Biochemical sensor using an LC resonant circuit and varactor diodes for detecting changes in electric potential caused by biochemical events. Modularized system with customizable sensing interfaces	Polymeric ion-selective membranes (ISMs) and aptamers	Ø1 mm	Magnetic coupled	—	Wireless-powered, inductive coupling	The sensor utilizes a magnetic coupling strategy between the sensing interface and the LC circuit through varactor diodes. It can be adaptable to different applications, including wearable accessories and miniaturized implantable probes.

Table 2.1 (*Continued*)

Ref.	Application	Device description	Biocompatible materials	Footprint	Communication link	Power budget	Powering	Comments
77	Spinal and muscular electrical stimulation	Implantable, battery-free device for multichannel, biphasic, current-controlled electrical stimulation	Parylene-C, Parylene P6	(4 × 2) cm	Infrared	76 mW	Wireless-powered, inductive coupling	Enables high voltage compliance of ±20 V, current-controlled stimulation ranging from 40 μA to 4.7 mA, and the capacity to independently control eight channels, thereby improving power transfer by 500% over previous systems
78	Epilepsy treatment and neuro-modulation	A split-ring resonator that generates an enhanced microwave field at its gap with sub-mm spatial precision. The SRR acts as an antenna with a perimeter of $\sim\lambda/2$.	Titanium alloy	Ø2-4 mm	Microwave	—	Wireless-powered, radiative	The device offers sub-mm spatial resolution for neuro-modulation, exceeding the capabilities of traditional microwave methods.
79	Monitoring inflammation, performing transdermal electrical stimulation, and drug delivery	Microneedle implant connected to a wearable smart device with reconfigurable electronics	Conducting polymer hydrogel	(12 × 12 × 2) mm	Wired (connection to a wearable module)	—	Battery and photovoltaic harvesting	Demonstrates the anti-inflammatory effect of electrical stimulation and synergistic treatment with electrical stimulation and drugs. Reconfigurable electronic hardware enables dynamic and flexible signal control.

80	Monitoring the water vapour transmission rate (WVTR) of thin-film encapsulations	Wireless platform using backscatter communication and magnesium (Mg)-based microsensors to monitor water permeation across thin-film encapsulation in real-time	Poliamide, Parylene C	(22 × 14 × 1) mm	RF backscattering	5 mW	Wireless-powered, inductive coupling	This platform utilizes the corrosion of Mg microsensors due to water permeation to change the resonant frequency of a backscatter circuit, enabling real-time WVTR monitoring. It is miniaturized for *in vivo* testing and adaptable to both rigid and flexible substrates.
81	Monitoring bladder function	Stretchable strain gauges (permanent and bioresorbable)	Bioresorbable materials: poly (octamethyl-enemaleate (anhydride) citrate) (POMaC) and tungsten particles	~Ø25 mm	Bluetooth	—	Battery	This system can be adapted to operate as a permanent or bioresorbable device.
82	Treatment of gastro-esophageal reflux disease by stimulating the lower esophageal sphincter (LES)	A battery-free electronic stent for non-invasive electrical stimulation of the LES endoscopically delivered through the mouth	Nitinol, Eutectic gallium-indium, Polydimethyl-siloxane	Ø0.8 mm	Magnetic coupled	32.96 mW	Wireless-powered, inductive coupling	Microneedles on the stent structure penetrate the esophageal mucosa to deliver electrical stimulation directly to the LES muscle.

Table 2.1 (*Continued*)

Ref.	Application	Device description	Biocompatible materials	Footprint	Communication link	Power budget	Powering	Comments
83	Alleviating the retention of the capsules within the GI tract	Programmable adaptive anti-retention module integrated into a capsule that is capable of detecting pressure and triggering vibration according to a pressure program	—	(25 × Ø11.7) mm	Bluetooth	—	Battery	A novel programmable adaptive anti-retention module with miniaturized dimensions is integrated into the ingestible capsule in order to address the capsule endoscope blockage in the GI tract. The novel module comprises a durable and highly sensitive pressure sensor and four different trigger logic circuits.
84	Imaging and localization within the GI tract	Lightweight wireless capsule endoscope system consisting of body channel communication transceiver integrated-circuits capable of VGA-resolution image transferring from the GI tract	—	(32 × Ø12) mm	Body channel communication	<1 mW	Battery	The device comprises four integrated cameras that can perform 360 degrees image capturing within the GI tract, which greatly reduces the risk of missing abnormalities. A dual-band pulse-shaping transmitter embedded within the capsule enables 80 Mb s^{-1} image transmission.

85	Treatment GI haemorrhage	Wireless capsule comprising a balloon that can be inflated. When the balloon reaches a predetermined pressure level, it can stop at a bleeding site in the intestines, hence achieving haemostasis by tamponade effect	Silicone	(60 × Ø14) mm	—	—	Wireless-trigger	The therapeutic capsule endoscope design helps with the treatment of GI haemorrhage using a silicone balloon that can stop at the bleeding site in the bowel. When the bleeding is detected in the tract, the capsule can send an alert to a transponder on the body, which forwards this alert to user's mobile phone *via* Bluetooth.
87	Monitoring the physiological signals of the GI tract	Wireless capsule consisting primarily of a wireless capsule, a portable data recorder, and a workstation. The capsule features temperature, pH, and pressure sensors, an RF transceiver, and a controlling and processing application specific integrated circuit	—	(26 × Ø11) mm	RF	—	Battery	The sensors equipped within the capsule monitor the variations in the corresponding physiological parameters within the GI tract. The application-specific integrated circuit helps control and process the recorded data.

Table 2.1 (*Continued*)

Ref.	Application	Device description	Biocompatible materials	Footprint	Communication link	Power budget	Powering	Comments
89	Biomedical telemetry	In-body biotelemetry platform that is robust against the EM changes in the tissues. The robustness is achieved with the novel tissue-independent antenna integrated into the platform	PVC	(17.7 × Ø8.9) mm	RF	13 dBm	Battery	The platform comprises a novel antenna, a microcontroller, and a matching circuit. The platform is also suitable for active implantable medical devices featuring therapeutic actuation.

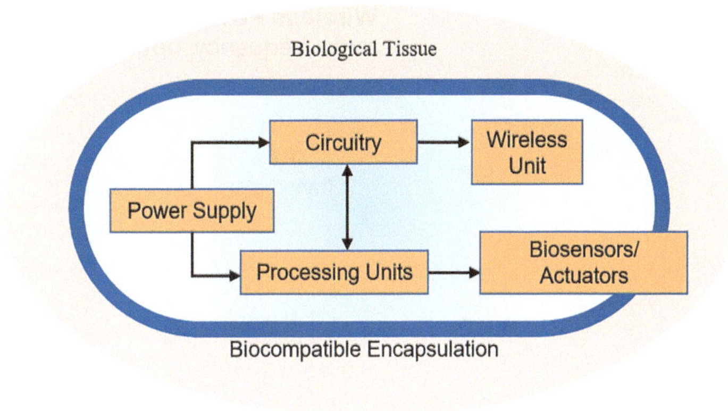

Figure 2.1 Electronic components of a typical implantable or ingestible device: Power supplies, processing units, electric and RF circuitry, biosensors/ actuators, and wireless unit. The device, except for its biosensing and actuator parts, is covered with a biocompatible encapsulation layer.

of multiple electronic and sensing components that function harmoniously within a compact form to perform their tasks efficiently. Such coordinated interaction between the components is essential for the successful operation and effectiveness of the devices. The integrated electronic components are typically protected by a biocompatible encapsulation layer, which serves two primary purposes: Firstly, it prevents direct contact of the components with biological tissues, thereby reducing the risk of adverse reactions. Secondly, it preserves the compact structure of the device. This protective layer is crucial for ensuring the long-term functionality and biocompatibility of the device within the body. The specific components needed for a given device may vary depending on the target application; however, there are several fundamental elements that are commonly essential for the functionality of an implantable or ingestible device. As shown in Figure 2.1, these key components include:[35]

(1) Power supplies,
(2) Signal and data processing units,
(3) RF circuits,
(4) Biosensors and actuators,
(5) Wireless unit.

2.2.1 Power Supplies

Power supplies are essential components of implantable and ingestible devices, as they provide the energy required for all of the device's functions like data processing, signal transmission, and sensing operations. The power supply in most existing devices comes in the form of batteries (rechargeable or not), which are frequently the largest component in the device.[36] Energy harvesting methods, which capture and convert small amounts of ambient

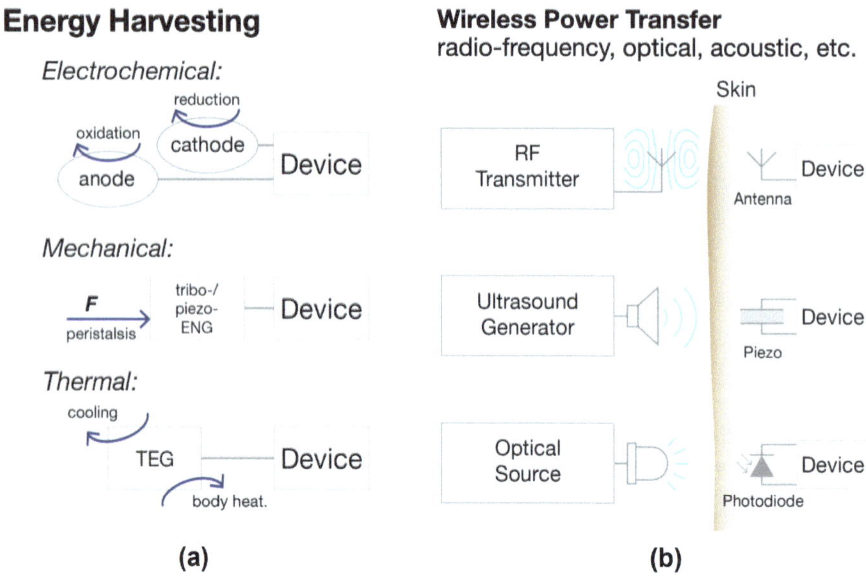

Figure 2.2 Energy harvesting methods and examples of wireless units than can be integrated into an implantable or ingestible device. (a) Electrochemical, mechanical, and thermal energy harvesting. (b) An antenna enables wireless communication by transmitting and receiving radio frequency signals between devices. A piezoelectric unit generates electrical energy by converting mechanical pressure from ultrasound waves. A photodiode converts incoming light energy into electrical energy.

energy from the surrounding environment into usable electrical power, are increasingly used in research-oriented implantable and ingestible technologies to enable battery-free operation. As seen in Figure 2.2(a), examples of these methods include electrochemical,[37] mechanical,[38,39] and thermal[40] energy harvesting. The implementation of reliable power supplies is crucial for ensuring the longevity of the device, since frequent replacement or recharging can be impractical and inconvenient in many clinical scenarios.

2.2.2 Signal and Data Processing Units

Signal and data processing units integrated into the devices coordinate various functions of the device. Their primary objective is to serve as the central processing unit that handles the processing of the data collected by other components such as sensors. Therefore, they control the overall operation of the device. Moreover, they may feature algorithms that can be executed for real-time analysis of the collected data.[41] Based on the interpretation of the collected data, these units enable the device to make informed decisions related to its next action, such as triggering alarms when necessary. Furthermore, they coordinate the communication between different components of the device as well as with the external systems.[42]

Therefore, signal and data processing units function as the 'brain' of the device, and their efficiency and performance are critical for an optimal operation.

2.2.3 Electrical Circuits

Circuits are an essential part of implantable and ingestible devices, serving various purposes within the overall system. Examples of most common circuits include RF circuits to interface between the processing and wireless units. These include, for instance, impedance matching circuits and power dividers/circulators/baluns.[43] Impedance matching circuits ensure maximum power transfer to the antenna to minimize the energy loss and improve overall efficiency. As for the power dividers, circulators, and baluns, they enable simultaneous communication between the components by distributing the RF signal to multiple paths. The selection of the RF circuits for a given device depends on the specific requirements of the application such as frequency range, power requirements, and data transmission rates. Meticulous design of RF circuits can help optimize the performance and reliability of implantable devices and ensure effective communication.[44]

2.2.4 Biosensors and Actuators

Biosensors and actuators play critical roles in monitoring, diagnostics, and therapy delivery with implantable and ingestible technologies. Biosensors can detect and measure a wide range of physiological parameters such as glucose levels and neural activity, providing real-time data related to the internal state of the body.[45] Actuators are able to respond to the signals processed by the signal and data processing units, delivering therapeutic actions such as electrical stimulation and drug release at a specific location. Miniaturization and long-term operation are the most critical general challenges in the design of biosensors and actuators.

2.2.5 Wireless Unit

The wireless unit in an implantable or ingestible device is the primary element enabling wireless communications and wireless powering/recharge features. It is responsible for the transmission and reception of signals and the energy between the in-body device and external systems. This wireless functionality is essential for real-time data exchange or activation of the device, providing the possibility to monitor the conditions within the body remotely in real time.[46] Depending on the target application, various types of wireless units can be integrated into implantable and ingestible devices. As shown in Figure 2.2(b), the key components of wireless systems include antenna systems for RF communication, piezoelectric units for acoustic communication, and photodiodes for optical communication. A wireless unit that is carefully designed to form the communication link plays an essential role in enhancing the signal quality and improving the overall efficiency of the device.[47] The wireless units with powering/recharge features are typically employed with rectifiers, electronic components that convert alternating to direct current. Careful design of rectifiers plays a critical role in stabilizing highly variable power from wireless energy transfer, enabling reliable operation of the other components.

2.3 Design Challenges

Implantable and ingestible device design presents several system-level challenges which must be addressed by the designers to ensure reliable performance. The main challenges encountered in developing these devices include biocompatibility of the materials used with the biological tissues, confined space available within the device, user acceptance and safety, restricted power availability and power consumption, and maintaining the quality of the communication link.[10,48] By effectively tackling these challenges, designers can improve both the efficacy and safety of implantable and ingestible technologies in healthcare.

2.3.1 Biocompatibility

Biocompatibility refers to the ability of a material to be safely and effectively integrated within the biological environment of the body without leading to any adverse reactions. This primarily includes the material's ability to be tolerated by the tissues and its interaction with cells and bodily fluids.[49] In the context of implantable and ingestible technologies, biocompatibility mainly concerns the encapsulation layer surrounding the device, since this layer is the only one that is in direct contact with the biological environment. A biocompatible encapsulation material is indispensable for implantable and ingestible devices in terms of ensuring patient safety and preventing inflammation and infection.[50] Biocompatible materials that are frequently used in current implantable and ingestible technologies include the following.

(1) **Titanium:** titanium is mainly used in implants for orthopaedic and dental applications due to its corrosion resistance nature.[51,52]
(2) **Polymers:** polymers such as polyvinyl chloride (PVC) and polylactic acid (PLA) are used due to their flexible structure.[53,54]
(3) **Ceramics:** ceramics, such as hydroxyapatite, are commonly utilized in implantable devices due to their high resistance to degradation in aqueous environments like biological tissues.[55,56]
(4) **Gold:** as it is resistant to corrosion, gold is a suitable biocompatible option for long-term implants.[57,58]

Furthermore, soft materials have been used both as encapsulation layers and as the entire platform for bioelectronic devices. For long-lasting biocompatibility, the Young's modulus of the implanted material should be like that of the targeted tissue.[59] This is a major challenge in current bioelectronics, as many implantable electrodes are much stiffer than the surrounding tissues.[59] This mechanical mismatch can lead to several negative consequences. For example, it can trigger a foreign body reaction (FBR), where the body's immune system attacks the implant. This reaction can be mitigated by employing soft materials.[60]

One example of a soft material commonly used for bioelectronic devices is the silicone elastomer (poly[dimethylsiloxane], PDMS). PDMS has a Young's modulus similar to that of the dura mater, the membrane that surrounds the brain and spinal cord. This similarity in mechanical properties makes PDMS a promising material for implantable electrodes designed to interface with the central nervous system. For example, in ref. 61, a soft, implantable electrode array called 'e-dura' has been developed using PDMS as the base material. The e-dura has been shown to be biocompatible and to cause minimal tissue damage when implanted in the epidural space of the spinal cord.

Other materials such as hydrogels and conducting polymers are also being explored for use in soft bioelectronics. Hydrogels can be designed to have a Young's modulus that is similar to that of soft tissues. Conducting polymers, such as the poly(3,4-ethylenedioxythiophene) polystyrene sulfonate (PEDOT:PSS), offer mixed electronic and ionic conductivity, which can improve charge transfer between the electrode and the tissue.[59]

2.3.2 Shape and Size Restrictions

Implantable and ingestible devices must be sufficiently small in order to fit within the specific locations in the body or travel through narrow passageways such as blood vessels or the GI tract without causing discomfort or obstruction.[62] The confined space available for these devices presents a significant challenge in the integration of all necessary components, which necessitates miniaturization of the devices. However, miniaturization must be achieved without compromising the functionality of the device, as the reduction of size often leads to limitations on battery capacity, component sizes, and communication range.[63] Moreover, implantable devices must have a shape that can promote seamless fitting into the target anatomical locations. As for the ingestible devices, they have generally a capsule-shaped structure to promote easy swallowing and smooth passage through the GI tract. Hence, any irregularities in shape can lead to discomfort of the patient or adverse reactions and prevent successful operation or transit through the GI tract.

2.3.3 User Acceptance and Safety

When using implantable and ingestible devices, both patients and healthcare professionals should feel comfortable. These devices must be designed to minimize discomfort for the patient and impose minimal restrictions on their daily activities. Furthermore, since these devices rely on wireless communication links through body tissues, user safety in terms of electromagnetic (EM) radiation exposure is a critical concern for these devices. Designers must carefully consider the potential effects of EM radiation on human tissues and ensure that the devices comply with international safety regulations to protect users from harmful exposure levels.[64,65]

2.3.4 Restricted Power Availability and Power Consumption

Restricted power availability is a major challenge in the design of implantable and ingestible devices. The miniaturized size of these devices restricts the capacity of batteries or other sources powering the device. This constraint influences the longevity and performance of the devices and turns energy efficiency into one of the major concerns.[66] Since frequent replacement or recharging of power sources is impractical in the majority of clinical cases, designers must optimize power consumption by utilizing energy-efficient components.[67] Furthermore, ensuring safety and adherence to regulatory standards regarding power management is essential for the efficient development of these devices.

2.3.5 Quality of the Communication Link

Maintaining a high-quality communication link between implantable and ingestible devices and external ones presents a significant challenge. For radiofrequency-based wireless devices, this challenge is primarily related to the design and performance of the antenna, since the antenna is the key element forming the wireless communication link. These devices transmit signals through lossy and dispersive body tissues that absorb electromagnetic waves, resulting in signal attenuation and degradation of link quality.[68] Additionally, variability in tissue composition among different individuals and the movement of the device within the body – particularly for ingestible devices that can have random orientations as they traverse the GI tract – further complicate the establishment of a robust communication link.[69] Consequently, antennas integrated into implantable and ingestible devices must operate efficiently, ensuring a safe, robust, and reliable communication link.

2.4 Implantable Technologies

Implantable bioelectronics has been a critical part of medical practice for over 60 years, ever since the successful implantation of the first pacemaker in 1958.[70] These devices have evolved rapidly, becoming more miniaturized while offering increasingly sophisticated functionalities and expanding their potential applications. Today, implantable bioelectronics are widely used across various medical fields, delivering more precise diagnostics and robust treatments. This section will explore state-of-the-art design strategies for implantable devices across various applications, focusing on how novel approaches are addressing the challenges of ultra-miniaturization, low-power consumption, material selection, and wireless connectivity to advance the development of next-generation bioelectronics.

Cardiovascular medicine is often the first application that comes to mind when discussing implantable devices, with pacemakers and stents being long-standing solutions in this field. However, the next generation of implantable devices is becoming increasingly miniaturized. Traditional pacemakers, which rely on electrodes wired to a subcutaneous pulse generator, are gradually replaced by ultra-miniature, leadless pacemakers implanted directly inside the heart's ventricles.

One innovative example is a microtubular pacemaker proposed by Wang *et al.*[71] that combines both electrical stimulation of the myocardium and wireless powering in a single device (Figure 2.3). This design features a resonant coil antenna built on a flexible substrate encapsulated between layers of a styrene–ethylene–butylene–styrene (SEBS) polymer. This material ensures biocompatibility and provides resistance to water and corrosion, which is essential for long-term use inside the body. A significant advantage of this proposed pacemaker is its minimally invasive implantation process. Unlike current leadless pacemakers, which require open-chest surgery, this new device is designed to be rolled up and deployed through the vascular system, eliminating the need for major surgery. In terms of performance, the

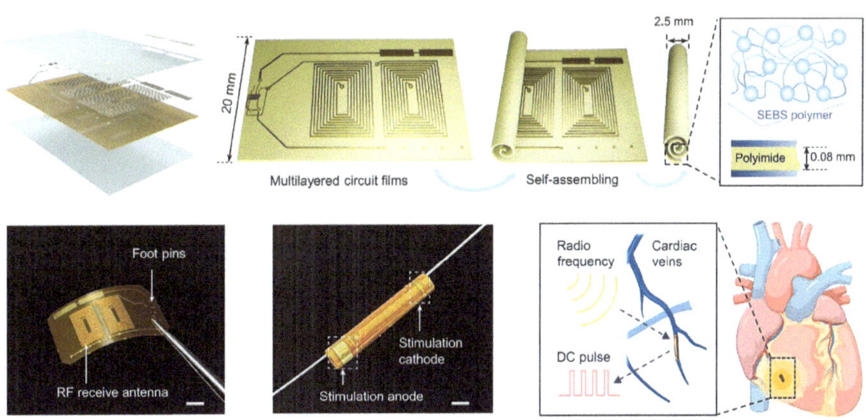

Figure 2.3 Self-assembled implantable microtubular pacemaker proposed by Wang *et al.*[71] The pacemaker's circuit film comprises several layers of SEBS (styrene–ethylene–butylene–styrene) encapsulation, a polyimide film, embedded circuits, another SEBS encapsulation, and gold electrodes for stimulation. The implanted pacemaker wirelessly receives RF energy and converts it into DC pulses for heart stimulation. Reproduced from ref. 71, https://doi.org/10.1126/sciadv.adj0540, under the terms of the CC BY 4.0 license, https://creativecommons.org/licenses/by/4.0/.

device demonstrated an impressive energy-harvesting efficiency of nearly 4%, even under misalignment conditions between the external transmitter and the internal receiver.

Another promising application in cardiovascular medicine involves the development of ultra-miniaturized electronic stents and interfaces designed to anchor inside blood vessels, providing precise measurements of arterial cross-section variations and pressure. One example is a device featuring a piezoelectric sheath made from polyvinylidene fluoride (PVDF) nanofibers, which conform to the external circumference of the artery (Figure 2.4).[72] As the artery's shape changes, these variations induce a change in the potential

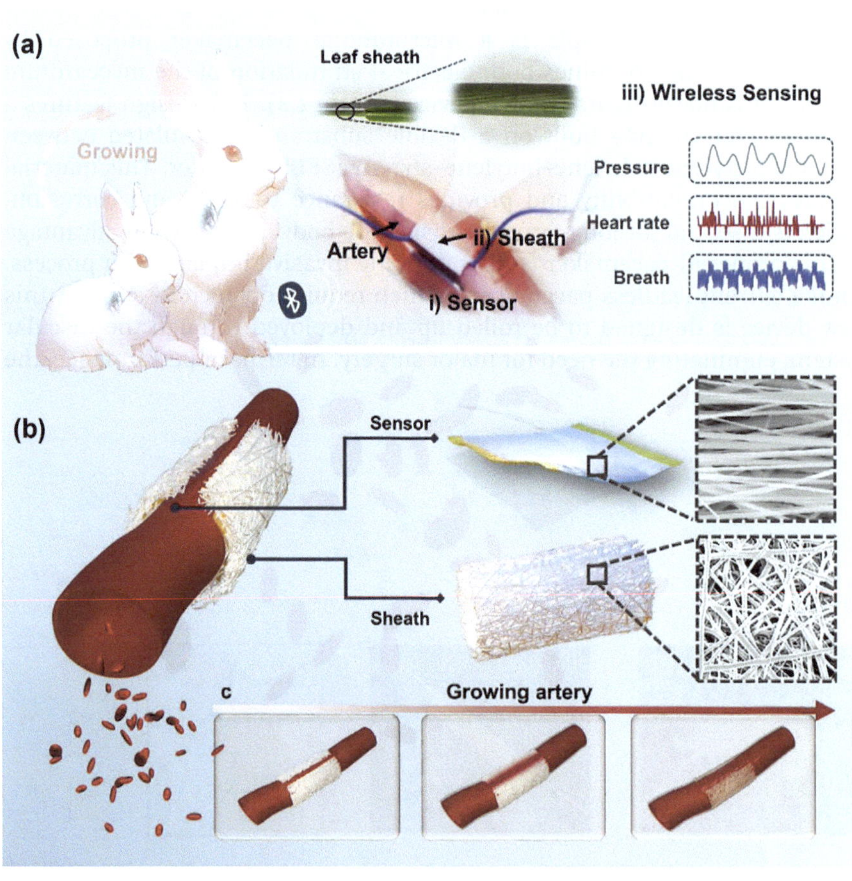

Figure 2.4 A hemodynamic bioelectronic sensor was proposed by Tang *et al.*[72] (a) The implantable electronic system comprises a flexible piezoelectric sensor designed for continuous hemodynamic monitoring, a growable sheath that anchors the sensor in the artery, and a wireless module for transmitting and receiving data. (b) A sheath made of polyvinylidene fluoride nanofibers anchors the sensor around the artery, conforming to the artery shape. Reproduced from ref. 72 with permission from John Wiley & Sons, Copyright 2023 Wiley-VCH GmbH.

of the piezoelectric material, enabling real-time monitoring of arterial dynamics.

An alternative approach employs a crown-shaped resonant loop that is anchored inside the blood vessel.[73] In this design, changes in the arterial cross-section affect the inductance of the loop, causing shifts in its resonant frequency. By tracking these frequency shifts, the device can provide real-time measurements of arterial cross-section variations, offering another method for continuous vascular monitoring.

Additionally, other sensor and actuator topologies have been proposed for measuring hemodynamic parameters or delivering cardiac stimulation. These designs are based on flexible, conformal arrays of microelectrodes[74] or micro-LEDs,[75] which can adapt to the morphology of the target area, enhancing their functionality and integration with the surrounding tissues. These advancements in flexible, adaptable implantable devices represent a significant step forward in precision cardiovascular monitoring and treatment.

Biotelemetry is another area where the use of miniature implantable devices is reviving growing interest. These devices are essential for monitoring various biomarkers within the body, providing valuable data for diagnostics and treatment. For instance, Liu *et al.*[76] proposed an implantable sensor that detects biomarkers such as glucose, ions, and neurotransmitters in body fluids (Figure 2.5). The device comprises a stretchable biochemical sensing interface and a loop antenna magnetically coupled to an external module. In this design, changes in the concentration of a specific biomarker alter the potential difference between electrodes located at the ends of the sensing interface. This voltage variation modulates the capacitance of a varactor within the circuit, which, in turn, shifts the resonant frequency of the loop antenna. The external module detects this frequency shift, providing

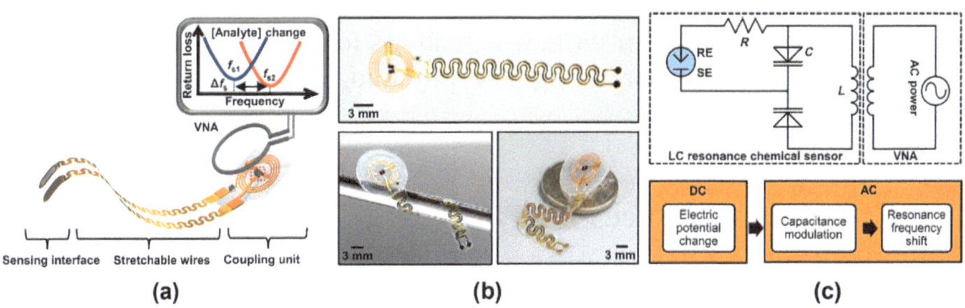

Figure 2.5 A bioelectronic implantable device proposed by Liu *et al.*[76] for biomarker sensing. (a) and (b) A stretchable, battery-free sensor comprises a sensing interface with a varactor diode connected to a coupling unit. (c) An equivalent circuit diagram and a flowchart describing the signal conversion and transmission process within the sensor system. Reproduced from ref. 76, https://doi.org/10.1126/sciadv.abo7049, under the terms of the CC BY 4.0 license, https://creativecommons.org/licenses/by/4.0/.

real-time information on biomarker levels. Moreover, this sensor topology can be tailored to detect a wide range of target biomarkers, such as aptamers or enzymes, by tuning the electronic circuit accordingly. This flexibility makes it a versatile tool for continuous health monitoring, offering significant potential in personalized medicine and chronic disease management.

Implantable bioelectronics also find applications in muscular and neurostimulation,[77] neuromodulation,[78] and drug delivery,[79] offering innovative treatments for various medical conditions. For instance, in one proposal, a split-ring resonator (SRR) made from a biocompatible titanium alloy is designed for neuromodulation with sub-millimetric spatial precision.[78] In this system, an external microwave radiation source excites the SRR, generating an electric field within the resonator's gap. This precisely targeted electric field stimulates specific brain regions, providing a potential treatment for conditions like epilepsy.

In another example, a microneedle implant has been proposed for treating inflammation, particularly in cases of rheumatoid arthritis.[79] This implant combines both electrical stimulation and drug delivery, allowing for more effective management of the condition. The microneedle system is controlled by a wearable smart device, enabling real-time adjustments to treatment protocols and providing a non-invasive, patient-friendly solution for chronic disease management.

These recent advancements demonstrate the potential of implantable bioelectronics for achieving more precise, adaptable, and multifunctional medical interventions, addressing complex conditions with improved precision and less invasive methods.

2.5 Ingestible Technologies

This section provides an overview of recent developments in the field of ingestible devices, with a focus on their primary applications. These devices are most commonly used in WCE, which allows for non-invasive imaging of the GI tract, and in monitoring key physiological parameters through sensing and biotelemetry.

2.5.1 Ingestible Devices for WCE

Ingestible devices form a critical part of WCE systems, with several recent designs specifically presented for this application.[3,83–85] One notable example is the wireless magnetically actuated capsule endoscope (MACE) introduced by Zhang *et al.*,[3] as shown in Figure 2.6(a). This system is part of an automated stomach monitoring setup and incorporates an EM actuation mechanism along with a wireless capsule that transmits real-time video. The MACE enables comprehensive mapping of the stomach by detecting and capturing specific gastric features. It features an LED-camera for image capture, three button batteries connected in series for power, and an AV

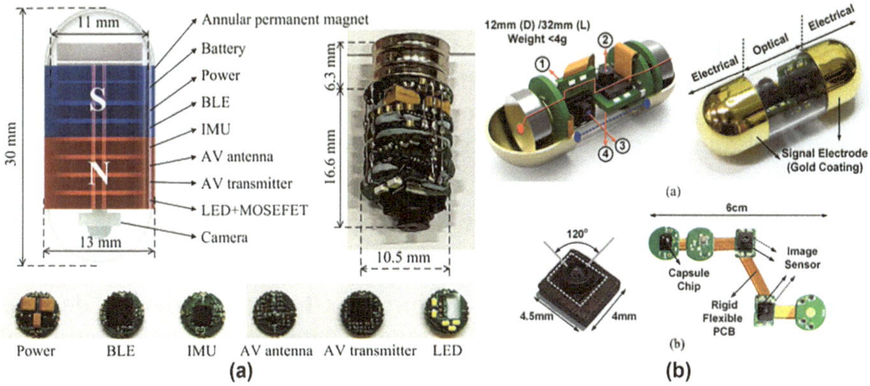

Figure 2.6 Examples of recent ingestible technologies developed for the capsule endoscopes. (a) MACE including an LED-camera for image capture, three button batteries connected in series for power, and an AV antenna for data transmission in an automated stomach monitoring setup presented by Zhang *et al.*[3] Reproduced from ref. 3 with permission from IEEE, Copyright 2024. (b) WCE device using four integrated cameras to capture images of the GI tract in 360 degrees designed by Jang *et al.*[84] Reproduced from ref. 84 with permission from IEEE, Copyright 2018.

antenna for data transmission, all contained within a 30 × Ø13 mm compact capsule.

Additionally, Zhou *et al.*[83] developed a WCE equipped with a programmable adaptive anti-retention module designed to minimize retention issues in the GI tract. This capsule integrates four modules, including a highly sensitive pressure sensor and four distinct trigger logic circuits, which detect internal pressure and prevent capsule retention when necessary. The developed pressure sensor is able to sense the changes in pressure in a wide interval, ranging from 0.39 to 1060 kPa. This broad sensing range helps mitigate the challenge posed by the large and unpredictable pressure differences experienced across individuals and different regions of the GI tract. The device includes lithium batteries to supply the power, and the components fit inside a 25 × Ø11.7 mm capsule.

Moreover, Jang *et al.*[84] presented a WCE device that can perform VGA resolution, low miss rate image capturing, and capsule location tracking inside the GI tract, as seen in Figure 2.6(b). The proposed device uses four integrated cameras to capture images of the GI tract in 360 degrees, significantly reducing the risk of missing abnormalities. Additionally, a transmitter embedded alongside the cameras allows data transmission at a rate of 80 Mb s^{-1} with less than 1 mW power consumption. The components are integrated into a 32 × Ø12 mm capsule.

Lastly, Leung *et al.*[85] proposed a therapeutic WCE designed for the treatment of GI hemorrhage. This capsule features a balloon that inflates through an endothermic acid–base reaction. Once the balloon reaches a predefined pressure level, it halts at the bleeding site in the bowel, achieving hemostasis *via* a balloon tamponade effect. When the capsule detects bleeding, it sends an alert to a transponder on the body, which in turn forwards the alert to a mobile phone *via* Bluetooth. Users can then activate the therapeutic module through a mobile interface to initiate the therapeutic module.

2.5.2 Ingestible Devices for Sensing and Biotelemetry

The second key application of ingestible devices is the sensing and monitoring of physiological conditions within the GI tract. The devices designed for this purpose can collect a wide range of data, such as temperature, pH levels, and pressure, which are transmitted wirelessly for biotelemetry purposes.[86–89] In that way, these devices enable continuous and real-time monitoring of the GI tract, offering valuable insights for diagnosing disorders and tracking disease progression. For instance, Hafezi *et al.*[86] developed an ingestible integrated-circuit microsensor designed for real-time, continuous monitoring of medication adherence. This sensor enables both patients and healthcare providers to track medication ingestion and adherence patterns, correlate pharmaceutical compliance with critical physiological metrics, and implement necessary interventions based on patient adherence trends and relevant health indicators.

Moreover, Xu *et al.*[87] proposed a wireless capsule system comprising a wireless capsule, a portable data recorder, and a workstation. The capsule is equipped with temperature, pH, and pressure sensors to monitor variations in these physiological parameters within the GI tract. Additionally, it incorporates an application-specific integrated circuit (ASIC) to facilitate the control and processing of the recorded data. All the components are integrated into a 26 × Ø11 mm capsule.

Furthermore, an ingestible device for sensing in the GI tract was developed by Kalantar-Zadeh *et al.*[88] This device detects various gases – specifically oxygen, hydrogen, and carbon dioxide – within the gut, monitoring changes in their concentrations. To accomplish this, it integrates thermal conductivity and semiconductor sensors, with heating elements calibrated to enhance selectivity and sensitivity to specific gases. The device has demonstrated accuracy and safety in monitoring dietary effects and holds potential as a diagnostic tool for GI health.

Nikolayev *et al.*[89] presented a wireless capsule platform operating at 434 MHz for biotelemetry in the GI tract. This platform features a novel antenna that is robust against the detuning throughout the GI tract. Moreover, it comprises a microcontroller with sensor readout capabilities and a matching circuit to transform the antenna impedance. The capsule dimensions are 17.7 × Ø8.9 mm, and its shell thickness is 0.65 mm. The authors indicated that the proposed platform may also be utilized in active implantable medical devices featuring therapeutic actuation and could be adapted for applications in non-medical fields where high robustness to the surrounding environment is required.

Cil *et al.*[90] studied the feasibility of using ingestible antennas as multiplexed biosensors to sense the GI tissues. For this purpose, they investigated impedance detuning throughout the GI tract for real-time localization of the capsule within the tract. This detuning is caused by the variations in the electromagnetic properties of the surrounding environment (changing GI tissues in this case). With this study, the authors showed that the variations in the antenna input impedance during travel in the GI tract can be used to locate the capsule in the tract, which can help diagnose specific GI diseases such as gastroparesis. Such multiplexed biosensor antennas can eliminate the need for complex sensor systems within the capsules.

Lastly, several key companies in the market are supplying ingestible devices to medical professionals for the diagnosis and management of various GI conditions. For instance, the PillCam SB3 capsule by Medtronic is a WCE device used for imaging the small intestine. CapsoVision's CapsoCam Plus capsule is widely utilized for visualizing the intestinal mucosa, while the NeoMom capsule, developed by Asept InMed, is designed for imaging the large intestine. Additionally, BodyCAP has developed the e-Celsius capsule for monitoring core body temperature in humans and the Anipill capsule for tracking core body temperature in animals. Comprehensive summaries of commercial ingestible capsules can be found in the literature.[91,92]

2.6 Injectable Technologies

One of the primary applications of injectable technologies is the health and physiological monitoring of animals. For example, Ahmmed *et al.*[29] designed a miniaturized, subcutaneously injectable multimodal device that can measure the heart and breathing rates and subcutaneous temperature of animals. For this purpose, the device features sensors with several modalities such as electrocardiography, accelerometry, and thermometry. Furthermore, Zhao *et al.*[93] reported an injectable device with a novel wireless applications specific integrated circuit that includes physiological sensing modalities such as body temperature monitoring and photoplethysmography. To monitor various physiological signs of animals, the device incorporates, besides the integrated circuit, two electrodes, a light-emitting diode, a photodetector pair, and a receiver coil.

As for the human applications, Chen *et al.*[94] proposed an ultralow-power injectable device for electrocardiography monitoring. This device is to be injected under the skin near the heart and designed primarily for atrial fibrillation arrhythmia monitoring. Moreover, Malik *et al.*[95] studied a novel approach that enables the development of passive injectable sensing devices. The devices designed with this approach have a thread-like shape with a diameter less than 1 mm and comprise a thin elongated body with two electrodes at their opposite ends and a simple electrical circuit. Zealear *et al.*[96] presented an injectable microstimulator device consisting of silicon and glass for reanimation of the paralyzed larynx.

2.7 Future Perspectives for Implantable and Ingestible Technologies

This section outlines various perspectives currently shaping the future of implantable and ingestible technologies. One of the primary perspectives in implantable and ingestible technologies lies in powering the devices. Many commercial devices still rely on bulky batteries consuming valuable space in the confined volume that could be used for additional sensors or actuators. Furthermore, batteries require periodic replacement, often through invasive surgeries, which increases patient risk and healthcare costs. Addressing this issue involves the development of alternative power strategies that can deliver reliable, long-term energy without the need for frequent maintenance. To this end, wireless power transfer and energy harvesting techniques are the focus of extensive research efforts.[97–99]

The second critical perspective involves material selection for implantable and ingestible technologies. Materials used in these devices must meet strict biocompatibility standards and perform well in terms of electrical, mechanical, and thermal properties. Enhanced biocompatibility of the materials used in these technologies can minimize the risk of adverse reactions or rejection of the device, ensuring safer integration within the human body. Moreover, these materials must be durable enough to withstand abrasion and other variations in the body's operating environment while maintaining their performance over time. Balancing these requirements is a significant hurdle for in-body devices. Advances in the materials for implantable and ingestible technologies focus on meticulously engineered materials, electrophysical, electrochemical, and electromagnetic systems that enable these devices with mechanical properties closely aligned to those of the target organ, thereby improving the functionality of in-body sensors and stimulators.[100,101] Furthermore, research into soft and deployable materials for future implantable and ingestible technologies promises enhanced biocompatibility and flexibility, which can reduce discomfort and lower the risk of tissue damage. Increasing use of soft materials in implantable and ingestible technologies is expected to improve the long-term stability and functionality of devices within dynamic biological environments.

Moreover, increasing the wireless connectivity of implantable and ingestible technologies is a substantial perspective for these devices. Wireless connectivity is essential for transmitting data collected by the implantable and ingestible devices to external monitoring systems or controlling the devices from outside the body. This capability allows for real-time monitoring and adjustments, which are crucial for patient safety and treatment efficacy. Increasing the wireless transmitting efficiency of implantable and ingestible technologies for ultra-low-power consumption is a vital research area, as it can lead to improved overall performance of these devices. By focusing on these advancements, future implantable and ingestible devices can be designed to be more sustainable and efficient, providing enhanced solutions for diagnostic and therapeutic applications within the human body.[102]

Lastly, the integration of artificial intelligence and machine learning into implantable and ingestible technologies presents a vital perspective for these devices. These advanced computational techniques could be used to process vast amounts of data collected from sensors in real time, allowing for more accurate monitoring of physiological parameters and timely identification of abnormalities in the organs. Furthermore, AI algorithms can facilitate predictive analytics, which can help healthcare providers notice potential health issues before they become critical. As research in this domain progresses, the integration of AI into biomedical engineering is expected to revolutionize the landscape of healthcare, paving the way for smarter and more efficient medical devices.[103,104]

2.8 Conclusions

Implantable and ingestible medical devices play a crucial role in current diagnostic and therapeutic applications within the human body. This chapter has delineated the fundamental concepts associated with these technologies, beginning with an overview of implantable and ingestible devices and their classifications. It provided insights into the general structure of these devices, elaborating on the functions of essential components necessary for their operation. The chapter then explained the challenges encountered in the design of such devices, such as the necessary biocompatibility of the devices, size and shape limitations, and user safety issues. Additionally, it included a review of recent innovations in implantable and ingestible devices currently employed in medical applications. Finally, the chapter discussed significant future perspectives for these technologies, underscoring the ongoing research endeavors aimed at advancing the field.

Acknowledgements

This study was supported by the French Agence Nationale de la Recherche (ANR) through the Project MedWave under Grant ANR-21-CE19-0045.

References

1. J. Y. Lucisano, T. L. Routh, J. T. Lin and D. A. Gough, *IEEE Trans. Biomed. Eng.*, 2017, **64**, 1982.
2. B. Bonaz, *Bioelectron. Med.*, 2024, **10**, 9.
3. H. Zhang, Y. Li, X. Shi, Y. Sun, J. Li and Y. Huang, *et al.*, *IEEE Trans. Med. Robot. Bionics*, 2024, **6**, 512.
4. N. A. White, S. L. Kroft, K. E. A. Bogt, T. J. C. O. Vrielink, C. Camenzuli and J. Calleja-Agius, *et al.*, *IEEE Trans. Biomed. Eng.*, 2024, **71**, 2379.
5. I. F. Akyildiz, A. Kak and S. Nie, *IEEE Access*, 2020, **8**, 133995.
6. N. Wu, R. Jiang, X. Wang, L. Yang, K. Zhang and W. Yi, *et al.*, *IEEE Commun. Mag.*, 2024, **62**, 144.

7. L. Berkelmann and D. Manteuffel, *IEEE Trans. Antennas Propag.*, 2022, **70**, 2593.
8. H. Cao, V. Leung, C. Chow and H. Chan, *IEEE Commun. Mag.*, 2009, **47**, 84.
9. M. Patel and J. Wang, *IEEE Wireless Commun.*, 2010, **17**, 80.
10. M. Mohamed, B. J. Maiseli, Y. Ai, K. Mkocha and A. Al-Saman, *IEEE Electromagn. Compat. Mag.*, 2021, **10**, 47.
11. B. Cornet, H. Fang, H. Ngo, E. W. Boyer and H. Wang, *IEEE Network*, 2022, **36**, 76.
12. S. K. Islam, A. Fathy, Y. Wang, M. Kuhn and M. Mahfouz, *IEEE Microwave Mag.*, 2014, **15**, S25.
13. D. P. Tobón, T. H. Falk and M. Maier, *IEEE Wireless Commun.*, 2013, **20**, 30.
14. A. Kiourti and K. S. Nikita, *IEEE Trans. Biomed. Eng.*, 2017, **64**, 1422.
15. A. K. Teshome, B. Kibret and D. T. H. Lai, *IEEE Rev. Biomed. Eng.*, 2019, **12**, 88.
16. M. Yin, D. A. Borton, J. Aceros, W. R. Patterson and A. V. Nurmikko, *IEEE Trans. Biomed. Circuits Syst.*, 2013, 7, 115.
17. U. Anwar, O. A. Ajijola, K. Shivkumar and D. Marković, *IEEE Trans. Biomed. Eng.*, 2022, **69**, 3074.
18. L. Dong, Z.-A. Qi, X.-F. Gao, L.-F. Wang and Q.-A. Huang, *IEEE Sens. J.*, 2024, **24**, 16774.
19. S. J. Bensmaia, D. J. Tyler and S. Micera, *Nat. Biomed. Eng.*, 2023, **7**, 443.
20. E. Losanno, M. Mender, C. Chestek, S. Shokur and S. Micera, *Nat. Rev. Bioeng.*, 2023, **1**, 390.
21. H. Jiang, Y. Guo, Z. Wu, C. Zhang, W. Jia and Z. Wang, *IEEE Trans. Biomed. Circuits Syst.*, 2018, **12**, 1076.
22. F. Del Bono, A. Bontempi, A. Dentis, N.-D. Trani, D. Demarchi and A. Grattoni, *et al.*, *IEEE Sens. J.*, 2024, **24**, 7345.
23. S. Fontana, S. D'agostino, A. Paffi, P. Marracino, M. Balucani and G. Ruocco, *et al.*, *IEEE Open J. Antennas Propag.*, 2024, **5**, 1282.
24. A. Kiourti, K. A. Psathas and K. S. Nikita, *Bioelectromagnetics*, 2014, **35**, 1.
25. C. Steiger, A. Abramson, P. Nadeau, A. P. Chandrakasan, R. Langer and G. Traverso, *Nat. Rev. Mater.*, 2019, **4**, 83.
26. R. Del-Rio-Ruiz, D. R. R. Silva, H. Suresh, H. Creasey, C. Asci and D. M. Santos, *et al.*, *Device*, 2024, **2**, 100406.
27. A. Abdigazy, M. Arfan, G. Lazzi, C. Sideris, A. Abramson and Y. Khan, *Nat. Electron.*, 2024, 7, 102.
28. Y. Sui, C. Hou, Y. Wang, K. L. Chung, C. Wei and Z. Li, *et al.*, *IEEE Antennas Wireless Propag. Lett*, 2024, **23**, 945.
29. P. Ahmmed, J. Reynolds and A. Bozkurt, *IEEE Sens. J.*, 2024, **24**, 11205.
30. J. M. Valero-Sarmiento, J. Reynolds, A. Krystal and A. Bozkurt, *IEEE Sens. J.*, 2018, **18**, 798.
31. X. Li, W. A. Serdijn, W. Zheng, Y. Tian and B. Zhang, *Trends Biotechnol.*, 2015, **33**, 388.
32. F. Cellesi and N. Tirelli, in *Injectable Biomaterials Science and Applications*, ed. B. Vernon, Woodhead Publishing, Cambridge, 2011, vol. 1, ch. 12, pp. 298–322.

33. K. Patch, *Nat. Biotechnol.*, 2021, **39**, 255.
34. D. De Ridder, M. A. Siddiqi, J. Dauwels, W. A. Serdijn and C. Strydis, *Neuromodulation*, 2024, **27**, 711.
35. X. Chen, X. Zhang, L. Zhang, X. Li, N. Qi and H. Jiang, *et al.*, *IEEE Trans. Biomed. Circuits Syst.*, 2009, **3**, 11.
36. C. McCaffrey, O. Chevalerias, C. O'Mathuna and K. Twomey, *IEEE Pervasive Comput.*, 2008, **7**, 23.
37. S. Bandyopadhyay, P. P. Mercier, A. C. Lysaght, K. M. Stankovic and A. P. Chandrakasan, *IEEE J. Solid-State Circuits*, 2014, **49**, 2812.
38. A. G. Fowler, S. O. R. Moheimani and S. Behrens, *J. Microelectromech. Syst.*, 2014, **23**, 1454.
39. M. Han, Q. Yuan, X. Sun and H. Zhang, *J. Microelectromech. Syst.*, 2014, **23**, 204.
40. J. Selvarathinam and A. Anpalagan, *IEEE Potentials*, 2016, **35**, 6.
41. X. Duan, Y. Chen, M. Li, Y. Rong, R. Xie and J. Han, *IEEE Trans. Biomed. Circuits Syst.*, 2023, **17**, 633.
42. F. Xu, G. Yan, K. Zhao, L. Lu, J. Gao and G. Liu, *IEEE Trans. Biomed. Circuits Syst.*, 2014, **8**, 871.
43. B. Chi, J. Yao, S. Han, X. Xie, G. Li and Z. Wang, *IEEE Trans. Biomed. Eng.*, 2007, **54**, 1291.
44. B. Chi, J. Yao, P. Chiang and Z. Wang, *IEEE Trans. Circuits Syst. II: Express Briefs*, 2009, **56**, 275.
45. S. P. Mohanty and E. Kougianos, *IEEE Potentials*, 2006, **25**, 35.
46. A. Kiourti and K. S. Nikita, *IEEE Antennas Propag. Mag.*, 2012, **54**, 210.
47. A. Basir, Y. Cho, I. A. Shah, S. Hayat, S. Ullah and M. Zada, *et al.*, *IEEE Antennas Propag. Mag.*, 2023, **65**, 70.
48. M. J. Christoe, N. Phaoseree, J. Han, A. Michael, S. Atakaramians and K. Kalantar-Zadeh, *IEEE Access*, 2021, **9**, 91874.
49. L. Mertz, *IEEE Pulse*, 2013, **4**, 16.
50. R. Das, F. Moradi and H. Heidari, *IEEE Trans. Biomed. Circuits Syst.*, 2020, **4**, 343.
51. A. Pandey and S. Sahoo, *IEEE Sens. J.*, 2023, **23**, 31171.
52. M. Langenmair, M. Hofert, M. Eickenscheidt and T. Stieglitz, *IEEE Trans. Compon. Packag. Manuf. Technol.*, 2021, **11**, 2046.
53. D. Nikolayev, M. Zhadobov, L. Le Coq, P. Karban and R. Sauleau, *IEEE Trans. Antennas Propag.*, 2017, **65**, 6107.
54. D. da Silva, M. Kaduri, M. Poley, O. Adir, N. Krinsky and J. Shainsky-Roitman, *et al.*, *J. Chem. Eng.*, 2018, **340**, 9.
55. K. Shen and M. M. Maharbiz, *IEEE Trans. Biomed. Eng.*, 2020, **67**, 223.
56. T.-F. Chien, C.-M. Cheng, H.-C. Yang, J.-W. Jiang and C.-H. Luo, *IEEE Antennas Wirel. Propag. Lett.*, 2010, **9**, 599.
57. D. H. Shin, K. W. Seong, H. H. Nakajima, S. Puria and J.-H. Cho, *IEEE Access*, 2020, **8**, 137947.
58. A. Qusba, A. K. RamRakhyani, J.-H. So, G. J. Hayes, M. D. Dickey and G. Lazzi, *IEEE Sens. J.*, 2014, **14**, 1074.
59. S. Lacour, G. Courtine and J. Guck, *Nat. Rev. Mater.*, 2016, **1**, 16063.

60. Z. Zongman, Z. Zhongtai, Z. Pengcheng, Z. Yunfan, Y. Jiawei and H. Hossam, *et al.*, *ACS Nano*, 2023, **17**, 17634.
61. G. Schiavone, F. Fallegger, X. Kang, B. Barra, N. Vachicouras and E. Roussinova, *et. al.*, *Adv. Mater.*, 2020, **32**, 1906512.
62. M. K. Magill, G. A. Conway and W. G. Scanlon, *IEEE Trans. Antennas Propag.*, 2017, **65**, 4406.
63. N. A. Malik, P. Sant, T. Ajmal and M. Ur-Rehman, *IEEE J. Electromag. RF Microwave Med. Biol.*, 2021, **5**, 84.
64. IEEE standard for safety levels with respect to human exposure to electric, magnetic, and electromagnetic Fields, 0 Hz to 300 GHz, IEEE Std. C95.1-2019, 2019, 1.
65. J. J. Smith and J. A. Henderson, *IEEE Sens. J.*, 2008, **8**, 52.
66. D. Trilla, C. Hernandez, J. Abella and F. J. Cazorla, *IEEE Trans. Sustainable Comput.*, 2021, **6**, 522.
67. A. Kim, M. Ochoa, R. Rahimi and B. Ziaie, *IEEE Access*, 2015, **3**, 89.
68. A. K. Skrivervik, M. Bosiljevac and Z. Sipus, *IEEE Trans. Antennas Propag.*, 2019, **67**, 4978.
69. H. Chu, P.-J. Wang, X.-H. Zhu and H. Hong, *IEEE Access*, 2019, **7**, 35.
70. U. F. Tesler, *A History of Cardiac Surgery: An Adventurous Voyage from Antiquity to the Artificial Heart*, Cambridge Scholars Publishing, Newcastle upon Tyne, 2020, vol. 1.
71. S. Wang, Q. Cui, P. Abiri, M. Roustaei, E. Zhu and Y.-R. Li, *et al.*, *Sci. Adv.*, 2023, **9**, eadj0540.
72. C. Tang, Z. Liu, Q. Hu, Z. Jiang, M. Zheng and C. Xiong, *et al.*, *Small*, 2024, **20**, 2304752.
73. W. S. Sheridan, F. Wetterling, J. M. Testani, B. A. Borlaug, M. Fudim and K. Damman, *et al.*, *Eur. J. Heart Fail.*, 2023, **25**, 754.
74. F. Wang, Y. Xue, X. Chen, P. Zhang, L. Shan and Q. Duan, *et al.*, *Adv. Funct. Mater.*, 2024, **34**, 2314471.
75. J. Ausra, M. Madrid, R. T. Yin, J. Hanna, S. Arnott and J. A. Brennan, *et al.*, *Sci. Adv.*, 2022, **8**, eabq7469.
76. T.-L. Liu, Y. Dong, S. Chen, J. Zhou, Z. Ma and J. Li, *Sci. Adv.*, 2022, **8**, eabo7049.
77. A. Burton, Z. Wang, D. Song, S. Tran, J. Hanna and D. Ahmad, *et al.*, *Nat. Commun.*, 2023, **14**, 7887.
78. C. Marar, Y. Jiang, Y. Li, L. Lan, N. Zheng and G. Chen, *et al.*, *Sci. Adv.*, 2024, **10**, eado5560.
79. Y. Liu, W. Xie, Z. Tang, Z. Tan, Y. He and J. Luo, *et al.*, *Sci. Adv.*, 2024, **10**, eadj0604.
80. M. Mariello, J. D. Rosenthal, F. Cecchetti, M. Gao, A. K. Skrivervik and Y. Leterrier, *et al.*, *Nat. Commun.*, 2024, **15**, 7443.
81. J. Kim, M. I. Bury, K. Kwon, J.-Y. Yoo, N. V. Halstead and H.-S. Shin, *et al.*, *Proc. Natl. Acad. Sci. U.S.A.*, 2024, **121**, e2400868121.
82. C. Zhang, C. Pan, K. F. Chan, J. Gao, Z. Yang and K. K. C. Leung, *et al.*, *Sci. Adv.*, 2023, **9**, eade8622.

83. C. Zhou, W. Wang, J. Jiang, W. Wang, N. Tang and Y. Liu, *et al.*, *IEEE Sens. J.*, 2024, **24**, 167.
84. J. Jang, J. Lee, K.-R. Lee, J. Lee, M. Kim and Y. Lee, *et al.*, *IEEE J. Solid-State Circuits*, 2019, **54**, 538.
85. B. H. K. Leung, C. C. Y. Poon, R. Zhang, Y. Zheng, C. K. W. Chan and P. W. Y. Chiu, *et al.*, *IEEE Trans. Biomed. Eng.*, 2017, **64**, 1106.
86. H. Hafezi, T. L. Robertson, G. D. Moon, K.-Y. Au-Yeung, M. J. Zdeblick and G. M. Savage, *IEEE Trans. Biomed. Eng.*, 2015, **62**, 99.
87. F. Xu, G. Yan, K. Zhao, L. Lu, J. Gao and G. Liu, *IEEE Trans. Biomed. Circuits Syst.*, 2014, **8**, 871.
88. K. Kalantar-Zadeh, K. J. Berean, N. Ha, A. F. Chrimes, K. Xu and D. Grando, *et al.*, *Nat. Electron.*, 2018, **1**, 79.
89. D. Nikolayev, M. Zhadobov and R. Sauleau, *IEEE Trans. Biomed. Circuits Syst.*, 2019, **13**, 403.
90. E. Cil, I. V. Soares, D. Renaudeau, R. Lucas, S. Dumanli and R. Sauleau, *et al.*, *IEEE Trans. Antennas Propag*, 2023, **71**, 1977.
91. P. A. Thwaites, C. K. Yao, E. P. Halmos, J. G. Muir, R. E. Burgell and K. J. Berean, *et al.*, *Aliment. Pharmacol. Ther.*, 2024, **59**, 459.
92. S. V. Hoffmann, J. P. O'Shea, P. Galvin, V. Jannin and B. T. Griffin, *Eur. J. Pharm. Sci.*, 2024, **203**, 106911.
93. L. Zhao, R. G. Stephany, Y. Han, P. Ahmmed, T.-P. Huang and A. Bozkurt, *et al.*, *IEEE Trans. Biomed. Circuits Syst.*, 2024, **18**, 1037.
94. Y.-P. Chen, D. Jeon, Y. Lee, Y. Kim, Z. Foo and I. Lee, *et al.*, *IEEE J. Solid-State Circuits*, 2015, **50**, 375.
95. S. Malik, Q. Castellvi, L. Becerra-Fajardo, M. Tudela-Pi, A. Garcia-Moreno and M. S. Baghini, *et al.*, *IEEE Trans. Biomed. Circuits Syst.*, 2020, **14**, 867.
96. D. L. Zealear, K. C. Garren, R. J. Rogriguez, J. H. Reyes, S. Huang and M. R. Dokmeci, *et al.*, *IEEE Trans. Biomed. Eng.*, 2001, **48**, 890.
97. G. L. Barbruni, P. M. Ros, D. Demarchi, S. Carrara and D. Ghezzi, *IEEE Trans. Biomed. Circuits Syst.*, 2020, **14**, 1160.
98. I. V. Soares, M. Gao, E. Cil, Z. Sipus, A. K. Skrivervik and J. S. Ho, *et al.*, *IEEE Trans. Microwave Theory Tech.*, 2023, **71**, 2680.
99. I. V. Soares, Z. Sipus, A. K. Skrivervik, R. Sauleau, A. Al'u and J. S. Ho, *et al.*, *IEEE Antennas Propag. Mag.*, 2025, **67**, 74.
100. A. Bhatia, J. Hanna, T. Stuart, K. A. Kasper, D. M. Clausen and P. Gutruf, *Chem. Rev.*, 2024, **124**, 2205.
101. Henry Royce Institute, Materials for bioelectronics in healthcare: Strategy and action plan.
102. A. Mohan and N. Kumar, *BioMed. Eng. Online*, 2024, **23**, 87.
103. P. A. Catherwood, D. D. Finlay and J. A. D. McLaughlin, *IEEE Tech. Soc. Mag.*, 2016, **35**, 73.
104. S. J. M. Yazdi, H.-J. Park, C.-S. Son and J.-H. Lee, *IEEE Access*, 2021, **9**, 14.

CHAPTER 3

Energy Solutions for Implantable and Ingestible Electronics

A. ABDIGAZY[a,b] AND Y. KHAN*[a,b]

[a] Department of Electrical and Computer Engineering, University of Southern California, Los Angeles, CA 90089, USA; [b] The Institute for Technology and Medical Systems (ITEMS), Keck School of Medicine, University of Southern California, Los Angeles, CA 90033, USA
*Email: yasser.khan@usc.edu

3.1 Introduction

Implantable and ingestible medical devices provide a convenient and reliable way to diagnose, treat, and monitor a wide range of health conditions in real time. Implantable technologies, such as pacemakers, glucose monitoring sensors, neurostimulators, and drug delivery systems, are designed to be placed inside the body, providing continuous therapeutic monitoring and necessary interventions. Ingestibles, on the other hand, equipped with cameras, sensors, and actuators traverse the gastrointestinal (GI) tract, enabling non-invasive diagnostics, and delivering medication directly to the target site of the GI tract. Both technologies can monitor vital signs, detect abnormalities, apply therapeutic actions, and provide valuable insights into patients' health, paving the way for personalized treatment plans. Over the past decade, this field has advanced significantly, with numerous startups and several FDA-approved devices emerging on the market.[1–3] As of 2024, the global market for implantable and ingestible devices was valued at approximately $93 billion and is expected to exceed $142 billion by 2030.[4–6]

Detection Science Series No. 33
Energy Materials for Ingestible and Implantable Healthcare Devices
Edited by Santanu Patra and Anja Boisen

Published by the Royal Society of Chemistry, www.rsc.org

Pacemakers, along with advancements in neural and gastric stimulation implants, are poised to be major drivers of future growth. In the ingestible devices sector, alongside capsule endoscopy, targeted drug delivery systems are gaining traction and are set to become a key component of the emerging smart pill market.

Depending on the technology and application, the average power consumption of these devices ranges from nanowatts to tens or hundreds of milliwatts (see Table 3.1).[6–8] Existing implantable and ingestible devices employ various powering methods: conventional lithium-ion or silver oxide, flexible, or transient batteries for reliable power supply; acoustic and radiofrequency (RF) power transfer for relatively high-power but battery-free devices; thermal energy harvesting and optical power transfer methods for low-power near-skin implants; and local energy harvesting using galvanic cells, piezoelectric, and triboelectric transducers for ultra-low-power devices.

This chapter examines a range of implantable and ingestible technologies, with a focus on their power requirements and sources. It offers an overview of their operational principles, emphasizing the specific power levels each device requires, and describes various powering methods used to sustain the effective operation of these devices.

3.2 Powering Methods for Implantable and Ingestible Electronics

The power source is one of the most crucial elements determining the operational life of the implantable or ingestible device. Key considerations when selecting a power source include its output power, capacity, size, and safety. As shown in Table 3.1, a wide variety of implantable and ingestible electronic devices exist, with power consumption ranging from nanowatts (nW) to tens of milliwatts (mW), depending on the application. Batteries remain the most common power source for several reasons. First, some sensitive devices such as those involved in therapeutic interventions like electrical stimulation require reliable and stable power. Second, implementing wireless power transfer can be challenging, especially for devices located deep within the body, where signals are significantly attenuated by surrounding tissues. Additionally, dynamic motion, such as in the case of ingestible pills, makes continuous power delivery difficult due to the constant change of the device's location and orientation. When the application allows, a radiofrequency (RF) and acoustic power transfer are increasingly used, particularly for devices positioned closer to the skin or in static locations, where power can be delivered more easily. However, these methods have limitations, including the potential for tissue heating and damage if excessive power is transferred. Power transfer *via* thermal and optical modalities is also emerging as a potential method for near-skin implantable devices. Local energy harvesting through chemical and mechanical means is being actively explored as it enables battery-free, self-sustaining operation.

Table 3.1 Summary of existing implantable and ingestible devices, their power consumption range, and power source employed for their operation.

Device type	Description	Power range and power source	Some examples of pre-clinically or clinically tested, and commercial devices
Retinal implants	• Device that electrically stimulates neurons in the retina and partially restores sight to people who have lost their photoreceptors due to retinal diseases	• Sub-mW to few mW depending on the number of pixels and stimulation frequency[9–11] • RF power transfer and battery are the most common power sources	• IMIE 256 by Golden Eye Bionic and IntelliMicro Medical • EPI-RET 3 by RWTH Aachen University • NR600 System by Nano Retina • Argus II by Second Sight Medical Products, and more[12]
Cochlear implants	• Device that electrically stimulates the auditory nerve to restore hearing in people with profound hearing loss	• A few to tens of mW depending on the device type, number of channels and stimulation frequency[13–16] • RF power transfer and battery are the most common power sources	• Nucleus, Osia, and Baha Systems by Cochlear • Naida CI M and Sky CI Marvel implants by Advanced Bionics • Sonata, Synchrony, and Concerto implants by MED-EL
Pacemakers	• Device implanted in the chest to regulate a slow heart rate by sending electrical impulses to the heart	• A few hundreds of μW to mW[17,18] • Battery is the most common power source • RF power transfer and mechanical energy harvesting methods are being explored	• Micra, Advisa, and Azure pacemakers by Medtronic • Accolade, Essentio, Ingenio, and Vitalio pacemakers by Boston Scientific • Aveir, Assurity, and Endurity pacemakers by Abbott
Bladder implants	• Device that sends mild electrical pulses to the sacral nerves to help treat bladder control problems, such as overactive bladder and urinary retention	• A few hundreds of μW to several mW[19–22] • Battery and RF power transfer are the most common power sources	• InterStim System by Metronic • eCoin implant by Valencia Technologies • F15 and R20 implants from Axonics

Table 3.1 (*Continued*)

Device type	Description	Power range and power source	Some examples of pre-clinically or clinically tested, and commercial devices
Drug delivery implants	• Device that stores and delivers small, precise doses of therapeutic drugs or medicines into the bloodstream or to specific tissue sites	• A few to tens of mW[23–26] • Battery is the most common power source • RF and acoustic power transfer, and optical excitation are being explored	• Synchromed and MiniMed implants by Medtronic • Prometra implants by Flowonix, and more[27]
Glucose sensing implants	• Device that is inserted under the skin to continuously monitor glucose levels for several months	• A few to hundreds of μW[28–31] • RF power transfer is the most common power source • Acoustic power transfer is being explored	• Eversense E3 CGM implant by Eversense • Subcutaneously implantable device by UNIST[32]
Neural implants	• Device implanted into the brain that records neural activity and stimulates various brain regions	• A few to tens of mW depending on the number of channels[33–36] • Battery is the most common power source • RF and acoustic power transfer, and optical methods are being explored	• N1 implant by Neuralink • Brain–Machine Interfaces by Axoft, Synchron, Blackrock Neurotech, BrainGate, and Precision Neuroscience
Capsule endoscopy	• Ingestible device with multiple cameras used to monitor and diagnose the GI tract in real time	• A few to tens of mW depending on the image acquisition and communication frequency[6,37,38] • Batteries remain the only reliable power source	• PillCam by Medtronic • NaviCam by AnX Robotica • MicroCam by Intro-Medic • EndoCapsule EC-10 by Olympus Medical
Ingestible pills	• Ingestible device with various sensors and actuators used to diagnose and treat GI disorders in real time	• A few μW to tens of mW depending on the application[6,39–42] • Battery is the most common power source	• RaniPill drug delivery system by Rani Therapeutics • Atmo Gas Capsule by Atmo Biosciences

Table 3.1 (*Continued*)

Device type	Description	Power range and power source	Some examples of pre-clinically or clinically tested, and commercial devices
		• RF and acoustic power transfer for devices with power consumption up to hundreds of μW • Mechanical and chemical energy harvesting for ultra-low-power devices in sub-μW to tens of μW range	• SmartPill for pH, temperature, and pressure sensing by Medtronics • HemoPill for acute bleeding detection by Ovesco • Vibrant Capsule for chronic constipation treatment by Vibrant, and more[6]

However, these approaches are typically limited to ultra-low-power devices, operating in the sub-μW to tens of μW range.

Figure 3.1 offers a summary and comparison of the existing powering methods for implantable and ingestible devices. The following subsections provide a more in-depth analysis of these techniques.

3.2.1 Energy Harvesting

Energy harvesting is one way of powering implantable and ingestible devices. For instance, flexible piezoelectric materials can be used to sense peristaltic movements of the GI tract or expansion of the chest cavity due to breathing, and to convert that mechanical energy into electric power (see Figure 3.1g).[51] Triboelectric energy harvesting is another mechanical method that is used to harvest energy to power the device (see Figure 3.1h).[42,52] Recent *in vivo* experiments of the self-powered intracardiac pacemaker demonstrated that an open circuit voltage and short circuit current of about 6 V and 0.2 μA can be achieved, which paves the way for future development of these devices.[49] Galvanic cells, a type of electrochemical energy harvesting, have also been used to power ingestible capsules using gastric acid as an electrolyte,[53] including clinically approved Proteus Digital[54] (Mg–CuCl couple) and EtectRx[55] (MgCl–AgCl couple) medication adherence capsules. Also, nutrients such as glucose[40] and microorganisms in the gut[56,57] can be exploited for sub-mW-level power generation (see Figure 3.1i). Thermal energy harvesting is another method that utilizes thermoelectric materials to detect the temperature difference across it, and convert it to energy to power the device. This technique is a subject of ongoing research which needs further optimization of thermoelectric

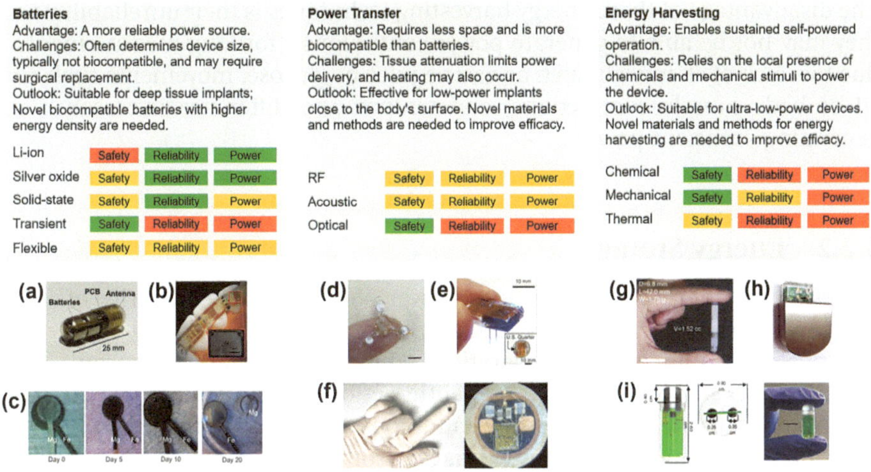

Figure 3.1 Summary of powering methods and existing technologies. (a) Battery-powered ingestible pill for monitoring vital signs. Reproduced from ref. 43 with permission from Elsevier, Copyright 2023. (b) Flexible and biocompatible solid-state micro-battery for implantable orthodontic system. Reproduced from ref. 44, https://doi.org/10.1038/s41528-017-0008-7, under the terms of the CC BY 4.0 license, http://creativecommons.org/licenses/by/4.0/. (c) Biodegradable magnesium/iron battery for transient implantable devices. Reproduced from ref. 45, https://doi.org/10.1038/micronano.2015.24, under the terms of the CC BY 4.0 license, http://creativecommons.org/licenses/by/4.0/. (d) Ultrasonically powered, wireless battery-free eye modulation patch for high myopia therapy. Reproduced from ref. 46, https://doi.org/10.1038/s41467-024-46049-6, under the terms of the CC BY 4.0 license, http://creativecommons.org/licenses/by/4.0/. (e) RF-powered, wirelessly rechargeable, fully implantable, soft optoelectronic system for optogenetics. Reproduced from ref. 47, https://doi.org/10.1038/s41467-020-20803-y, under the terms of the CC BY 4.0 license, http://creativecommons.org/licenses/by/4.0/. (f) RF-powered, implantable, wireless, battery-free system for tactile pressure sensing. Reproduced from ref. 48, https://doi.org/10.1038/s41378-023-00602-3, under the terms of the CC BY 4.0 license, http://creativecommons.org/licenses/by/4.0/. (g) Mechanically self-powered intracardiac pacemaker. Reproduced from ref. 49, https://doi.org/10.1038/s41467-023-44510-6, under the terms of the CC BY 4.0 license, http://creativecommons.org/licenses/by/4.0/. (h) Mechanically self-rechargeable cardiac pacemaker system. Reproduced from ref. 50, https://doi.org/10.1038/s41467-021-24417-w, under the terms of the CC BY 4.0 license, http://creativecommons.org/licenses/by/4.0/. (i) Galvanically self-powered ingestible wireless biosensing system for real-time *in situ* monitoring of gastrointestinal tract metabolites. Reproduced from ref. 40, https://doi.org/10.1038/s41467-022-35074-y, under the terms of the CC BY 4.0 license, http://creativecommons.org/licenses/by/4.0/.

materials, and is suitable only for ultra-low-power devices operating in the μW or sub-μW range.[58] The transducers for these energy harvesting methods are typically custom designed and optimized for the target device specifications.

The disadvantage of these energy harvesting techniques is their unreliability as they may not be able to generate power continuously for long measurements due to a lack of chemicals such as gastric juice or glucose, movements such as GI or body muscle contractions, and temperature difference necessary for generating sufficient power.

3.2.2 Energy Storage

Batteries are the most widely used power sources as the majority of implantable and ingestible devices require a continuous and stable supply of electrical power for reliable operation. This is particularly important for critical devices that perform therapeutic actions such as electrical stimulation and drug delivery. Capsule endoscopy pills are another example as they must continuously transmit images from the GI tract to an external reader. However, batteries tend to occupy significant space and often determine the overall size of the device. This can be particularly problematic for ingestibles because a larger pill size increases the risk of its retention in the GI tract. There are various types of batteries available for implantable and ingestible devices: some are commercially available off-the-shelf, while others are still under research and development. Standard lithium-ion batteries offer high electrical capacity and can output up to several hundred mW of power (see Figure 3.1a). However, they are vulnerable to issues such as thermal runaway, explosions, and toxic electrolyte leakage.[59] Silver-oxide batteries, which are not prone to thermal runaway but still contain toxic metals and caustic electrolytes, are preferred as an on-board power source and the only type of battery approved for clinical use to power capsule endoscopes.[60–62] They typically generate up to 100 mW of power, which, while less than standard lithium-ion batteries, is generally sufficient for most applications. Solid-state batteries are also attractive due to their reduced risk of thermal runaway, flammability, and leakages (see Figure 3.1b).[63,64] Transient batteries made from biodegradable materials may eliminate the risks associated with toxicity and minimize GI obstruction through partial or complete decomposition, which also makes them attractive for implants and ingestibles designed to self-degrade over time (see Figure 3.1c). However, they have much lower capacity than traditional batteries, typically thousands of times smaller, which could only be suitable for devices operating in the μW-level power range.[65–67] Similarly, flexible batteries made from biocompatible materials provide another option to minimize the device's volume.[44,68] Implantable and ingestible devices can also benefit from new technologies such as fully 3D-printed batteries made from disposable papers[69] and gel-based micro-batteries consisting of organic and organometallic molecules.[70] Most regular coin cells, and some solid-state and flexible batteries, can be purchased off-the-shelf. However, transient, and most solid-state and flexible batteries remain part of ongoing research studies.

3.2.3 Power Transfer

Remote powering of implantable and ingestible devices is also possible using acoustic,[10,71–73] RF,[74,75] and optical[76] power transfer. Due to the low propagation loss of sound waves, the first method can deliver several mWs of power with the implant placed under 6 cm of animal tissue. Typically, acoustic power transfer relies on transmitting an ultrasonic wave at sub- or MHz-level frequency, which is picked up by a piezoelectric material on the device and further converted to a stable voltage by power management circuits (see Figure 3.1d). Similarly, RF power transfer is implemented by transmitting electromagnetic waves at one of the industrial, scientific, and medical (ISM) frequency bands, which is detected by coils on the implant or ingestible device, and converted to a stable direct current (DC) voltage (see Figure 3.1e and f). *In vivo* experiments in pigs showed the feasibility of this method delivering hundreds of μW to the implanted device. These numbers may be affected by frequency-dependent path loss, antenna size, and orientation of the device in the body. When utilizing this power delivery method, it is essential to ensure that the specific absorption rate (SAR) values are lower than the standard provided by the Institute of Electrical and Electronics Engineers (IEEE) (the standard limit for 1-g tissue is 1.6 $W\,kg^{-1}$ for an input power of 1 W).[77] Optical power transfer is an emerging technology used in wearable and implantable devices.[76,78] It uses an LED as a light source, and miniature solar cells or photodiodes on the device as optical receivers. It has been demonstrated that a few hundred of μW can be delivered to a device positioned beneath 3 mm of chicken skin.[79] This method remains as the least studied power delivery technique and is highly sensitive to misalignments between the external light source and the on-board photodiodes. Typically, the piezoelectric materials used in acoustic power transfer, LEDs, and photodiodes employed in optical power transfer, and circuits that convert an AC signal from the external transmitter to the stable DC voltage rail are available off-the-shelf, and they can be integrated into an implantable or ingestible device.

3.3 Summary

This chapter provides an overview of existing implantable and ingestible devices, focusing on the various powering methods used to sustain their operation. The goal of this chapter is not to delve deeply into each method, but rather to offer readers a broad understanding of the available approaches. Some methods such as utilizing batteries and certain applications of wireless power transfer are well-established with certain limitations. Others, such as optical power transfer and most of the energy harvesting techniques, are still the subjects of ongoing research. Generally, energy harvesting is the most appealing method for powering implantable and ingestible devices, as it offers a seamless and safe solution that is not constrained by batteries or external power delivery systems. However, due to its inherently unstable nature of relying on the presence of local chemicals or mechanical stimuli, this approach remains largely limited to specific applications such as pacemakers, where continuous heartbeats and breathing can drive piezoelectric or triboelectric generators, or ingestible devices designed for short-term monitoring of biomarkers using chemical modality for self-powering. If a more reliable and long-term powering solution is required for a near-skin device, wireless power transfer serves as a middle ground that enables consistent energy delivery. However, to ensure safe operation, it is essential that the device is designed to operate with minimal power consumption, thereby allowing lower power transfer and reducing heat generation. To further mitigate heating concerns, further development of acoustic and optical power transfer methods may offer safer alternatives. For devices located deeper within the body, the development of transient batteries becomes critical as they may provide a safe and reliable means of powering the implant over extended periods of time. More detailed descriptions of these methods will be provided in the following chapters of this book.

References

1. Y. U. Cho, S. L. Lim, J. H. Hong and K. J. Yu, Transparent neural implantable devices: a comprehensive review of challenges and progress, *npj Flexible Electron.*, 2022, **6**(1), 1–18.
2. Front-matter, in *Implantable Electronic Medical Devices*, D. Fitzpatrick, ed. Academic Press, Oxford, 2015, pp. i–iii.
3. S. Mukherjee, S. Suleman, R. Pilloton, J. Narang and K. Rani, State of the Art in Smart Portable, Wearable, Ingestible and Implantable Devices for Health Status Monitoring and Disease Management, *Sensors*, 2022, **22**(11), 4228.
4. Implantable Medical Devices Market Size Report, 2030, 2024. Available from: https://www.grandviewresearch.com/industry-analysis/implantable-medical-devices-market-report.
5. Smart Pills Market Size, Share, Trends, Growth Report, 2030, 2024. Available from: https://www.grandviewresearch.com/industry-analysis/smart-pills-market.
6. A. Abdigazy, M. Arfan, G. Lazzi, C. Sideris, A. Abramson and Y. Khan, End-to-end design of ingestible electronics, *Nat. Electron.*, 2024, **7**(2), 102–118.
7. S. Y. Yang, V. Sencadas, S. S. You, N. Z. X. Jia, S. S. Srinivasan and H. W. Huang, *et al.*, Powering Implantable and Ingestible Electronics, *Adv. Funct. Mater.*, 2021, **31**(44), 2009289.
8. S. Roy, A. N. M. W. Azad, S. Baidya, M. K. Alam and F. Khan, Powering Solutions for Biomedical Sensors and Implants Inside the Human Body: A Comprehensive Review on Energy Harvesting Units, Energy Storage, and Wireless Power Transfer Techniques, *IEEE Trans. Power Electron.*, 2022, **37**(10), 12237–12263.

9. T. Yousefi, G. Zoidl and H. Kassiri, 33.5 Closed-Loop 100-Channel Highly-Scalable Retinal Implant with 1.02μW Analog ED-Based Adaptive-Threshold Spike Detection and Poisson-Coded Temporally Distributed Optogenetic Stimulation, in 2024 IEEE International Solid-State Circuits Conference (ISSCC), 2024, pp. 550–552.
10. J. H. Park, J. S. Y. Tan, H. Wu and J. Yoo, 34.2 1225-Channel Localized Temperature-Regulated Neuromorphic Retinal-Prosthesis SoC with 56.3nW/Channel Image Processor, in 2020 IEEE International Solid-State Circuits Conference - (ISSCC), 2020, pp. 508–510.
11. K. Eom, M. Park, H. S. Lee, S. B. Ku, N. Kim, S. Cha, *et al.*, 32.2 A Stimulus-Scattering-Free Pixel-Sharing Sub-Retinal Prosthesis SoC with 35.8dB Dynamic Range Time-Based Photodiode Sensing and Per-Pixel Dynamic Voltage Scaling, in 2023 IEEE International Solid-State Circuits Conference (ISSCC), 2023, pp. 480–482.
12. Overview of Retinal Prosthesis and Future Directions, PentaVision, 2024. Available from: https://retinalphysician.com/issues/2024/may/overview-of-retinal-prosthesis-and-future-directions/.
13. Y. Park, S. T. Koh, J. Lee, H. Kim, J. Choi, S. Ha, *et al.*, 33.7 A Frequency-Splitting-Based Wireless Power and Data Transfer IC for Neural Prostheses with Simultaneous 115mW Power and 2.5Mb/s Forward Data Delivery, in 2021 IEEE International Solid-State Circuits Conference (ISSCC), 2021, pp. 472–474.
14. M. Yip, R. Jin, H. H. Nakajima, K. M. Stankovic and A. P. Chandrakasan, 18.2 A fully-implantable cochlear implant SoC with piezoelectric middle-ear sensor and energy-efficient stimulation in 0.18μm HVCMOS, in 2014 IEEE International Solid-State Circuits Conference Digest of Technical Papers (ISSCC), 2014, pp. 312–313.
15. H. Külah, H. Uluşan, S. Chamanian, A. Batu, M. B. Uğur, M. B. Yüksel, *et al.*, A Fully-Implantable Mems-Based Autonomous Cochlear Implant, in 2022 IEEE 35th International Conference on Micro Electro Mechanical Systems Conference (MEMS), 2022, pp. 396–399.
16. H. Uluşan, M. B. Yüksel, Ö Topçu, H. A. Yiğit, A. M. Yılmaz and M. Doğan, *et al.*, A full-custom fully implantable cochlear implant system validated *in vivo* with an animal model, *Commun. Eng.*, 2024, **3**(1), 1–11.
17. F. V. Y. Tjong and V. Y. Reddy, Permanent Leadless Cardiac Pacemaker Therapy, *Circulation*, 2017, **135**(15), 1458–1470.
18. C. Moerke, A. Wolff, H. Ince, J. Ortak and A. Öner, New strategies for energy supply of cardiac implantable devices, *Herzschrittmachertherapie Elektrophysiologie*, 2022, **33**(2), 224–231.
19. InterStim Therapy for Bladder Control | SSM Health, 2024. Available from: https://www.ssmhealth.com/services/urology/interstim-therapy-for-bladder-control-problems.
20. S. J. A. Majerus, S. L. Garverick, M. A. Suster, P. C. Fletter and M. S. Damaser, Wireless, Ultra-Low-Power Implantable Sensor for Chronic Bladder Pressure Monitoring, *ACM J. Emerg. Technol. Comput. Syst.*, 2012, **8**(2), 11.

21. R. Karam, S. J. A. Majerus, D. J. Bourbeau, M. S. Damaser and S. Bhunia, Tunable and Lightweight On-Chip Event Detection for Implantable Bladder Pressure Monitoring Devices, *IEEE Trans. Biomed. Circuits Syst.*, 2017, **11**(6), 1303–1312.
22. Neuspera Medical | Implantable Medical Device Technology | OAB, Neuspera, 2024. Available from: https://neuspera.com/.
23. D. Fitzpatrick, Chapter 14 - Implantable Drug Delivery Systems, in *Implantable Electronic Medical Devices*, ed. D. Fitzpatrick, Academic Press, Oxford, 2015, pp. 139–157.
24. M. L. Wang, P. Yeon, C. F. Chamberlayne, M. Mofidfar, H. Xu, J. P. Annes, *et al.* A Wireless Implantable Potentiostat for Programmable Electrochemical Drug Delivery, in 2021 IEEE Biomedical Circuits and Systems Conference (BioCAS), 2021, pp. 1–4.
25. H. Joo, Y. Lee, J. Kim, J. S. Yoo, S. Yoo and S. Kim, *et al.*, Soft implantable drug delivery device integrated wirelessly with wearable devices to treat fatal seizures, *Sci. Adv.*, 2021, 7(1), eabd4639.
26. Y. Zhang, F. Liu, Y. Zhang, J. Wang, D. D'Andrea and J. B. Walters, *et al.*, Self-powered, light-controlled, bioresorbable platforms for programmed drug delivery, *Proc. Natl. Acad. Sci. U. S. A.*, 2023, **120**(11), e2217734120.
27. A. Fayzullin, A. Bakulina, K. Mikaelyan, A. Shekhter and A. Guller, Implantable Drug Delivery Systems and Foreign Body Reaction: Traversing the Current Clinical Landscape, *Bioengineering*, 2021, **8**(12), 205.
28. M. Mujeeb-U-Rahman, M. H. Nazari, M. Sencan and W. V. Antwerp, A Novel Needle-Injectable Millimeter scale Wireless Electrochemical Glucose Sensing Platform for Artificial Pancreas Applications, *Sci. Rep.*, 2019, **9**(1), 17421.
29. H. Basaeri, D. Christensen, S. Roundy, Y. Yu, T. Nguyen, P. Tathireddy, *et al.* Ultrasonically powered hydrogel-based wireless implantable glucose sensor, in 2016 IEEE SENSORS, 2016, pp. 1–3.
30. H. Lee, H. Park, T. Kim, M. S. Nam, Y. J. Heo and S. Kim, A mm-sized implantable glucose sensor using a fluorescent hydrogel, in 2023 IEEE Radio Frequency Integrated Circuits Symposium (RFIC), 2023, pp. 101–104.
31. J. Malik, S. Kim, J. M. Seo, Y. M. Cho and F. Bien, Minimally Invasive Implant Type Electromagnetic Biosensor for Continuous Glucose Monitoring System: In Vivo Evaluation, *IEEE Trans. Biomed. Eng.*, 2023, **70**(3), 1000–1011.
32. S. Kim, J. Malik, J. M. Seo, Y. M. Cho and F. Bien, Subcutaneously implantable electromagnetic biosensor system for continuous glucose monitoring, *Sci. Rep.*, 2022, **12**(1), 17395.
33. S. Miziev, W. A. Pawlak and N. Howard, Comparative analysis of energy transfer mechanisms for neural implants, *Front. Neurosci.*, 2024, **17**, 1320441.
34. C. Lee, B. Kim, J. Kim, S. Lee, T. Jeon and W. Choi, *et al.*, A Miniaturized Wireless Neural Implant With Body-Coupled Power Delivery and Data Transmission, *IEEE J. Solid-State Circuits*, 2022, **57**(11), 3212–3227.
35. S. R. Nason, A. K. Vaskov, M. S. Willsey, E. J. Welle, H. An and P. P. Vu, *et al.*, A low-power band of neuronal spiking activity dominated by local

single units improves the performance of brain–machine interfaces, *Nat. Biomed. Eng.*, 2020, **4**(10), 973–983.
36. B. Chatterjee, M. Nath, G. Kumar K, S. Xiao, K. Jayant and S. Sen, Biphasic quasistatic brain communication for energy-efficient wireless neural implants, *Nat. Electron.*, 2023, **6**(9), 703–716.
37. A. Sahafi, Y. Wang, C. Rasmussen, P. Bollen, G. Baatrup and V. Blanes-Vidal, *et al.*, Edge artificial intelligence wireless video capsule endoscopy, *Sci. Rep.*, 2022, **12**(1), 13723.
38. TinyML for Gastroscopic Image Processing. Available from: https://docs.edgeimpulse.com/experts/prototype-and-concept-projects/tinyml-gastroscopic-image-processing.
39. M. Inda-Webb, M. Jimenez, Q. Liu, N. Phan, J. Ahn and C. Steiger, *et al.*, Sub-1.4 cm3 capsule for detecting labile inflammatory biomarkers *in situ*, *Nature*, 2023, **620**(7973), 386–392.
40. E. De la Paz, N. H. Maganti, A. Trifonov, I. Jeerapan, K. Mahato and L. Yin, *et al.*, A self-powered ingestible wireless biosensing system for real-time *in situ* monitoring of gastrointestinal tract metabolites, *Nat. Commun.*, 2022, **13**(1), 7405.
41. C. Dagdeviren, F. Javid, P. Joe, T. von Erlach, T. Bensel and Z. Wei, *et al.*, Flexible piezoelectric devices for gastrointestinal motility sensing, *Nat. Biomed. Eng.*, 2017, **1**, 807–817.
42. G. Yao, L. Kang, J. Li, Y. Long, H. Wei and C. A. Ferreira, *et al.*, Effective weight control *via* an implanted self-powered vagus nerve stimulation device, *Nat. Commun.*, 2018, **9**(1), 1–10.
43. G. Traverso, V. Finomore, J. Mahoney, J. Kupec, R. Stansbury and D. Bacher, *et al.*, First-in-human trial of an ingestible vitals-monitoring pill, *Device*, 2023, **1**(5), 100125.
44. A. T. Kutbee, R. R. Bahabry, K. O. Alamoudi, M. T. Ghoneim, M. D. Cordero and A. S. Almuslem, *et al.*, Flexible and biocompatible high-performance solid-state micro-battery for implantable orthodontic system, *npj Flexible Electron.*, 2017, **1**(1), 1–8.
45. M. Tsang, A. Armutlulu, A. W. Martinez, S. A. B. Allen and M. G. Allen, Biodegradable magnesium/iron batteries with polycaprolactone encapsulation: A microfabricated power source for transient implantable devices, *Microsyst. Nanoeng.*, 2015, **1**(1), 1–10.
46. T. Zhong, H. Yi, J. Gou, J. Li, M. Liu and X. Gao, *et al.*, A wireless battery-free eye modulation patch for high myopia therapy, *Nat. Commun.*, 2024, **15**(1), 1766.
47. C. Y. Kim, M. J. Ku, R. Qazi, H. J. Nam, J. W. Park and K. S. Nam, *et al.*, Soft subdermal implant capable of wireless battery charging and programmable controls for applications in optogenetics, *Nat. Commun.*, 2021, **12**(1), 535.
48. L. Du, H. Hao, Y. Ding, A. Gabros, T. C. E. Mier and J. Van der Spiegel, *et al.*, An implantable, wireless, battery-free system for tactile pressure sensing, *Microsyst. Nanoeng.*, 2023, **9**(1), 1–12.

49. Z. Liu, Y. Hu, X. Qu, Y. Liu, S. Cheng and Z. Zhang, *et al.*, A self-powered intracardiac pacemaker in swine model, *Nat. Commun.*, 2024, **15**(1), 507.
50. H. Ryu, H. M. Park, M. K. Kim, B. Kim, H. S. Myoung and T. Y. Kim, *et al.*, Self-rechargeable cardiac pacemaker system with triboelectric nanogenerators, *Nat. Commun.*, 2021, **12**(1), 4374.
51. C. Dagdeviren, F. Javid, P. Joe, T. von Erlach, T. Bensel and Z. Wei, *et al.*, Flexible piezoelectric devices for gastrointestinal motility sensing, *Nat. Biomed. Eng.*, 2017, **1**(10), 807–817.
52. L. Liu, S. Towfighian and Z. Jin, A cylindrical triboelectric energy harvester for capsule endoscopes, in 2015 IEEE Biomedical Circuits and Systems Conference (BioCAS), 2015, pp. 1–4.
53. S. Yoshida, H. Miyaguchi and T. Nakamura, Development of Ingestible Thermometer With Built-in Coil Antenna Charged by Gastric Acid Battery and Demonstration of Long-Time in Vivo Telemetry, *IEEE Access*, 2021, **9**, 102368–102377.
54. H. Hafezi, T. L. Robertson, G. D. Moon, K. Y. Au-Yeung, M. J. Zdeblick and G. M. Savage, An Ingestible Sensor for Measuring Medication Adherence, *IEEE Trans. Biomed. Eng.*, 2015, **62**(1), 99–109.
55. G. P. Flores, B. Peace, T. C. Carnes, S. L. Baumgartner, D. E. Buffkin Jr and N. R. Euliano, *et al.*, Performance, reliability, usability, and safety of the ID-Cap system for ingestion event monitoring in healthy volunteers: a pilot study, *Innovations Clin. Neurosci.*, 2016, **13**(9–10), 12.
56. K. Dong, B. Jia, C. Yu, W. Dong, F. Du and H. Liu, Microbial fuel cell as power supply for implantable medical devices: A novel configuration design for simulating colonic environment, *Biosens. Bioelectron.*, 2013, **41**, 916–919.
57. Y. Han, C. Yu and H. Liu, A microbial fuel cell as power supply for implantable medical devices, *Biosens. Bioelectron.*, 2010, **25**(9), 2156–2160.
58. W. Gao, Y. Wang and F. Lai, Thermoelectric energy harvesting for personalized healthcare, *Smart Med.*, 2022, **1**(1), e20220016.
59. Y. Chen, Y. Kang, Y. Zhao, L. Wang, J. Liu and Y. Li, *et al.*, A review of lithium-ion battery safety concerns: The issues, strategies, and testing standards, *J. Energy Chem.*, 2021, **59**, 83–99.
60. S. Le, L. Zhang, X. Song, S. He, Z. Yuan and F. Liu, *et al.*, status of zinc-silver battery, *J. Electrochem. Soc.*, 2019, **166**(13), A2980.
61. G. Ciuti, A. Menciassi and P. Dario, Capsule Endoscopy: From Current Achievements to Open Challenges, *IEEE Rev. Biomed. Eng.*, 2011, **4**, 59–72.
62. A. Abdigazy, M. Arfan, J. Shao, M. S. Islam, M. F. Hassan and Y. Khan, 3D gas mapping in the gut with AI-enabled ingestible and wearable electronics, *Cell Rep. Phys. Sci.*, 2024, **5**(7), 101990.
63. C. Li, Z. Y. Wang, Z. J. He, Y. J. Li, J. Mao and K. H. Dai, *et al.*, An advance review of solid-state battery: Challenges, progress and prospects, *Sustainable Mater. Technol.*, 2021, **29**, e00297.
64. A. Abdigazy, M. S. Islam, M. Arfan, M. F. Hassan, H. Hashemi and Y. Khan, A Self-Orienting Single-Chip Ingestible Pill for Electrochemical

Sensing in the GI Tract, in 2024 IEEE Biomedical Circuits and Systems (BioCAS). Xi'an, IEEE, China, 2024.
65. P. Kumar, E. Di Mauro, S. Zhang, A. Pezzella, F. Soavi and C. Santato, *et al.*, Melanin-based flexible supercapacitors, *J. Mater. Chem. C*, 2016, **4**(40), 9516–9525.
66. N. Mittal, A. Ojanguren, M. Niederberger and E. Lizundia, Degradation behavior, biocompatibility, electrochemical performance, and circularity potential of transient batteries, *Adv. Sci.*, 2021, **8**(12), 2004814.
67. Y. Chen, R. Jamshidi, K. White, S. Çınar, E. Gallegos and N. Hashemi, *et al.*, Physical–chemical hybrid transiency: A fully transient li-ion battery based on insoluble active materials, *J. Polym. Sci., Part B: Polym. Phys.*, 2016, **54**(20), 2021–2027.
68. N. Li, Z. Chen, W. Ren, F. Li and H. M. Cheng, Flexible graphene-based lithium ion batteries with ultrafast charge and discharge rates, *Proc. Natl. Acad. Sci. U. S. A.*, 2012, **109**(43), 17360–17365.
69. X. Aeby, A. Poulin, G. Siqueira, M. K. Hausmann and G. Nyström, Fully 3D printed and disposable paper supercapacitors, *Adv. Mater.*, 2021, **33**(26), 2101328.
70. F. N. Crespilho, G. C. Sedenho, D. De Porcellinis, E. Kerr, S. Granados-Focil and R. G. Gordon, *et al.*, Non-corrosive, low-toxicity gel-based microbattery from organic and organometallic molecules, *J. Mater. Chem. A*, 2019, **7**(43), 24784–24787.
71. T. C. Chang, M. L. Wang, J. Charthad, M. J. Weber and A. Arbabian, 27.7 A 30.5mm3 fully packaged implantable device with duplex ultrasonic data and power links achieving 95kb/s with <10−4 BER at 8.5cm depth, in 2017 IEEE International Solid-State Circuits Conference (ISSCC), 2017, pp. 460–461.
72. Y. Zhang and K. L. Shepard, A 0.6-mm2 Powering and Data Telemetry System Compatible with Ultrasound B-Mode Imaging for Freely Moving Biomedical Sensor Systems, in 2019 IEEE Custom Integrated Circuits Conference (CICC), 2019, pp. 1–4.
73. L. Zhao, M. Annayev, Ö Oralkan and Y. Jia, An Ultrasonic Energy Harvesting IC Providing Adjustable Bias Voltage for Pre-Charged CMUT, *IEEE Trans. Biomed. Circuits Syst.*, 2022, **16**(5), 842–851.
74. J. Koo, S. B. Kim, Y. S. Choi, Z. Xie, A. J. Bandodkar and J. Khalifeh, *et al.*, Wirelessly controlled, bioresorbable drug delivery device with active valves that exploit electrochemically triggered crevice corrosion, *Sci. Adv.*, 2020, **6**(35), eabb1093.
75. A. Abid, J. M. O'Brien, T. Bensel, C. Cleveland, L. Booth and B. R. Smith, *et al.*, Wireless power transfer to millimeter-sized gastrointestinal electronics validated in a swine model, *Sci. Rep.*, 2017, **7**(1), 1–6.
76. D. H. Nguyen, Optical Wireless Power Transfer for Implanted and Wearable Devices, *Sustainability*, 2023, **15**(10), 8146.
77. A. Iqbal, P. R. Sura, M. Al-Hasan, I. B. Mabrouk and T. A. Denidni, Wireless power transfer system for deep-implanted biomedical devices, *Sci. Rep.*, 2022, **12**(1), 13689.

78. S. R. Khan, S. K. Pavuluri, G. Cummins and M. P. Y. Desmulliez, Wireless Power Transfer Techniques for Implantable Medical Devices: A Review, *Sensors*, 2020, **20**(12), 3487.
79. M. Mujeeb-U-Rahman, D. Adalian, C. F. Chang and A. Scherer, Optical power transfer and communication methods for wireless implantable sensing platforms, *J. Biomed. Opt.*, 2015, **20**(9), 095012.

CHAPTER 4

Characteristics of Materials for Biomedical Energy Transfer

NASIM GOLAFSHAN*[a,b] AND MARYAM KAMAEI[c]

[a] Department of Health Technology, Technical University of Denmark, 2800, Kgs. Lyngby, Denmark; [b] The Danish National Research Foundation and Villum Foundation's Center for Intelligent Drug Delivery and Sensing Using Microcontainers and Nanomechanics (IDUN), Department of Health Technology, Technical University of Denmark, Ørsted Plads, 2800, Kgs. Lyngby, Denmark; [c] Nanotechnology and Advanced Materials Department, Materials & Energy Research Center (MERC), Karaj, Iran
*Email: Golafshan.na@gmail.com

4.1 Introduction

Materials are the main component of energy-transferring systems, so their selection is among the most critical aspects in designing systems.[1,2] Depending on the energy sources, such as mechanical, electrochemical, or thermal energy, the design and performance of the devices change based on the use of various materials. For example, piezoelectric and triboelectric materials are used in mechanical energy-transferring systems, while electrochemical energy storage relies on materials capable of reversible redox reactions. Also, thermal energy harvesting may involve thermoelectric materials or materials with high thermal conductivity.[3–5] The design of the device is directly linked to the selection of materials, whether it is a flexible sensor or a rigid implant.[6,7]

In biomedical energy systems, even more restrictions are applied due to the toxicity and ion leaching of materials in the human body.[6,7] The component materials must not be toxic and have no significant impact on the

Detection Science Series No. 33
Energy Materials for Ingestible and Implantable Healthcare Devices
Edited by Santanu Patra and Anja Boisen

Published by the Royal Society of Chemistry, www.rsc.org

immune systems in the body. Biocompatibility is the ability of materials to interact with biological systems without adverse effects, which is the most crucial feature of biomedical energy systems. Certain ions and their high concentrations can be toxic, and their interference with biological processes leads to inflammation, immune reactions, or other adverse effects.[8,9] Moreover, the stability of the materials during their operational lifetime within the body is another parameter to consider. Materials that degrade or corrode over time can release harmful byproducts, compromising biocompatibility and device functionality.[10]

Mechanical properties play an important role in energy systems' design, fabrication, and efficiency.[10] The materials should withstand the stresses applied to them during the system's operation. The specific mechanical properties and parameters depend highly on the intended applications.[11] For example, in load-bearing applications, the materials should be high in strength and resilience,[12] while for flexible applications, they should be super flexible and fatigue-resistant.[13]

We next discuss the main features of the energy systems, focusing on biocompatibility and mechanical properties. Following this, the characterization of the devices is described in terms of biocompatibility, miniaturization, and energy efficiency.

4.2 Main Features of Materials Used in Biomedical Energy Transfer Devices

Recent studies emphasize innovative materials for efficient energy transfer devices to strengthen functionality and reliability in various biomedical applications, such as targeted drug delivery,[14] cancer treatment,[15] diagnostics,[16] implants,[17] and tissue engineering.[18] These materials include nanomaterials and bio-inspired systems, which are selected to maximize biocompatibility and electrochemical stability under physiological conditions.[16] Biocompatible materials used in energy transfer devices improve targeting in therapy and reduce side effects, especially in applications like cancer treatment, thereby enhancing patient outcomes.[19] In the following, we explore the main features of materials used in biomedical energy transfer devices, divided into chemical and mechanical properties.

4.2.1 Chemical Properties of Energy-transferring Systems

With the increasing demand for energy-transferring devices in biomedical applications, the chemical properties of materials effectively play a critical role in their performance. Advanced materials are selected, evaluated, and applied based on their sufficient chemical stability and non-toxicity under physiological conditions.[20–23] Their interaction with biological components like biofluids and the cellular environment highlights the importance of their biocompatibility, biodegradation, and controlling ion release as key factors in their design. The following sections explore the combination of biocompatibility, chemical stability, and degradation behavior of materials used in biomedical energy transfer systems.

Biocompatibility is defined as the inert interaction of materials with biological systems without toxicity, which is crucial for integrating materials into medical devices.[24,25] To be more specific, biocompatible materials used in biomedical energy transfers neither release harmful ions nor degrade into toxic substances.[26] Various biocompatible materials can be integrated into biomedical energy systems, such as metallic ones (gold, titanium),[27,28] bioceramics (barium titanate, potassium sodium niobate),[29–31] and polymeric materials (cellulose, polyvinylidene fluoride).[32,33] For example, in mechanical energy harvesters such as piezoelectric and triboelectric generators, polyvinylidene fluoride (PVDF) has been extensively studied due to its biocompatibility.[34] Figure 4.1A shows that one study demonstrated biocompatible electrospun nanofibers used in triboelectric nanogenerators for wearable sensing systems by combining PVDF, carbon, and polyurethane (PU) nanofibers. This combination is attached to an arm as electronic skin.[35] Another research developed a flexible supercapacitor (SC) cell with biocompatible oxidized single-walled carbon nanotubes (SWCNTs) in body fluids for implantable electronic medical devices.[36] In Figure 4.1B, a schematic of this sensor and its implantation into mice is shown. The results demonstrated the biocompatibility of the sensor for use as an implantable device.[36]

Another approach in materials used for biomedical energy devices is related to a suitable combination of chemical stability and biodegradation. The stability of the materials used in biomedical energy transfer devices plays a critical role in ensuring optimal performance, as these materials are in contact with the human body. However, balancing controlled stability and appropriate biodegradation is essential to accelerate functionality. Thus, these materials must degrade at a controlled rate to maintain their integrity during the device's lifespan while preventing long-term adverse effects in biological applications.[37] Most recent research in biomedical energy transfer devices focuses on common degradable materials, including polylactic acid (PLA),[38] polylactic acid glycolic acid copolymer (PLGA),[39] polyvinyl alcohol (PVA),[40] and polycaprolactone (PCL).[41] Also, controlling the degradation rate of materials is necessary to improve the efficiency of energy-harvesting systems in physiological environments.

Addressing these issues involves using innovative materials to improve controlled biodegradation mechanisms to ensure timely and safe absorption.[38] In one study, the degradation rate of different materials was suggested by innovative layer-by-layer composites used in bioresorbable triboelectric nanogenerators (TENGs).[42] Developing more efficient fabrication techniques and innovative composites can help overcome these limitations and enhance the capability of materials used in these devices. Regarding efficient biodegradation, a study developed implantable TENGs using surface-microstructured carbonized cellulose films (M-CCFs), as

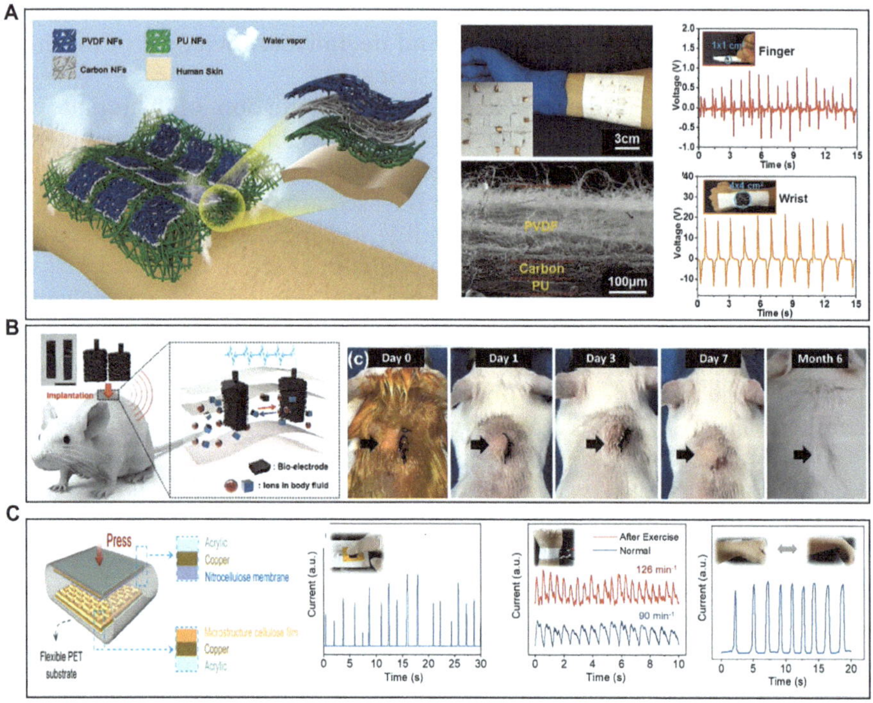

Figure 4.1 Chemical properties of materials used in energy transfer devices. (A) Schematic illustration and photograph of electronic skin. The SEM image shows a cross-section of the electronic skin and the voltage output of the electronic skin when it is attached to the finger and wrist. Reproduced from ref. 35 with permission from John Wiley & Sons, Copyright 2019 Wiley-VCH Verlag GmbH & Co. KGaA, Weinheim. (B) Schematic of the implantation of the flexible electrode materials into mice to show its biocompatibility and stability over 6 months. Reproduced from ref. 36 with permission from the Royal Society of Chemistry. (C) Illustration of biocompatible cellulose-based TENG with current output generated by finger and wrist movement. Reproduced from ref. 43 with permission from Elsevier, Copyright 2022.

shown in Figure 4.1C.[43] The biocompatibility and biodegradation of these materials make them suitable for implantable energy-harvesting devices.[44]

4.2.2 Mechanical Properties of Energy-transferring Systems

With the acceleration of the demand for high-performance biomedical energy systems, the mechanical properties of advanced materials play a key role in ensuring the structural integrity of the device under physiological stresses and cyclic deformation. While chemical features often get primary attention, mechanical integrity becomes another substantial element in energy device efficacy. In biomedical energy devices, materials are subjected to repetitive mechanical stresses due to cyclic loading induced by volume changes during loading and unloading cycles.[45] Additionally, body movements and deformations, as other contributing factors, are major factors in developing flexible biomedical devices as a sufficient solution to enhance durability and adaptability.[46]

Although ceramics-based materials are among the advanced approaches in energy transfer systems, their brittle nature limits their applications under dynamic applications. There are challenges like brittleness in ceramic generators, which reduce biomedical energy transfer effectiveness.[47] Addressing these issues paves the way for developing the mechanical capability to withstand cyclic loading without failure in biomedical energy devices. For instance, stretchable organic solar cells (OSCs) with various concentrations of polymer acceptors as additives have been designed to experience repetitive mechanical deformation with strong mechanical resilience in biomedical energy devices (Figure 4.2A).[48] Accordingly, the effect of the additional polymer acceptor on the tensile test was analyzed. The results show the intrinsically stretchable OSCs+20 wt% P_{VH} maintained 84% of their initial performance after 100 stretching–releasing cycles at 15% strain. Thus, compared to other reported intrinsically stretchable OSCs, the mechanical flexibility and durability OSCs incorporated with 20 wt% of P_{VH} are exceptional.[48] These characteristics make them well-suited for sustained energy harvesting under mechanical loading and they are highly recommended for mechanical flexibility applications such as dynamic wearable devices.

Another example of ceramic-based materials in energy systems is traditional piezoelectric nanogenerators, often characterized by their inherent brittleness under mechanical loading.[49] In response to this challenge, stretchable piezoelectric devices have been designed with materials such as hybrid nanocomposite fibers composed from poly(vinylidene fluoride) (PVDF), reduced graphene oxide (rGO), and barium-titanium oxide (BTO), which exhibit high stretchability that can endure mechanical deformation while maintaining functionality.[50] Moreover, polymeric materials like poly(vinylidene fluoride-*co*-trifluoro ethylene) (PVDF-TrFE) and their composites with ceramics are used due to their mechanical properties for integrating energy-transferring systems. For example, recent research focusing on PVDF-TrFE-$BaTiO_3$ piezoelectric nanofibers exhibits improved crystallinity by adding $BaTiO_3$, enhancing sensitivity. It confirms that PVDF-TrFE suits flexible, sensitive piezoelectric devices in wearable applications.[51,52]

Polymers like polyurethane (PU) are commonly used due to their excellent elasticity and toughness. For example, the elongation at break of PU was reported to be 1114.6% and the toughness as 126.4 MJ m^{-3}, confirming its ability to withstand deformation without failure.[53] In recent research, composite designs have been changed to develop flexibility in a composite-based piezoelectric generator. Accordingly, the optimum content of PMN-0.3PT in $0.7Pb(Mg_{1/3}Nb_{2/3}O_3)$-$0.3PbTiO_3$ (PMN-0.3PT) incorporated into polydimethylsiloxane (PDMS) based 2–2 has been investigated, and two different reinforcement configurations have been examined. The results confirm superior mechanical flexibility attributed to 0.2-PMN-0.3PT/PDMS 2-2 composite with parallel connectivity.[54]

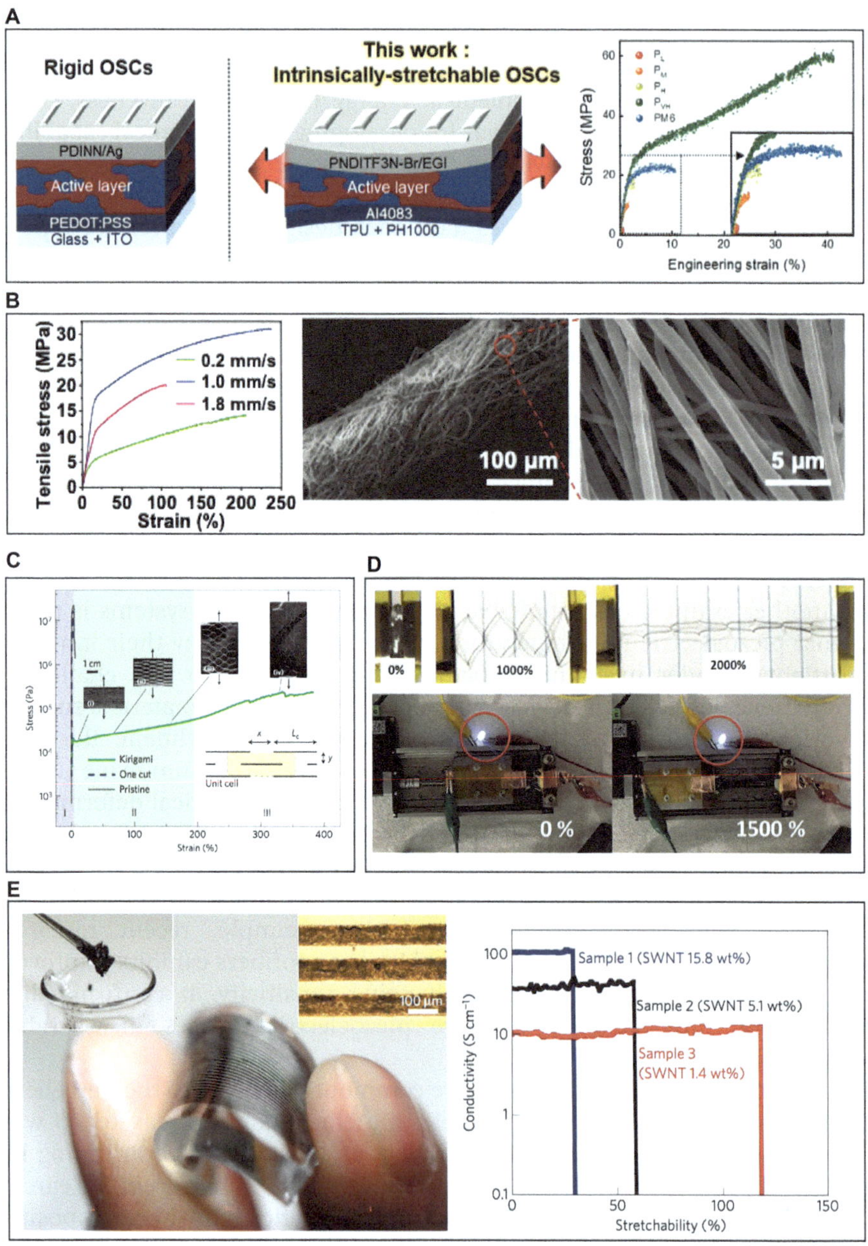

To improve the tensile strain in piezoelectric systems, recent research has been developed on a piezoelectric nanoyarn using PVDF. The strength of nanoyarn PVDF is 313 MPa, and the tensile strength is 46 MPa, as shown in Figure 4.2B.[55] Research on enhancing both capacitance and mechanical strength demonstrated that structural supercapacitors by load bearing and energy storage functions promise to expand energy storage.[56] To improve pseudocapacitive capacitance and tensile strength, a carbon-fiber coating was used to reinforce structural electrodes.[56] The mechanical properties of the structural supercapacitor with the gradient electrolyte were compared to two other structures, one laminate was bonded by epoxy only (maximum mechanical strength), and the other was with the PEO electrolyte (low structural rigidity). The tensile and flexural stress have shown that the mechanical property of the structural supercapacitor is between that of the two other groups.[56]

There is another mechanical feature to consider before materials selection and designing for energy transfer systems. Friction, another crucial factor for developing and operating biomedical energy-harvesting systems, can occur between moving parts. While high friction usually can lead to energy loss, reduced efficiency, and increased heat generation, it plays a critical role in generating electricity in triboelectric nanogenerators.[60] In frictional energy harvesting, as energy is harvested through contact electrification and materials are subjected to repeated frictional forces, materials selection, contact pressure, and friction direction can modify energy transfer efficiency.[61] In regard to using human skin as a friction material in energy harvesting and storage systems (EHSS), a recent study employed a combination of novel electronics and skin-EHSS. This system is used as an energy harvester through contact with objects and a sensor for detecting the electrostatic properties of clothing.[62]

Figure 4.2 Mechanical properties of materials used in energy transfer devices. (A) Molecular structure of materials used to make OSC stretchable. The tensile stress–strain diagram shows the mechanical properties of pristine film. Reproduced from ref. 48 with permission from John Wiley & Sons, Copyright 2022 Wiley-VCH GmbH. (B) Stress–strain curve of nanoyarn PVDF and SEM images of nanoyarn. Reproduced from ref. 55 with permission from John Wiley & Sons, Copyright 2022 Wiley-VCH GmbH. (C) The stress–strain diagram for a sheet with a kirigami pattern, single notch, and a non-patterned sheet. Reproduced from ref. 57, under the terms of the CC BY 4.0 license. (D) Schematic of stretching up to 2000% strain of the kirigami conductor and lighting up a bulb by the conductor under 0% and 1500% tensile strain. Reproduced from ref. 58 with permission from American Chemical Society, Copyright 2019. (E) Conductivity of the SWCNT conductors as a function of tensile strain. Reproduced from ref. 59 with permission from Springer Nature, Copyright 2009.

Mechanical abrasion caused by friction originating from solid–solid interactions in TENGs causes instability. Also, it deteriorates output voltage at high frequencies due to heat generation in the sliding motion of layers.[63] To overcome this obstacle, a biomimetic composite film inspired by snake scales synthesized by PVA and keratin on triboelectric nanogenerators (TENGs) indicates improvement in mechanical properties and wear resistance of the contact layer. Also, the tooth patterns of the film not only reduce friction but also improve charge transfer.[63]

Considering other mechanical properties in energy devices is essential. As often these devices operate under fluctuating temperatures and dynamic mechanical loads, the material's fracture toughness—capability to withstand crack propagation—is critical in these environments. Double network hydrogels like PAM/CA/$EtOH_{30}$ demonstrate remarkable stretchability and toughness for use as electrodes in TENG applications.[64] About stretchability, another study focusing on recent developments in creating stretchable conductors demonstrated that stable conductivity under significant strain is essential for flexible energy devices and is extremely affected by synthesized methods. Also, to improve stretchability, exceeding 10-fold and even 20-fold stretchable structures can be designed based on origami/kirigami patterns. The stress–strain function of a kirigami pattern is shown in Figure 4.2C.[57] Also, in Figure 4.2D, the results of lighting up a bulb in a kirigami structure can be seen.[58] In this work, two-dimensional transition-metal dichalcogenide (2D TMD), because of its limitations in mechanical properties, is converted to 2D TMD-based kirigami electrical conductors by kirigami patterning. This developed structure is synthesized by integrating 2D platinum diselenide (2D $PtSe_2$) layers into a flexible polyimide (PI) substate. Finally, 2D $PtSe_2$/PI-based kirigami patterns show high stretchability of ~2000%. Moreover, Figure 4.2E shows a conductor composed of SWCNTs, ionic liquid (IL), and fluorinated copolymer on polydimethylsiloxane (PDMS) sheets, and its conductivity as a tensile function.[59] These results are optimized by the jet milling process, for example, which can change the structure of the conductor into net patterns leading to uniform dispersion. This process led to an elastic conductor with 102 $S\,cm^{-1}$ conductivity and 29% stretchability.[59]

So far, the main characterization of materials used energy-transferring systems has been discussed. However, the energy materials need to be integrated into electronical components to make the device applicable. In the following section, we will discuss the characterization of the energy-harvesting device as a unique unit for energy-transferring systems.

4.3 Characterization of Energy Transfer Devices for Biomedical Applications

Energy transfer devices as a kind of medical device need to fulfill the requirements for their applications in the healthcare system. Besides the properties and specific characteristics that the components of the devices should have to be proven in biomedical applications (Section 4.1), there are some features that need to be considered before developing an energy transfer device for biomedical applications. The characterization of these devices will ensure their biocompatibility, efficiency, and safety for the human body.[65,66] Alongside that, an energy transfer device needs to be optimized performance for specific medical needs. Therefore, the characterization and design of energy transfer devices require strict standards in order to address clinical challenges. The performance of these devices is linked to their chemical and mechanical properties.[67] These properties highlight their interaction with the human body and their efficiency to carry out their specific task in the human body.

In the following, the most important features of energy transfer devices will be discussed in detail and the most recent advancements in the field will highlight the potential of such devices in biomedical science.

4.3.1 Biocompatibility of Energy-transferring Devices

Apart from the period that a device needs to stay in contact with the body, biocompatibility is one of the most important features of energy-transferring devices. As discussed previously in Section 4.1, the chemical composition and biocompatibility of the materials in biomedical energy-transferring devices are of great importance as well. However, biomedical devices also need to prove their biocompatibility to ensure that the device does not cause irritation, damage, or toxicity through leaching in biological systems; animal studies to examine the biocompatibility behavior of devices serve this purpose.

Biocompatible materials have recently been used widely in the field of biomedical research. Piezoelectric and triboelectric materials are among the most important biomechanical energy harvesters. Recently, a biodegradable piezoelectric device made of aligned crystals of γ-glycine/polyvinyl alcohol was developed for use in ultrasound electrotherapy devices.[68] In another study, an organic piezoelectric material, poly-L-lactic acid (PLLA) with inorganic piezoelectric nanowires (potassium sodium niobate coated with polydopamine, KNN@PDA), was used to fabricate PLLA/KNN@PDA piezoelectric composite fibers (Figure 4.3A).[69] The PLLA/KNN@PDA piezoelectric composite fibers are used as biodegradable scaffolds in regenerative tissue engineering, serving as versatile scaffolds for broader tissue engineering applications. Also, these composite fibers are used as platforms for electrical stimulation to improve growth, proliferation, and neuronal differentiation in the dorsal root ganglion axon.[69] In recent research, an implantable ultrasound-driven TENG (IU-TENG) was fabricated based on polyetherether ketone (PEEK) to achieve remarkable chemical stability in physiological conditions in the human body, with acoustic impedance agreeable to human organs.[70]

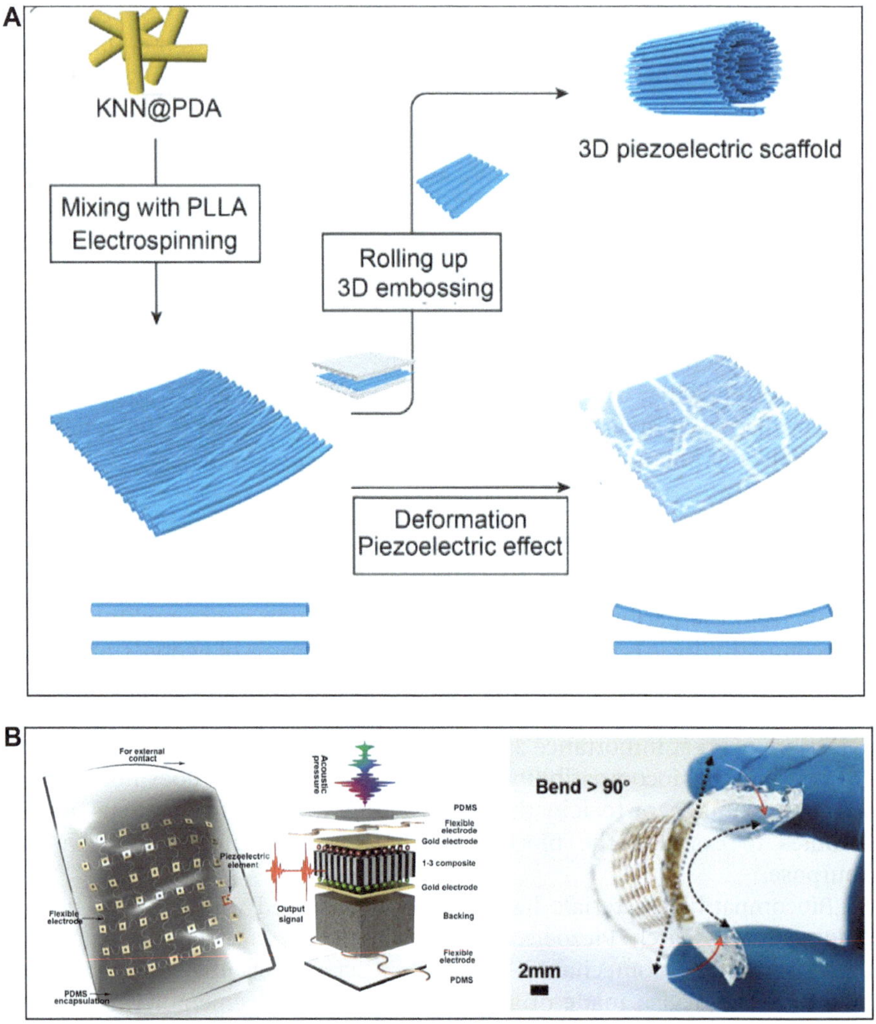

Figure 4.3 The application of biocompatible materials for energy-transferring devices. (A) The fabrication process of biocompatible piezoelectric material nanofibers for neural regeneration. Reproduced from ref. 69 with permission from Elsevier, Copyright 2024. (B) Schematic of the wireless generator made of PZT, and the design of the device to ensure biocompatibility. Reproduced from ref. 71 with permission from Elsevier, Copyright 2019.

Furthermore, as some of these materials have great functions as piezoelectric materials, they are toxic to the human body. Still, the design and architecture of the device can be modulated to ensure the safety of such materials for biomedical applications. A study developed a flexible piezoelectric ultrasonic energy harvester (PUEH) by synthesizing piezoelectric elements integrated into flexible electrodes in elastomer membranes.[71]

Finally, this study selected piezoelectric lead-based zirconate titanate (PZT)/epoxy 1–3 composite as the energy harvester's active element. Although this composite contains more than 60% toxic leads, the PUEH is designed to encapsulate PZT safely to avoid ions leaching into the physiological human body (Figure 4.3B)[71].

4.3.2 Miniaturization of Energy-transferring Devices

Another important characterization of energy-transferring devices is miniaturization.[72] Miniaturization is necessary for medical devices to ensure the creation of relatively small and minimally invasive devices.[73] For example, miniaturization of implantable sensors can enhance patient comfort and reduce the recovery time.[74] Although the benefits of miniaturization are clear, there remain some significant challenges in this field. Miniaturizing the device needs smaller components, resulting in new requirements in materials handling, processing, and designing.[75] Therefore, this area is of great interest to scientists who want to explore new materials and processes for the miniaturization of devices.

In one research work, a heart-worn pacemaker strategy allowed battery-less powering and leadless pacing. This device was made of piezoelectric materials and designed to be miniaturized for *in vivo* systems in swine, as shown in Figure 4.4A.[76] The implanted piezoelectric transducer obtained a record *in vivo* output current of 30 μA in an adult swine, which could effectively drive a commercial cardiac pacemaker.[76] In another research, a mm-sized acoustic wireless implantable neural stimulator based on a piezoelectric micromachined ultrasound transducer was developed.[77] The wireless implantable neural stimulator in tissue accessed 3.24 μW of output power.[77] A miniaturized wireless sensor was created, as shown in Figure 4.4B.[78] The sensor was implanted in a rat's calf muscle, showing a marked increase in longitudinal relaxation time after circulation to the leg was restricted. The arrow indicates the time at which the cuff was tightened. This sensor can be implanted during a biopsy and return chemical information from within the body.[78] In another study, a miniaturized self-powered trinity triboelectric nanogenerator with double-spacer structure harvested energy for therapeutics and diagnosis in one cardiac patch.[79] This device can also wirelessly monitor the electro cardio signal for diagnosis *via* a mobile device, as shown in Figure 4.4C.[79]

4.3.3 Energy Efficiency in Energy-transferring Devices

The energy efficiency of miniaturized medical devices is defined as their ability to transfer energy to the target tissue with minimal losses. This phrase also includes controlling the focus of the energy, ensuring it is not wasted, as well as the ability to modulate the energy output over time to achieve precise outcomes. In the following section, we will discuss these

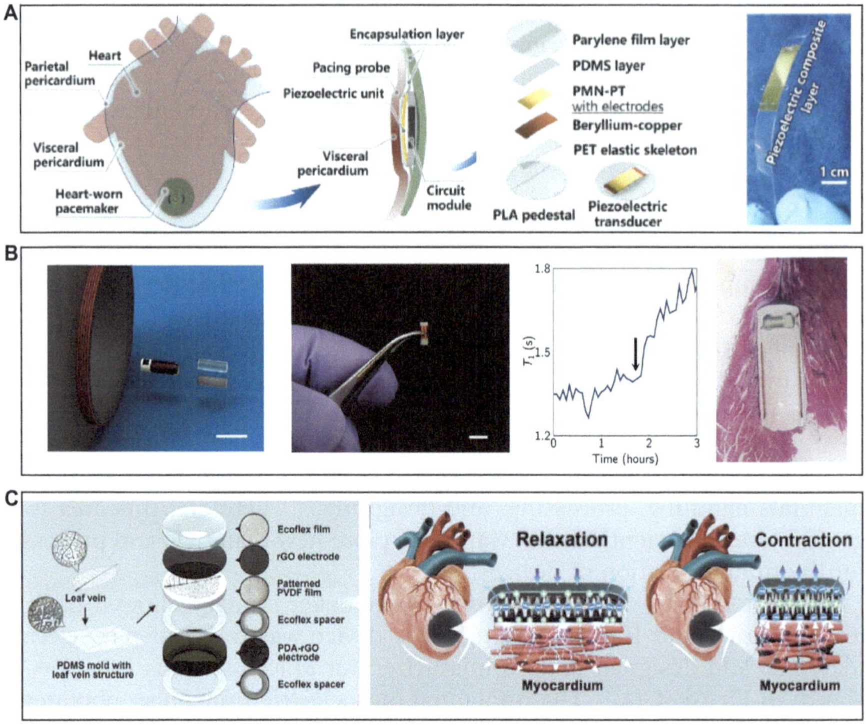

Figure 4.4 The miniaturized architecture of energy-transferring devices. (A) Schematic of the heart-worn pacemaker and the component of the piezoelectric transducer. Reproduced from ref. 76 with permission from John Wiley & Sons, Copyright 2020 Wiley-VCH Verlag GmbH & Co. KGaA, Weinheim. (B) Photographs of the miniaturized sensor with 28 mm diameter. The longitudinal relaxation time of a sensor and the tissue section of the sensor in muscle. Reproduced from ref. 78 with permission from the Royal Society of Chemistry. (C) A schematic of the preparation of the triboelectric nanogenerator, and the electrical mechanism of acquisition of an open circuit voltage between electrodes and the ground using the heart activity. Reproduced from ref. 79, under the terms of the CC BY 4.0 license.

factors and explore strategies to enhance energy efficiency in energy-transferring medical devices.

In electronic medical devices, energy can be dissipated through heat, scattering, or even absorption by the various components of the device.[80] These forms of energy dissipation should be carefully considered during the design and miniaturization phases. The dissipation of energy can be harmful to the human body, and the device can be reduced in functionality. To overcome these challenges, advanced materials with excellent thermal and electrical properties have been introduced in biomedical research.[81] A great group of biocompatible materials, such as graphene and ceramics, can be a good option to prevent the energy dissipation of energy-transferring devices.[81] Furthermore, innovative fabrication methods and strategies like optimizing the device design, insulators, and energy-recycling mechanisms are promising approaches to enhance performance.[82]

In medical devices, precise control over the focus of the energy is essential.[83] This action can ensure that the energy is concentrated on the intended site of the body, spreading to the adjacent tissues. By controlling the focus of the energy, clinicians can deliver energy more accurately while minimizing any collateral damage.[84] This control over the focus can be applied by energy delivery systems such as ultrasound or laser systems.[85]

The other essential aspect of energy efficiency for these devices is the ability to control energy output in response to various medical procedures.[86] Some medical tasks, such as delivery and sensing, need more energy than others. Therefore, the device can adapt its energy output in real time using advanced methods such as machine learning algorithms or feedback loops. This leads to operation at optimal efficiency through the medical procedure.

In conclusion, biocompatible and miniaturized energy transfer devices can be a great option that opens avenues in healthcare systems. However, energy efficiency in miniaturized devices remains a big challenge, however this can be significantly improved with the application of advanced materials, innovative strategies, and intelligent systems.

4.4 Conclusion and Future Outlook

The development of energy-transferring systems requires a high demand for materials selection in biomedical applications. Selection of materials is necessary first to enhance the functional performance of the devices and then for their biocompatibility. As discussed in this chapter, materials must be non-toxic and mechanically stable with the specific energy-transferring systems. Moreover, the integration of these materials into energy-transferring devices is a great challenge in the field of bioelectronics and biomedical applications, as this can influence the clinical translation and longevity of the devices. In the near future, advancements in materials development and selection will open new avenues into the field of energy-transferring systems and self-powered medical devices. Smart materials such as stimuli-responsive and self-healing polymers have a great opportunity to enhance the efficiency and safety of energy-transferring medical devices.

References

1. Z. Tang, C. He, H. Tian, J. Ding, B. S. Hsiao and B. Chu, *et al.*, Polymeric nanostructured materials for biomedical applications, *Prog. Polym. Sci.*, 2016, **60**, 86–128.
2. S. Jayalath and A. Khan, Design, challenges, and trends of inductive power transfer couplers for electric vehicles: A review, *IEEE J. Emerg. Sel. Top. Power Electron.*, 2021, **9**(5), 6196–6218.

3. R. Hinchet and S. W. Kim, Wearable and Implantable Mechanical Energy Harvesters for Self-Powered Biomedical Systems, *ACS Nano*, 2015, **9**(8), 7742–7745.
4. B. Shi, Z. Li, Y. Fan, B. Shi, Y. Fan and Z. Li, Implantable Energy-Harvesting Devices, *Adv. Mater.*, 2018, **30**(44), 1801511.
5. D. Jiang, B. Shi, H. Ouyang, Y. Fan, Z. L. Wang and Z. Li, Emerging Implantable Energy Harvesters and Self-Powered Implantable Medical Electronics, *ACS Nano*, 2020, **14**(6), 6436–6448.
6. A. Umapathi, M. Kumawat and H. K. Daima, Engineered nanomaterials for biomedical applications and their toxicity: a review, *Environ. Chem. Lett.*, 2021, **20**(1), 445–468.
7. K. M. Patel and C. P. Devatha, Investigation on leaching behaviour of toxic metals from biomedical ash and its controlling mechanism, *Environ. Sci. Pollut. Res.*, 2019, **26**(6), 6191–6198.
8. M. Summer, R. Ashraf, S. Ali, H. Bach, S. Noor and Q. Noor, *et al.*, Inflammatory response of nanoparticles: Mechanisms, consequences, and strategies for mitigation, *Chemosphere*, 2024, **363**, 142826.
9. M. Jaishankar, T. Tseten, N. Anbalagan, B. B. Mathew and K. N. Beeregowda, Toxicity, mechanism and health effects of some heavy metals, *Interdiscip. Toxicol.*, 2014, **7**(2), 60–72.
10. W. Wu, R. Xia, G. Qian, Z. Liu, N. Razavi and F. Berto, *et al.*, Mechanostructures: Rational mechanical design, fabrication, performance evaluation, and industrial application of advanced structures, *Prog. Mater. Sci.*, 2023, **131**, 101021.
11. M. Benedetti, A. du Plessis, R. O. Ritchie, M. Dallago, N. Razavi and F. Berto, Architected cellular materials: A review on their mechanical properties towards fatigue-tolerant design and fabrication, *Mater. Sci. Eng., R*, 2021, **144**, 100606.
12. J. Zhang, J. Yan, Y. Zhao, Q. Zhou, Y. Ma and Y. Zi, *et al.*, High-strength and machinable load-bearing integrated electrochemical capacitors based on polymeric solid electrolyte, *Nat. Commun.*, 2023, **14**(1), 1–12.
13. M. Wang, L. Li and T. Zhang, Hysteresis-free, fatigue-resistant and self-adhesive conductive hydrogel electronics towards multimodal wearable application, *Nano Energy*, 2024, **126**, 109586.
14. M. Abbas, R. Marzouki, H. F. M. Ameen, A. Dilsora, J. Younis and A. Akgül, Performance of magnetic dipole contribution on electromagnetic Ellis tetra hybrid nanofluid with the applications of surface tension gradient: A Xue model exploration, *Int. J. Thermofluids*, 2024, **24**, 100951.
15. F. A. Rahim, K. Niyas, R. Vivek, S. Pathan and P. A. Rasheed, An overview of the use of non-titanium MXenes for photothermal therapy and their combinatorial approaches for cancer treatment, *Nanoscale Adv.*, 2024, **7**(4), 963–983.
16. B. D. Kevadiya, J. Machhi, J. Herskovitz, M. D. Oleynikov, W. R. Blomberg and N. Bajwa, *et al.*, Diagnostics for SARS-CoV-2 infections, *Nat. Mater.*, 2021, **20**(5), 593–605.

17. Y. J. Kim, J. Lee, J. H. Hwang, Y. Chung, B. J. Park and J. Kim, *et al.*, High-Performing and Capacitive-Matched Triboelectric Implants Driven by Ultrasound, *Adv. Mater.*, 2024, **36**(2), 2307194.
18. Z. Shariatinia and Z. Karimzadeh, Perovskite oxides as efficient bioactive inorganic materials in tissue engineering: A review, *Mater. Today Chem.*, 2024, **35**, 101846.
19. M. Al Ali and M. Shimoda, Designing piezoelectric stretcher through multiphysics topology for enhanced photodynamic therapy in cancer treatment, *Adv. Eng. Software*, 2024, **191**, 103621.
20. S. Y. Yang, V. Sencadas, S. S. You, N. Z. X. Jia, S. S. Srinivasan and H. W. Huang, *et al.*, Powering Implantable and Ingestible Electronics, *Adv. Funct. Mater.*, 2021, **31**(44), 2009289.
21. M. Narayanan, G. Sołowski and F. A. Özdemir, Green Synthesis and Applications of Nanoparticles in Biomedical Chemistry, *ChemistrySelect*, 2024, **9**(35), e202402472.
22. R. Ajdary, N. Z. Ezazi, A. Correia, M. Kemell, S. Huan and H. J. Ruskoaho, *et al.*, Multifunctional 3D-Printed Patches for Long-Term Drug Release Therapies after Myocardial Infarction, *Adv. Funct. Mater.*, 2020, **30**(34), 2003440.
23. N. Zanjanizadeh Ezazi, R. Ajdary, A. Correia, E. Mäkilä, J. Salonen and M. Kemell, *et al.*, Fabrication and Characterization of Drug-Loaded Conductive Poly(glycerol sebacate)/Nanoparticle-Based Composite Patch for Myocardial Infarction Applications, *ACS Appl. Mater. Interfaces*, 2020, **12**(6), 6899–6909.
24. Biocompatibility - an overview | ScienceDirect Topics, 2025. Available from: https://www.sciencedirect.com/topics/materials-science/biocompatibility.
25. M. Wang and B. Duan, Materials and their biomedical applications, *Encycl. Biomater. Biomed. Eng.*, 2019, **1–3**, 135–152.
26. D. F. Williams, Biocompatibility pathways and mechanisms for bioactive materials: The bioactivity zone, *Bioact. Mater.*, 2022, **10**, 306–322.
27. T. Hanawa, Biocompatibility of titanium from the viewpoint of its surface, *Sci. Technol. Adv. Mater.*, 2022, **23**(1), 457–472.
28. S. Khan, M. Ul-Islam, M. W. Ullah, M. Israr, J. H. Jang and J. K. Park, Nano-gold assisted highly conducting and biocompatible bacterial cellulose-PEDOT:PSS films for biology-device interface applications, *Int. J. Biol. Macromol.*, 2018, **107**(PartA), 865–873.
29. C. K. Jeong, Toward bioimplantable and biocompatible flexible energy harvesters using piezoelectric ceramic materials, *MRS Commun.*, 2020, **10**(3), 365–378.
30. S. W. Ibrahim, T. I. Hamad and J. Haider, Biological properties of polycaprolactone and barium titanate composite in biomedical applications, *Sci. Prog.*, 2023, **106**(4), 00368504231215942.
31. A. Sood, M. Desseigne, A. Dev, L. Maurizi, A. Kumar and N. Millot, *et al.*, A Comprehensive Review on Barium Titanate Nanoparticles as a Persuasive Piezoelectric Material for Biomedical Applications: Prospects and Challenges, *Small*, 2023, **19**(12), 2206401.

32. P. Wang, Y. Yin, L. Fang, J. He, Y. Wang and H. Cai, *et al.*, Flexible cellulose/PVDF composite films with improved breakdown strength and energy density for dielectric capacitors, *Composites, Part A*, 2023, **164**, 107325.
33. X. Wang, C. Yao, F. Wang, Z. Li, P. X. Wang and C. Yao, *et al.*, Cellulose-Based Nanomaterials for Energy Applications, *Small*, 2017, **13**(42), 1702240.
34. W. Zhang, G. Wu, H. Zeng, Z. Li, W. Wu and H. Jiang, *et al.*, The Preparation, Structural Design, and Application of Electroactive Poly(vinylidene fluoride)-Based Materials for Wearable Sensors and Human Energy Harvesters, *Polymers*, 2023, **15**(13), 2766.
35. Z. Li, M. Zhu, J. Shen, Q. Qiu, J. Yu and B. Ding, *et al.*, All-Fiber Structured Electronic Skin with High Elasticity and Breathability, *Adv. Funct. Mater.*, 2020, **30**(6), 1908411.
36. W. S. Cho, Y. Suk Huh, K. Chul Roh, J. Su Chae, H. Lee and S. H. Kim, *et al.*, A durable high-energy implantable energy storage system with binder-free electrodes useable in body fluids, *J. Mater. Chem. A*, 2022, **10**, 4611–4620.
37. C. Li, C. Guo, V. Fitzpatrick, A. Ibrahim, M. J. Zwierstra and P. Hanna, *et al.*, Design of biodegradable, implantable devices towards clinical translation, *Nat. Rev. Mate.*, 2019, **5**(1), 61–81.
38. K. Rajaram, S. M. Yang and S. W. Hwang, Transient, Biodegradable Energy Systems as a Promising Power Solution for Ecofriendly and Implantable Electronics, *Adv. Energy Sustainability Res.*, 2022, **3**(9), 2100223.
39. L. Lu, Z. Yang, K. Meacham, C. Cvetkovic, E. A. Corbin and A. Vázquez-Guardado, *et al.*, Biodegradable Monocrystalline Silicon Photovoltaic Microcells as Power Supplies for Transient Biomedical Implants, *Adv. Energy Mater.*, 2018, **8**(16), 1703035.
40. M. Teodorescu, M. Bercea and S. Morariu, Biomaterials of PVA and PVP in medical and pharmaceutical applications: Perspectives and challenges, *Biotechnol. Adv.*, 2019, **37**(1), 109–131.
41. M. Boffito, P. Sirianni, A. M. Di Rienzo and V. Chiono, Thermosensitive block copolymer hydrogels based on poly(ε-caprolactone) and polyethylene glycol for biomedical applications: State of the art and future perspectives, *J. Biomed. Mater. Res., Part A*, 2015, **103**(3), 1276–1290.
42. M. Kang, D. M. Lee, I. Hyun, N. Rubab, S. H. Kim and S. W. Kim, Advances in Bioresorbable Triboelectric Nanogenerators, *Chem. Rev.*, 2023, **123**(19), 11559–11618.
43. Y. Chen, D. Li, Y. Xu, Z. Ling, H. Nawaz and S. Chen, *et al.*, Surface-microstructured cellulose films toward sensitive pressure sensors and efficient triboelectric nanogenerators, *Int. J. Biol. Macromol.*, 2022, **208**, 324–332.
44. S. G. Ramaraj, D. Elamaran, H. Tabata, F. Zhang and X. Liu, Biocompatible triboelectric energy generators (BT-TENGs) for energy harvesting and healthcare applications, *Nanoscale*, 2024, **16**(39), 18251–18273.

45. Y. Cheng, X. Xiao, K. Pan and H. Pang, Development and application of self-healing materials in smart batteries and supercapacitors, *Chem. Eng. J.*, 2020, **380**, 122565.
46. L. Wang, Z. Lou, K. Jiang and G. Shen, Bio-Multifunctional Smart Wearable Sensors for Medical Devices, *Adv. Intell. Syst.*, 2019, **1**(5), 1900040.
47. R. Bai, Q. Sun, Y. He, L. Peng, Y. Zhang and L. Zhang, *et al.*, Ceramic Toughening Strategies for Biomedical Applications, *Front. Bioeng. Biotechnol.*, 2022, **10**, 840372.
48. J. W. Lee, G. U. Kim, D. J. Kim, Y. Jeon, S. Li and T. S. Kim, *et al.*, Intrinsically-Stretchable, Efficient Organic Solar Cells Achieved by High-Molecular-Weight, Electro-Active Polymer Acceptor Additives, *Adv. Energy Mater.*, 2022, **12**(28), 2200887.
49. S. Siddiqui, D Il Kim, E. Roh, L. T. Duy, T. Q. Trung and M. T. Nguyen, *et al.*, A durable and stable piezoelectric nanogenerator with nanocomposite nanofibers embedded in an elastomer under high loading for a self-powered sensor system, *Nano Energy*, 2016, **30**, 434–442.
50. F. Mokhtari, H. Y. Nam, A. Ruhparwar, R. Raad, J. M. Razal and R. J. Varley, *et al.*, Highly stretchable nanocomposite piezofibers: a step forward into practical applications in biomedical devices,*J. Mater. Chem. B*, 2024, **12**(38), 9727–9739.
51. X. Wan, Z. Wang, X. Zhao, Q. Hu, Z. Li and Z. Lin Wang, *et al.*, Flexible and highly piezoelectric nanofibers with organic–inorganic coaxial structure for self-powered physiological multimodal sensing, *Chem. Eng. J.*, 2023, **451**, 139077.
52. Q. He, H. Heidari, D. A. Zeze and E. Hosseini, Highly Sensitive Flexible Pressure Sensor Based on PVDF-TrFE-BaTiO3 Piezoelectric Nanofibers, *IEEE Sens. Lett.*, 2024, **8**(8), 1–4.
53. S. Wang, X. Chen, L. Guo, S. Wang, F. Dong and H. Liu, *et al.*, A human muscle-inspired, high strength, good elastic recoverability, room-temperature self-healing, and recyclable polyurethane elastomer based on dynamic bonds, *Compos. Sci. Technol.*, 2024, **248**, 110457.
54. A. Kumar, K. Das and A. Roy, Improving the energy density and flexibility of PMN-0.3PT based piezoelectric generator by composite designing, *Sens. Actuators, A*, 2024, **376**, 115609.
55. W. Fan, R. Lei, H. Dou, Z. Wu, L. Lu and S. Wang, *et al.*, Sweat permeable and ultrahigh strength 3D PVDF piezoelectric nanoyarn fabric strain sensor, *Nat. Commun.*, 2024, **15**(1), 1–11.
56. L. Yao, K. Zheng, N. Koripally, N. Eedugurala, J. D. Azoulay and X. Zhang, *et al.*, Structural pseudocapacitors with reinforced interfaces to increase multifunctional efficiency, *Sci. Adv.*, 2023, **9**(25), eadh0069.
57. T. C. Shyu, P. F. Damasceno, P. M. Dodd, A. Lamoureux, L. Xu and M. Shlian, *et al.*, A kirigami approach to engineering elasticity in nanocomposites through patterned defects, *Nat. Mater.*, 2014, **14**(8), 785–789.
58. E. Okogbue, S. S. Han, T. J. Ko, H. S. Chung, J. Ma and M. S. Shawkat, *et al.*, Multifunctional Two-Dimensional PtSe2-Layer Kirigami

Conductors with 2000% Stretchability and Metallic-to-Semiconducting Tunability, *Nano Lett.*, 2019, **1911**, 7598–7607.

59. T. Sekitani, H. Nakajima, H. Maeda, T. Fukushima, T. Aida and K. Hata, *et al.*, Stretchable active-matrix organic light-emitting diode display using printable elastic conductors, *Nat. Mater.*, 2009, **8**(6), 494–499.
60. W. G. Kim, D. W. Kim, I. W. Tcho, J. K. Kim, M. S. Kim and Y. K. Choi, Triboelectric Nanogenerator: Structure, Mechanism, and Applications, *ACS Nano*, 2021, **15**(1), 258–287.
61. W. Lin, Y. Wei, X. Wang, K. Zhai, X. Ji and W. Lin, *et al.*, Study on Human Motion Energy Harvesting Devices: A Review, *Machines*, 2023, **11**(10), 977.
62. F. Liu, S. Chang, L. Xia, H. Zhou, C. Zhang and Y. Tong, *et al.*, Experimental study and application on a novel skin energy harvesting and storage system, *J. Energy Storage*, 2024, **88**, 111610.
63. M. Z. Huang, P. Parashar, A. R. Chen, S. C. Shi, Y. H. Tseng and K. C. Lim, *et al.*, Snake-scale stimulated robust biomimetic composite triboelectric layer for energy harvesting and smart health monitoring, *Nano Energy*, 2024, **122**, 109266.
64. K. Ou, M. Wang, C. Meng, K. Guo, N. Shariar Emon and J. Li, *et al.*, Enhanced mechanical strength and stretchable ionic conductive hydrogel with double-network structure for wearable strain sensing and energy harvesting, *Compos. Sci. Technol.*, 2024, **255**, 110732.
65. B. Huzum, B. Puha, R. M. Necoara, S. Gheorghevici, G. Puha, A. Filip, P. D. Sirbu and O. Alexa, Biocompatibility assessment of biomaterials used in orthopedic devices: An overview (Review), *Experimental and Therapeutic Medicine*, 2021, **22**, 1315.
66. M. Bernard, E. Jubeli, M. D. Pungente and N. Yagoubi, Biocompatibility of polymer-based biomaterials and medical devices – regulations, *in vitro* screening and risk-management, *Biomater. Sci.*, 2018, **6**(8), 2025–2053.
67. Z. Gao, Y. Zhou, J. Zhang, J. Foroughi, S. Peng and R. H. Baughman, *et al.*, Advanced Energy Harvesters and Energy Storage for Powering Wearable and Implantable Medical Devices, *Adv. Mater.*, 2024, **36**(42), 2404492.
68. H. Xue, J. Jin, Z. Tan, K. Chen, G. Lu and Y. Zeng, *et al.*, Flexible, biodegradable ultrasonic wireless electrotherapy device based on highly self-aligned piezoelectric biofilms, *Sci. Adv.*, 2024, **10**(22), 31.
69. J. Zhang, Q. Wang, X. Tang, M. Chai, N. Liu and Z. Jiang, *et al.*, A biodegradable piezoelectric scaffold promotes spinal cord injury nerve regeneration, *Nano Energy*, 2024, **132**, 110382.
70. S. Jeon, X. Meng, N. Rubab, D. Kim, H. Mo and X. Xiao, *et al.*, Ultrasound-Driven Highly Stable Implantable Triboelectric Nanogenerator with Human-Tissue Acoustic Impedance-Matched Polyether Ether Ketone, *Adv. Mater. Technol.*, 2024, **9**(21), 2400317.
71. L. Jiang, Y. Yang, R. Chen, G. Lu, R. Li and D. Li, *et al.*, Flexible piezoelectric ultrasonic energy harvester array for bio-implantable wireless generator, *Nano Energy*, 2019, **56**, 216–224.

72. W. Lian, Y. Xuan and Q. Li, Characterization of miniature automatic energy transport devices based on the thermomagnetic effect, *Energy Convers. Manage.*, 2009, **50**(1), 35–42.
73. H. Dinis, I. Colmiais and P. M. Mendes, Extending the Limits of Wireless Power Transfer to Miniaturized Implantable Electronic Devices, *Micromachines*, 2017, **8**(12), 359.
74. P. Li, G. H. Lee, S. Y. Kim, S. Y. Kwon, H. R. Kim and S. Park, From Diagnosis to Treatment: Recent Advances in Patient-Friendly Biosensors and Implantable Devices, *ACS Nano*, 2021, **15**(2), 1960–2004.
75. J. Niemann, S. Härter, C. Kästle and J. Franke, Challenges of the Miniaturization in the Electronics Production on the example of 01005 Components, *Tagungsband des 2 Kongresses Montage Handhabung Industrieroboter*, 2017, pp. 113–123.
76. Z. Yi, F. Xie, Y. Tian, N. Li, X. Dong and Y. Ma, *et al.*, A Battery- and Leadless Heart-Worn Pacemaker Strategy, *Adv. Funct. Mater.*, 2020, **30**(25), 2000477.
77. Y. Ning, A. C. Lombarte, Y. Gong, S. Sun, Z. Rong and S. Middya, *et al.*, A mm-sized acoustic wireless implantable neural stimulator based on a piezoelectric micromachined ultrasound transducer, *Sens. Actuators, B*, 2024, **405**, 135382.
78. C. C. Vassiliou, V. H. Liu and M. J. Cima, Miniaturized, biopsy-implantable chemical sensor with wireless, magnetic resonance readout, *Lab Chip*, 2015, **15**(17), 3465–3472.
79. R. Qiu, X. Zhang, C. Song, K. Xu, H. Nong and Y. Li, *et al.*, E-cardiac patch to sense and repair infarcted myocardium, *Nat. Commun.*, 2024, **15**(1), 1–20.
80. P. D. Wolf, Thermal Considerations for the Design of an Implanted Cortical Brain–Machine Interface (BMI), *Indwelling Neural Implants: Strategies for Contending with the in Vivo Environment*, 2008, pp. 63–86.
81. V. R. Raphey, T. K. Henna, K. P. Nivitha, P. Mufeedha, C. Sabu and K. Pramod, Advanced biomedical applications of carbon nanotube, *Mater. Sci. Eng.: C*, 2019, **100**, 616–630.
82. A. Chauhan, B. D. Y. Sunil, V. Revathi, B. Manideep, N. Singh and A. Prakash, Innovative Approaches to Energy-Efficient Insulation in Building Construction, E3S Web of Conferences, 2024, p. 596.
83. M. M. H. Shuvo, T. Titirsha, N. Amin and S. K. Islam, Energy Harvesting in Implantable and Wearable Medical Devices for Enduring Precision Healthcare, *Energies*, 2022, **15**(20), 7495.
84. B. J. O'Daly, E. Morris, G. P. Gavin, J. M. O'Byrne and G. B. McGuinness, High-power low-frequency ultrasound: A review of tissue dissection and ablation in medicine and surgery, *J. Mater. Process. Technol.*, 2008, **200**(1–3), 38–58.
85. S. K. S. Huang and J. M. Miller, *Catheter Ablation of Cardiac Arrhythmias*, 2019.
86. T. Rault, A. Bouabdallah and Y. Challal, Energy efficiency in wireless sensor networks: A top-down survey, *Comput. Networks*, 2014, **67**, 104–122.

CHAPTER 5

Biodegradable Mechanical Energy Harvesting for Medical Devices: The Promise of Triboelectric Nanogenerators

AISWARYA BABURAJ,[a,b] AKSHAYA KUMAR ALIYANA,*[c] S. K. NAVEEN KUMAR*[a] AND GEORGE K. STYLIOS*[c]

[a] Department of Electronics, Mangalore University, Mangalore, India; [b] PES University, 100 Feet Ring Road, BSK III Stage, Bangalore – 560085, India, Email: aiswaryababuraj@pes.edu; [c] Research Institute for Flexible Materials, School of Textiles and Design, Heriot-Watt University, UK
*Emails: a.aliyana@hw.ac.uk; nave12@gmail.com; g.stylios@hw.ac.uk

5.1 Introduction

Implantable electronic devices, powered by the Internet of Things (IoT) and artificial intelligence (AI), are transforming healthcare by enabling precise monitoring of vital signs, targeted treatments, and neural stimulation. From life-saving pacemakers to innovative brain implants, these devices are transforming lives.[1] The rapid advancements in miniaturization, biocompatibility, and energy efficiency have led to a surge in the development of diverse implantable devices. These include pH sensors,[2] intelligent stomach sensors,[3] cardiac pacemakers,[4] cochlear implants,[5] implanted cardioverter defibrillators,[6,7] and deep brain, nerve,[7] and bone stimulators.[8] As these devices become increasingly sophisticated, they promise to further improve patient outcomes and quality of life. The implantable medical device market

Detection Science Series No. 33
Energy Materials for Ingestible and Implantable Healthcare Devices
Edited by Santanu Patra and Anja Boisen

Published by the Royal Society of Chemistry, www.rsc.org

is steadily growing, driven by an aging global population and the rising prevalence of chronic diseases like heart disease, diabetes, and neurological disorders. This increasing demand for effective disease management solutions fuels the development of advanced devices.[9] Technological advancements, such as miniaturization, controlled drug delivery, neurostimulation, biomedical imaging, smart sensors, energy harvesting, biocompatible materials, data analytics, and AI integration, are creating more sophisticated and efficient devices.[10,11] The market, valued at approximately USD 167 billion in 2021, is projected to exceed USD 335 billion by 2030, with a compound annual growth rate (CAGR) of 7.2%.[12,13] The projected growth of the implantable medical device market is driven by increasing demand for innovative solutions due to aging populations, rising chronic diseases, and advancements in healthcare technologies. Biodegradable materials, gaining traction in research and development, offer a promising avenue for improving the biocompatibility and sustainability of implantable devices, aligning with the growing focus on eco-friendly and patient-centered healthcare.[14] When designing implantable medical devices, engineers focus on several critical aspects, including packaging, wireless communication, biocompatibility, structural design, and power management.[15] Power management is essential for ensuring sustained operation and longevity within the body. Efficient power management systems optimize energy usage, extend the device's lifespan, and reduce the need for repairs or replacements.[12,13,16,17] Researchers are dedicated to developing innovative power solutions that balance energy efficiency and the power requirements of the implanted device, thereby enhancing its overall reliability and effectiveness.[18] A short battery life leads to frequent replacements, which increases patient risks and costs. Additionally, size constraints can impede miniaturization and comfort, while potential deterioration raises concerns about long-term effectiveness.[19–22]

Currently, implant power sources include piezoelectric nanogenerators (PENGs),[23] rechargeable batteries,[24] and radiofrequency (RF) harvesting.[25] However, piezoelectric technologies require significant movement for energy generation, RF power is often limited and can interfere with surrounding tissues, and batteries need to be surgically replaced.[26] A promising alternative to these methods is biodegradable triboelectric nanogenerators (B-TENGs),[27] which are made from biocompatible and flexible materials.[28–30] These devices are autonomous, require no maintenance, and generate energy from subtle bodily movements such as breathing or the pulse. This innovation has the potential to extend the lifespan of implanted devices and enhance patient comfort.[26] Dr. Zhong Lin Wang pioneered TENG technology in 2012, revolutionizing the field of energy harvesting by converting mechanical energy into electrical power. These devices harness mechanical energy from various sources, such as motion, vibrations, and environmental elements, to generate electricity.[31–33] TENGs provide a consistent and sustainable power supply for self-powered medical devices, removing the need for conventional batteries and external charging. This breakthrough greatly enhances patient care and compliance by enabling the continuous operation of essential medical devices like pacemakers, implanted sensors, and wearables, all of which improve patient comfort and mobility.[34–36] Additionally, TENGs have significant potential in remote and resource-constrained environments where electricity access is limited. The versatility and broader applications of this technology make it a game changer in energy harvesting, shaping the future of self-powered medical devices and beyond.[37–39]

There is an increasing demand for inherently biodegradable B-TENGs that can power implantable or deployable medical devices while also serving as self-powered, precise sensing tools. However, this presents challenges, as these devices need to be adaptable, conformable, lightweight, and capable of integrating into various surfaces for real-time healthcare applications.[40–42] Achieving this level of flexibility and comfort without sacrificing functionality, durability, and energy generation capacity is a significant hurdle. Innovative engineering of biodegradable materials and device structures is crucial to ensuring these devices can withstand repeated mechanical deformations without degradation.[43–45] This chapter examines the transformative potential of B-TENGs in advancing implantable medical devices. It highlights the diverse biodegradable materials used in their construction, including active layers, electrodes, and associated circuitry, and provides a detailed analysis of how these materials influence device performance and enable self-sustaining medical solutions. The discussion also delves into the critical issue of material lifespan and its impact on B-TENG functionality, while addressing the challenges that hinder commercialization. Finally, the chapter outlines a forward-looking roadmap for future research and development to realize the full potential of B-TENGs in healthcare.

5.2 Transitioning from Traditional to Biodegradable Materials in TENGs

TENGs are devices that convert mechanical energy into electrical energy. This conversion is based on the principles of triboelectric electrification and electrostatic induction. When two materials with different electronegativities come into contact and then separate, one material gains electrons while the other loses electrons, creating a potential difference.[46,47] This phenomenon, described by Maxwell's displacement current theory, generates an electric current as the materials move back into contact. The combination of these effects allows TENGs to efficiently convert mechanical energy into electrical energy, making them a promising technology for self-powered devices and energy harvesting applications.[32,48] The transfer of electric charge through the exchange of electrons is initiated by the relative motion or separation of materials, leading to a redistribution of charges on their surfaces.[48–50] TENGs operate through four primary modes, each leveraging distinct mechanical interactions to generate electrical energy (Figure 5.1). In the contact–separation (C–S) mode, materials with varying electron affinities come into contact, and upon separation, surface charges accumulate,

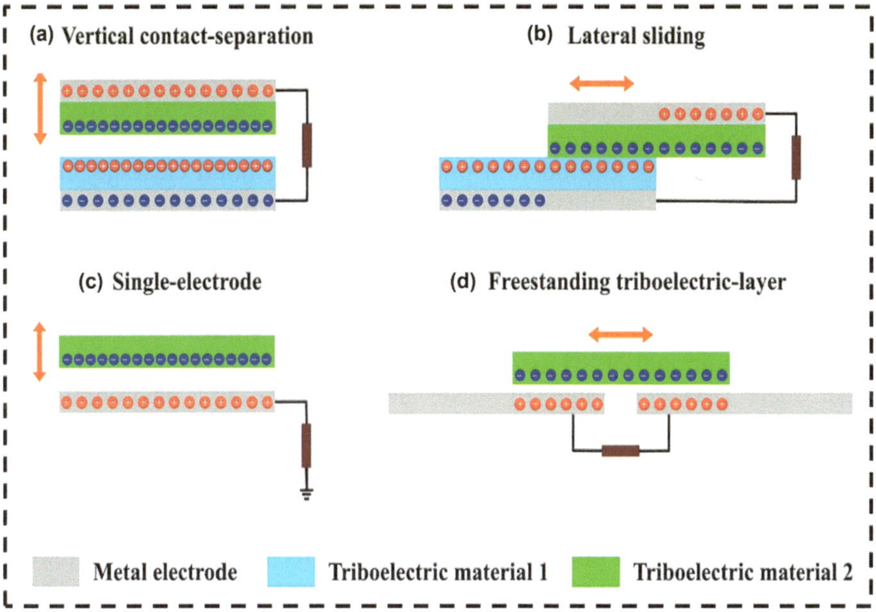

Figure 5.1 A schematic representation of the four primary configurations of TENGs: vertical contact–separation, lateral sliding, single-electrode, and freestanding triboelectric layer.[51] Reproduced from ref. 51, https://doi.org/10.1002/advs.202304232, under the terms of the CC BY 4.0 license, https://creativecommons.org/licenses/by/4.0/.

inducing a potential difference that drives electron flow through an external circuit. When the materials return to contact, the process reverses, producing an alternating current (AC) output. The lateral sliding (L-S) mode involves the sliding motion of materials against each other, creating changes in the contact area. This leads to charge accumulation on non-contacting electrode areas and generates a potential difference, which drives electron flow; as equilibrium is restored, the current direction reverses. The single-electrode (S-E) mode features a stationary electrode and a moving dielectric material. As the dielectric approaches and recedes from the electrode, charge transfer occurs between the electrode and the ground, providing flexibility in design but generally yielding lower output. Lastly, the freestanding triboelectric layer (FTL) mode employs a dielectric material moving between two grounded electrodes. This motion creates an asymmetric charge distribution, driving charge transfer through an external load, making this mode particularly effective for applications such as grating structures. Each mode offers unique advantages suited to specific energy harvesting scenarios.

The triboelectric series ranks materials based on their electron affinity.[50] Materials positioned further apart on the series generally exhibit better triboelectric performance.[52–55] This knowledge helps researchers select optimal material pairs to maximize energy output from TENGs. Environmental factors like temperature and humidity can influence TENG performance.[56,57]

Materials with high charge density and moisture resistance are preferred for consistent operation. To address environmental concerns, the development of B-TENGs has garnered significant attention. These devices employ biodegradable materials derived from natural sources or engineered to degrade efficiently, thereby reducing their environmental impact. Such materials undergo a multi-stage degradation process, including bond cleavage, chain scission, and eventual breakdown into environmentally assimilable components. The use of hydrophobic materials enhances moisture resistance, ensuring stable TENG performance in various conditions, while the introduction of micro-pores, nanostructures, or chemical modifications to polymers can optimize surface charge density and improve triboelectrification. In contrast to traditional TENG materials, often synthetic polymers derived from fossil fuels that contribute to waste accumulation and environmental contamination, biodegradable alternatives offer sustainable, eco-friendly solutions. These materials not only minimize waste and reliance on non-renewable resources but also align with increasing regulatory demands for environmental compliance. Furthermore, natural biodegradable materials are safer for applications in implantable and wearable technologies, posing fewer health risks compared to synthetic options. Their cost-effectiveness over time further incentivizes adoption, supporting green manufacturing practices and promoting long-term sustainability. By addressing environmental, health, and regulatory challenges, biodegradable TENGs pave the way for cleaner and more responsible energy harvesting solutions (Figure 5.2).

5.3 Self-powered Implantable Medical Devices: B-TENGs

The integration of B-TENGs in self-powered implantable medical devices opens new avenues for healthcare innovation. These devices can continuously monitor vital signs, deliver therapeutic interventions, and facilitate real-time communication with healthcare providers, all while minimizing the need for battery replacements or external power sources.[59–61] Traditional health monitoring devices often suffer from bulky designs and a tangle of wires, limiting their suitability for flexible and wearable applications. B-TENGs offer a promising solution by enabling the continuous monitoring of body signals without the need for external power sources.[62]

Their biodegradable nature and the wide range of materials available for their construction make them ideal for various applications, including sensors, health monitoring devices, and power generators for healthcare sensors. B-TENGs are composed of several key components, including the active layer, electrode, and power management circuit.[63] The choice of materials for the active layer is crucial, as it must be biocompatible and biodegradable. For instance, B-TENGs made of reed film and polylactic acid have been explored as power sources for implantable medical devices, demonstrating their potential for applications like cancer treatment.[64–66]

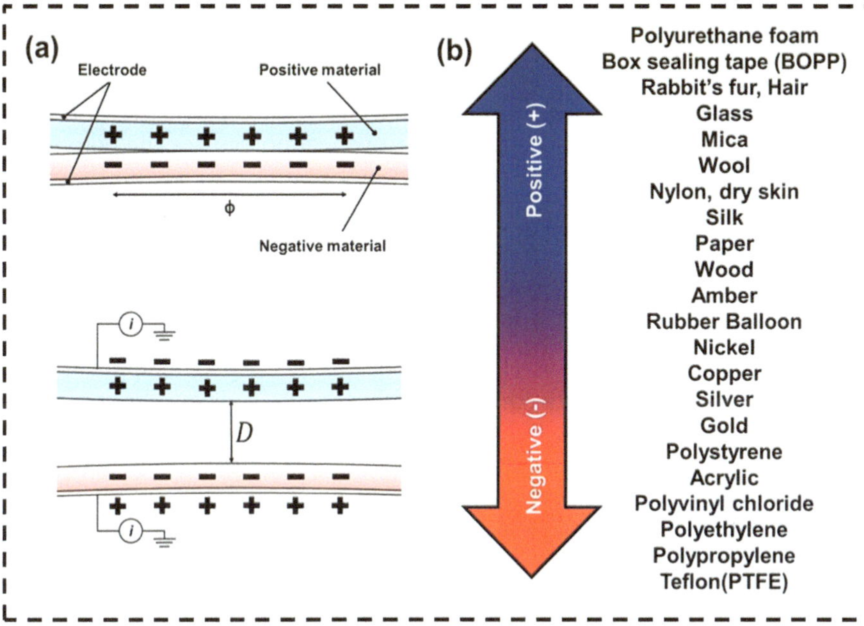

Figure 5.2 Schematic representation of a triboelectric nanogenerator, highlighting the role of material triboelectric series.[58] Reproduced from ref. 58 with permission from Elsevier, Copyright 2018.

Natural polymers like cellulose, starch, and proteins, and synthetic biodegradable polymers like PLGA, PPDO, PCL, and PBS, offer biocompatibility and eco-friendliness. Cellulose-based TENGs, in particular, have shown promise in various applications, including health monitoring. TENGs have found numerous applications in healthcare, including human motion monitoring, pulse sensing, and powering implantable devices like pacemakers. These devices can convert mechanical energy from body movements into electrical energy, eliminating the need for battery replacements.[72–75] By harnessing the power of the human body and utilizing biodegradable materials, TENGs offer a sustainable and innovative approach to powering wearable and implantable medical devices. Gelatin, chitin, and silk fibroin are among the materials that have shown promise in TENGs. Silk fibroin, for example, has been used in TENGs as an electron donor material. This TENG showed remarkable performance when used with MXene-integrated polyvinyl alcohol (PVA) nanofibers as the substrate, producing a power density of around 1087.6 $mW\,m^{-2}$ at a load resistance of 5 MΩ.[76] These gadgets have demonstrated promise for tracking human body motions in real time, underscoring the potential of biodegradable polymers in medical settings. Zhuo *et al.*[68] have developed a ground-breaking self-powered endocardial pressure sensor (SEPS) that offers significant advancements in cardiac monitoring. This miniaturized, flexible device, powered by a B-TENG, can be implanted minimally invasively through a surgical catheter. The SEPS exhibits exceptional sensitivity, enabling real-time monitoring of even subtle pressure variations. Its high linearity ($R^2 = 0.997$) and sensitivity of 1.195 $mV\,mmHg^{-1}$ ensure accurate and reliable measurements. Moreover, the SEPS can detect critical cardiac arrhythmias like ventricular fibrillation and premature contractions. This innovative technology has the potential to revolutionize cardiac care by providing continuous, precise, and self-powered monitoring of heart function (Figure 5.3). Silk-based hydrogels, combining silk with other polymers, offer biocompatibility and flexibility, making them suitable for stretchable and compressible TENGs. These devices can function as effective strain sensors, capable of detecting a wide range of deformations. Researchers are actively working on tailoring the degradation rate of these silk-based TENGs to meet specific application requirements. Chitosan, derived from crustacean shells, is another promising biodegradable material for TENGs. Its biocompatibility, bio-absorbability, and antibacterial properties make it attractive for various biomedical applications. Surface engineering techniques, such as laser processing, can further enhance the performance of chitosan-based TENGs by modifying their surface properties.[77] Gelatin has also found application in TENGs. Collagen, a major component of gelatin, has been utilized as a triboelectric layer. By combining collagen-based electrospun membranes with other materials like PVDF, researchers have achieved significant output voltages.[78,79] Researchers have explored the use of cellulose-based TENGs for various applications, including human motion monitoring and energy harvesting. Wood has also been utilized in TENGs. All-wood-based TENGs have been developed, demonstrating the potential of this natural material for energy harvesting applications.[72–75] Polylactic acid (PLA) is a biodegradable polymer gaining popularity in biomedical applications due to its biocompatibility and bio-absorbability. Its good mechanical properties and transparency make it suitable for implantable devices. Researchers have developed seawater-degradable TENGs using a blend of PLA and poly(lactic-*co*-glycolic acid) (PLGA), demonstrating their potential for energy harvesting from water waves. For wearable applications, TENGs made from Janus electrospun nanofibers of PLA and PVDF-HFP have been developed. These TENGs can generate significant output voltage and power density, making them suitable for powering various wearable devices (Table 5.1).[80]

5.4 Challenges and Future Prospects in the Lifespan of B-TENGs

The lifetime of a B-TENG is influenced by various factors, including the materials used, operating conditions, and intended application. Biodegradable materials, while advantageous for environmental impact, have a finite lifespan determined by their degradation rate. This rate can be influenced by factors like pH, temperature, and enzymatic activity.[85–88] Non-biodegradable components, such as electrodes or encapsulation materials,

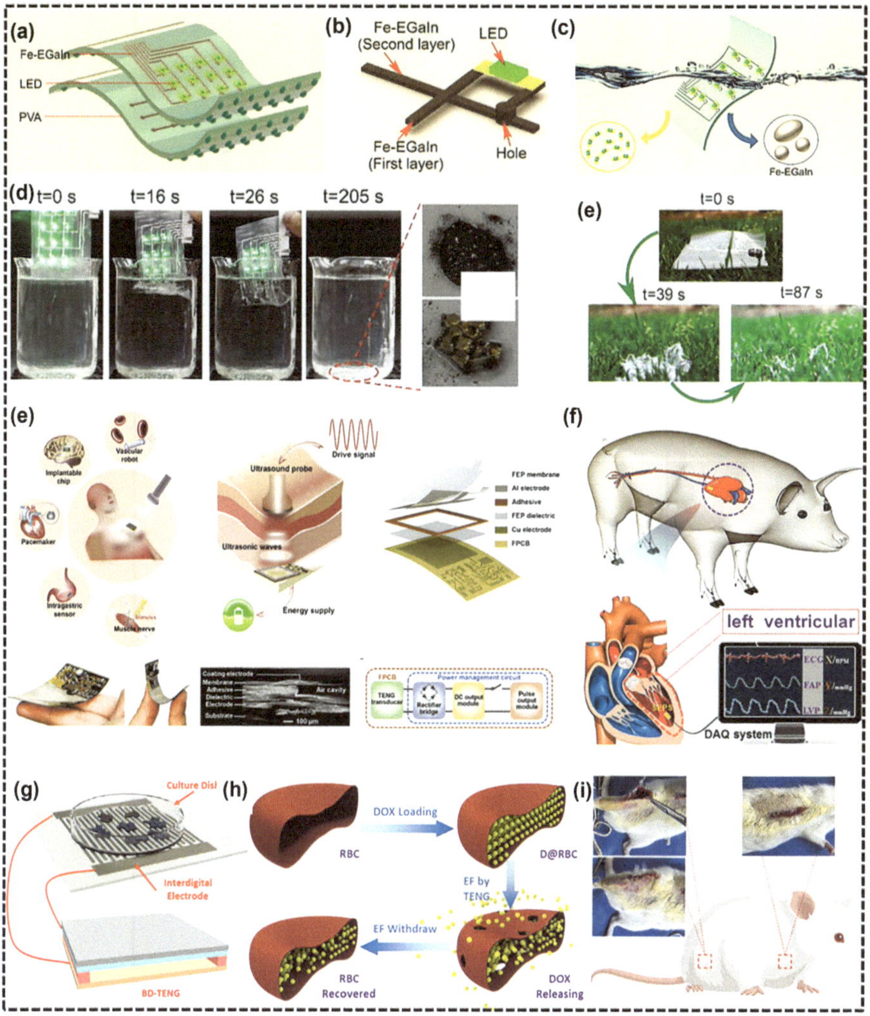

Figure 5.3 Biodegradable electronics and bio-integrated energy harvesting devices. (a) Structure of Fe-EGaIn-based electronics. (b) Schematic illustration of the structure of Fe-EGaIn lines on two layers. (c) Schematic illustration of the PVA/fructose-enabled degradable and recyclable electronics. (d) PVA/fructose enabled transient and recyclable electronics: photos of double-layer LED array and the dissolution and recycle process of the double-layer LED array. (e) Schematic diagram depicting the process of semaphore acquisition facilitated by the Sub-Epicardial Pacing System (SEPS) implanted within an adult Yorkshire swine's heart.[67] (f) BI-TENG-controlled constructs are used to control doxorubicin (DOX) release.[68] (g) Process of loading DOX into the BI-TENG-controlled constructs. (h) *In vivo* output properties and blood/tissue toxicity of FB-HNG.[69] (i) Schematic image and digital photographs of the implantations of FB-HNGs in the thigh and chest areas of a rat.[70] Reproduced from ref. 71, https://doi.org/10.1016/j.nanoen.2024.109785, under the terms of the CC BY 4.0 license, https://creativecommons.org/licenses/by/4.0/.

Table 5.1 Summary of biodegradable materials-based TENGs, along with their performance evaluations and applications.[71]

S. no.	Biodegradable active materials	Dimensions	Force/frequency	Output: voltage/current/power density	Applications	Ref.
1	Silk fibroin/MXene integrated PVA nanofibers	Diameter: 6 cm	10 N/10 Hz	118.4 V/—/1087.6 mW m^{-2}	Monitoring wrist movement, finger bending, and knee movement	76
2	Silk fibroin, polyacrylamide, graphene oxide, and poly(3,4-ethylene dioxythiophene): poly(styrene sulfonate)	2 × 2 cm^2	0.5–119.4 kPa/—/	12 V/0.4 μA/—	Pressure sensor, monitoring of wrist, leg, and arm movement	81
3	Chitosan and Kapton	2 × 2 cm^2	—	—/—/17.5 μW m^{-2}	Human activity monitoring	82
4	Fish gelatin (FG) film and poly(tetrafluoroethylene)/poly(dimethyl siloxane) (PTFE/PDMS) composite film	5 × 5 cm^2	—	130 V/0.35 μA/45.8 μW cm^{-2}	Real-time respiratory monitoring	78
5	Crepe cellulose paper and nitrocellulose membrane	6.25 cm^2	6 N/10 Hz	196.8 V, 31.5 μA, 16.1 W m^{-2}	LEDs, electroluminescence display, paper piano	83
6	Rice paper and PVC	3 × 3 cm^2	—/5 Hz	244 V/6 μA/37.64 μW cm^{-2}	LEDs	84
7	PLA and PVDF-HFP	2 × 2 cm^2	—/2 Hz	140 V/3.8 μA/0.75 mW	LEDs, calculators, energy harvesting from throat vibration and wrist movement	80

can also impact the overall device lifespan, particularly for implantable devices. Mechanical stress, such as repeated bending or impact, can accelerate material degradation and reduce the device's lifespan. Proper design and encapsulation can help mitigate these effects. Environmental factors like moisture, temperature fluctuations, and chemical exposure can also influence the device's performance and longevity. To extend the lifespan of B-TENGs, careful material selection, effective encapsulation, optimized design, and rigorous quality control are crucial. By considering these factors, researchers can develop B-TENGs with longer lifespans, ensuring their reliable and sustainable operation in various applications.[89,90] The lifespan of B-TENGs is significantly influenced by the degradation rate of their constituent materials. Biodegradable materials, while offering environmental benefits, undergo decomposition over time due to factors like pH, temperature, and moisture. Therefore, selecting materials with a degradation rate compatible with the desired application is crucial. Designing B-TENGs with optimal thickness, morphology, and protective coatings can extend their operational lifespan. Additionally, efficient energy management systems and power storage units can help mitigate the impact of degradation on overall performance.[95] Biocompatibility studies are essential to assess the potential adverse effects of B-TENG materials on biological tissues and cells.[96] Understanding the biocompatibility profile of materials is crucial for minimizing the risk of adverse reactions and ensuring the safe and effective use of B-TENGs in biomedical applications. The choice of biodegradable materials significantly impacts the lifespan of B-TENGs. Synthetic biodegradable polymers like PLA and PCL offer a wide range of degradation rates, allowing for tailored device lifespans. For example, PLA can degrade within months, while PCL can take longer.[97] Zheng *et al.* have presented a significant advancement in the field of transient electronics with their development of B-TENGs. These devices offer a promising solution for powering *in vivo* sensors and therapeutic devices without the need for external power sources. A key aspect of B-TENGs is their controlled degradation profile.

Researchers have demonstrated that the device's lifetime can be tuned by selecting appropriate materials and optimizing the design. By employing biodegradable polymers and metal electrodes, the B-TENG can be designed to degrade at a specific rate, ensuring its safe and efficient operation. The degradation process typically involves an initial stable phase followed by a period of accelerated degradation. The choice of encapsulation material significantly influences the overall degradation rate. While materials like PLGA can provide initial stability, they eventually degrade, leading to the complete dissolution of the B-TENG. The ability to tailor the degradation profile is crucial for various biomedical applications. For instance, short-lived B-TENGs can power temporary implants, while longer-lasting devices can be used for continuous monitoring (Figure 5.4) [91]. Natural biodegradable materials, such as cellulose and silk fibroin, often have shorter degradation times, typically ranging from hours to months. The degradation rate of these materials is influenced by factors like pH, temperature, and the presence of enzymes. Researchers can fine-tune the degradation rate to match the desired application by carefully selecting materials and controlling environmental conditions. For short-term applications like drug delivery, materials with rapid degradation rates are suitable, while for long-term applications like pacemakers, slower-degrading materials are preferred. In addition to material selection, the design and fabrication of B-TENGs play a crucial role in determining their lifespan. Optimizing the thickness,

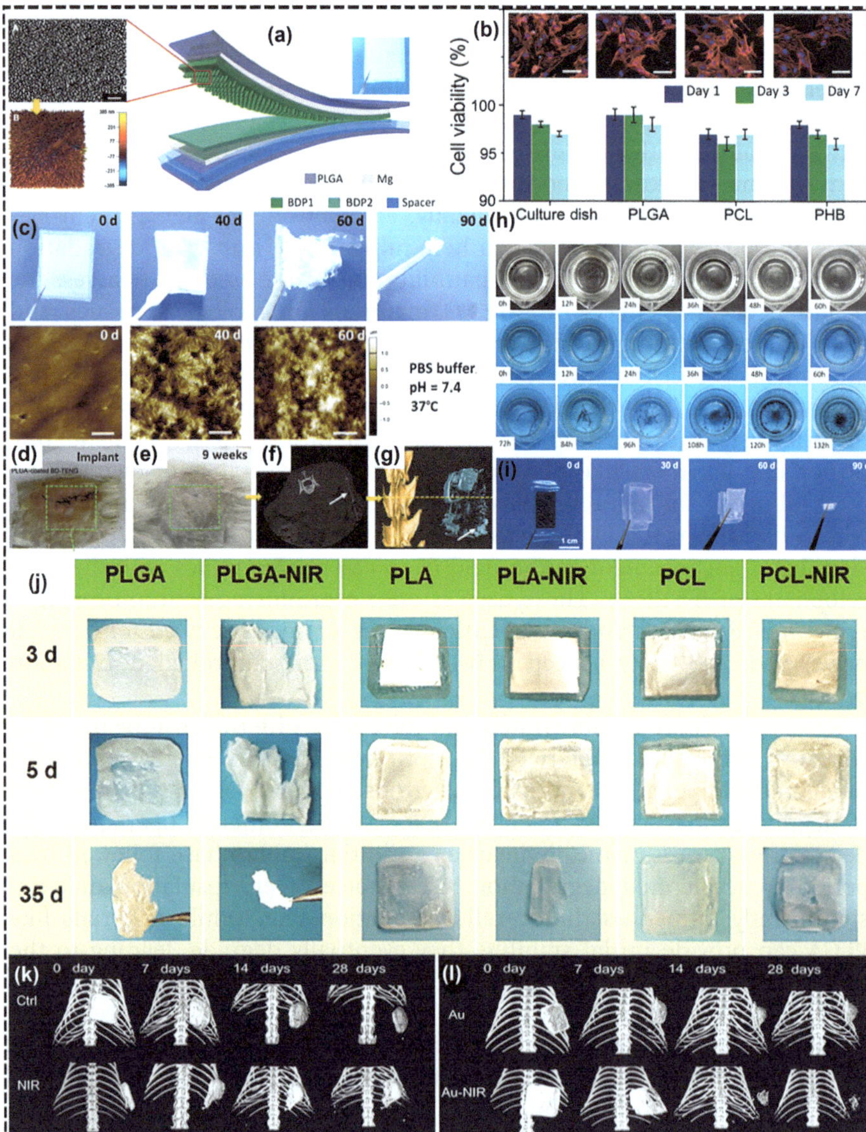

morphology, and protective coatings of the device can help prolong its operational life. Additionally, efficient energy management systems and power storage units can further enhance the device's performance and longevity (Table 5.2).[98]

The lifespan of B-TENGs, particularly for implantable devices, is significantly influenced by various factors. Material degradation, both in biodegradable (like silk, chitosan, and PLA) and non-biodegradable components, can limit device longevity. Mechanical stress, especially in wearable and implantable devices, accelerates material degradation, particularly at interfaces. Environmental factors, such as moisture, temperature fluctuations, and chemical agents, can further impact performance and lifespan. Additionally, inefficient energy harvesting, due to factors like material choice, triboelectric layer design, and energy conversion/storage mechanisms, can constrain operational time. Balancing the degradation rate with the desired device lifespan is crucial. Additionally, repeated mechanical stress can accelerate material degradation, especially at interfaces. Exposure

Figure 5.4 Lifetime and biodegradability of B-TENGs. (a) Device structure, typical output performance, and cytocompatibility of BD-TENG: scanning emission microscopy (SEM)and atomic force microscopy (AFM) images of the nanostructure on the biodegradable polymer (BDP) film (scale bar, 10 μm) with schematic diagram and photograph of BD-TENG. (b) Cell viability after cells were cultured with different BDP films for 7 days. (Inset: Fluorescence images of stained endothelial cells cultured on BDP films. Scale bar, 100 μm.) (c) Photographs from BD-TENG at various stages of the degradation timeline suggest that devices encapsulated in poly(lactic-*co*-glycolic acid) (PLGA) were initially resistant to mass degradation. However, after 40 days, significant mass loss and structure disintegration were initiated. Near-total mass loss was observed at 90 days, along with AFM images from the surface of BD-TENG at different degradation times demonstrating the destruction of the BDP structure throughout the degradation process (scale bar, 20 μm).[91] (d) Images of an implanted demonstration for BD-TENG located in the subdermal dorsal region of a Sprague-Dawley (SD) rat: Implant site right after suture. (e) Implant site after 9 weeks. (f) and (g) Micro-CT images of implanted BD-TENG (white arrow) after 9-week implantation demonstrate the degradation degree of BD-TENG *in vivo*.[91] (h) The degradation experiment of bacterial cellulose (BC) and BC/carbon nanotube/polypyrrole (BC/CNT/PPy) microfiber in cellulose solution. Photographs of macrofibers during degradation, BC macrofiber, and BC/CNT/PPy macrofiber.[92,93] (i) Photographs of flexible bioresorbable nanogenerators (FB-HNG) at various stages of degradation in phosphate-buffered saline (37 °C, pH = 7.4). (j) *In vitro* degradation evaluation of various bioresorbable implantable TENGs (BD-iTENG) (size, 1.2 cm × 1.2 cm; thickness, 0.65 mm) under different conditions. Biodegradation of the BD-iTENG made from PLGA, PLA, and polycaprolactone (PCL) in phosphate-buffered saline (PBS) at 37 °C with or without near-infrared (NIR) laser. (k) and (l) Micro-CT images of the implanted PLGA-iTENGs at various time points.[94] Reproduced from ref. 71, https://doi.org/10.1016/j.nanoen.2024.109785, under the terms of the CC BY 4.0 license, https://creativecommons.org/licenses/by/4.0/.

Table 5.2 The degradation behavior of biodegradable polymer materials.[71]

S. no.	Materials	Degradation rate	Degradation mechanism	Degradation by-products	Ref.
1	PLA (PLLA and PDLA)	100% in 12–16 month (*in vivo*)	Hydrolysis through the action of enzymes	Lactic acid	99, 100, and 101
2	PGA	100% in 6–12 months (*in vivo*)	Both enzymatic and non-enzymatic hydrolysis	Glycolic acid	102 and 103
3	PCL	50% in 4 years (*in vivo*)	Hydrolytic degradation	Caproic acid	104 and 105
4	PBS	5–10 wt% in 100 days (*in vitro*)	Enzymatic hydrolytic degradation	Succinic acid	106 and 107
5	PPC	6% in 200 days (*in vivo*)	Hydrolysis, or enzyme mediation	CO_2 and Water	99 and 104

to moisture, temperature fluctuations, and chemical agents can further impact device performance and longevity. Inefficient energy harvesting can also limit the device's operational time. To address these challenges and extend the lifespan of B-TENGs, researchers are exploring various strategies. Developing novel biodegradable materials with tailored degradation profiles can enhance device longevity. Improved encapsulation techniques can protect devices from environmental factors and mechanical stress. Optimizing the design of B-TENGs to improve energy harvesting efficiency and power management can maximize energy output and minimize power consumption. Rigorous biocompatibility and toxicity studies are essential to ensure the long-term safety of B-TENG materials. Incorporating sensors for real-time monitoring and predictive maintenance can optimize device maintenance and extend lifespan. Researchers can further extend the lifespan of B-TENGs by tackling these issues and looking into creative solutions, opening the door for their broad use in a range of applications, such as environmental monitoring, wearable technology, and healthcare.

5.5 Conclusions

This chapter highlights the innovative potential of B-TENGs as a sustainable solution for powering implantable medical devices. By leveraging biodegradable materials, B-TENGs not only address the growing need for eco-friendly energy solutions in healthcare but also set the stage for a future of energy-independent, self-powered medical technologies. The chapter delves into the principles, materials, and design innovations driving B-TENG development, offering valuable insights into their ability to balance performance with environmental responsibility. Crucially, the discussion tackles the challenges of material lifespan, durability, and real-world implementation, presenting viable strategies to overcome these obstacles and advance commercialization. By addressing regulatory, technical, and practical barriers, the chapter lays out a clear path for the integration of B-TENGs into mainstream medical practices. Looking ahead, B-TENGs promise to redefine healthcare by enabling self-sustaining, patient-focused medical devices that minimize their environmental impact and reduce dependence on traditional power sources. This innovation marks a significant step toward a more sustainable and technologically advanced future in healthcare.

Acknowledgements

This work was supported by the Engineering and Physical Sciences Research Council (EPSRC) through the NextGen T-TENG Standard Grant (Grant Ref: EP/V003380/1).

References

1. Q. Chen and G. A. Thouas, Metallic implant biomaterials, *Mater. Sci. Eng., R*, 2015, **87**, 1–57.
2. A. K. Aliyana, A. Baburaj, S. K. Naveen Kumar and R. E. Fernandez, Zinc Oxide-Based Miniature Sensor Networks for Continuous Monitoring of Aqueous PH in Smart Agriculture, *Sens. Technol. Real Time Monit. Water Qual.*, 2023, DOI: 10.1002/9781119775843.ch6.
3. Z. Liu, *et al.*, Endocardial Pressure Sensors: Transcatheter Self-Powered Ultrasensitive Endocardial Pressure Sensor (Adv. Funct. Mater. 3/2019), *Adv. Funct. Mater.*, 2019, **29**, 1807560.
4. Q. Zheng, *et al.*, In Vivo Self-Powered Wireless Cardiac Monitoring via Implantable Triboelectric Nanogenerator, *ACS Nano*, 2016, **10**, 6510–6518.
5. A. Kral, M. F. Dorman and B. S. Wilson, Annual Review of Neuroscience Neuronal Development of Hearing and Language: Cochlear Implants and Critical Periods, 2019, DOI: 10.1146/annurev-neuro-080317.
6. D. Jones and K. Jeffrey, *Machines in Our Hearts: The Cardiac Pacemaker, the Implantable Defibrillator, and American Health Care. 370 pp., figs., tables, bibl., index. Baltimore: Johns Hopkins University Press, 2001. $48 (cloth)*, 2002, vol. 93, p. 537.
7. Y. Long, J. Li, F. Yang, J. Wang and X. Wang, Wearable and Implantable Electroceuticals for Therapeutic Electrostimulations, *Adv. Sci.*, 2021, **8**, DOI: 10.1002/advs.202004023.
8. M. R. Neuman, *et al.*, Advances in medical devices and medical electronics, in *Proceedings of the IEEE*, Institute of Electrical and Electronics Engineers Inc., vol. 100, 2012, 1537–1550.
9. C. Li, *et al.*, Design of biodegradable, implantable devices towards clinical translation, *Nat. Rev. Mater.*, 2020, **5**, 61–81.
10. A. J. T. Teo, *et al.*, Polymeric Biomaterials for Medical Implants and Devices, *ACS Biomater. Sci. Eng.*, 2016, **2**, 454–472.

11. P. Turon and B. Braun, *Trends and Challenges in Textiles Associated with Medical Devices Used in Surgery*, 2017. https://www.researchgate.net/publication/321669807.
12. Z. Che, *et al.*, Implantable Triboelectric Nanogenerators for Self-Powered Cardiovascular Healthcare, *Small*, 2023, **19**, DOI: 10.1002/smll.202207600.
13. A. Khan, *et al.*, Piezoelectric and triboelectric nanogenerators: Promising technologies for self-powered implantable biomedical devices, *Nano Energy*, 2024, **119**, DOI: 10.1016/j.nanoen.2023.109051.
14. A. Kim, M. Ochoa, R. Rahimi and B. Ziaie, New and Emerging Energy Sources for Implantable Wireless Microdevices, *IEEE Access*, 2015, **3**, 89–98.
15. Engineering, B, Energy Harvesting: A Review of Novel Power Sources for Medical Devices.
16. S. Panda, *et al.*, Sustainable Solutions for Oral Health Monitoring: Biowaste-Derived Triboelectric Nanogenerator, *ACS Appl. Mater. Interfaces*, 2023, **15**, 36096–36106.
17. X. Chen, *et al.*, Advances in Healthcare Electronics Enabled by Triboelectric Nanogenerators, *Adv. Funct. Mater.*, 2020, **30**, DOI: 10.1002/adfm.202004673.
18. Y. H. Joung, Development of implantable medical devices: From an engineering perspective, *Int. Neurourol. J.*, 2013, **17**, 98–106.
19. D. Godwinraj and S. C. George, Recent advancement in TENG polymer structures and energy efficient charge control circuits, *Adv. Ind. Eng. Polym. Res.*, 2021, **4**, 1–8.
20. R. Zhang, *et al.*, Cellulose-Based Fully Green Triboelectric Nanogenerators with Output Power Density of 300 W m^{-2}, *Adv. Mater.*, 2020, **32**, 2002824.
21. Y. Mi, *et al.*, Biodegradable Polymers in Triboelectric Nanogenerators, *Polymers*, 2023, **15**, DOI: 10.3390/polym15010222.
22. W. Wang, *et al.*, Applications of nanogenerators for biomedical engineering and healthcare systems, *InfoMat*, 2022, **4**, DOI: 10.1002/inf2.12262.
23. S. C. Mukhopadhyay, Wearable sensors for human activity monitoring: A review, *IEEE Sens. J.*, 2015, **15**, 1321–1330.
24. S. Peng, Y. Yu, S. Wu and C. H. Wang, Conductive Polymer Nanocomposites for Stretchable Electronics: Material Selection, Design, and Applications, *ACS Appl. Mater. Interfaces*, 2021, **13**, 43831–43854.
25. M. A. Craven, J. W. Gardner and P. N. Bartlett, Electronic noses – development and future prospects, *TrAC, Trends Anal. Chem.*, 1996, **15**, 486–493
26. J. Lee, D. Kim, H. Y. Ryoo and B. S. Shin, Sustainable wearables: Wearable technology for enhancing the quality of human life, *Sustainability*, 2016, **8**, 466.
27. S. Chao, H. Ouyang, D. Jiang, Y. Fan and Z. Li, Triboelectric nanogenerator based on degradable materials, *EcoMat*, 2021, **3**, DOI: 10.1002/eom2.12072.

28. A. Adapa, F. F. H. Nah, R. H. Hall, K. Siau and S. N. Smith, Factors Influencing the Adoption of Smart Wearable Devices, *Int. J. Hum. Comput. Interact.*, 2018, **34**, 399–409.
29. Y. Yang, *et al.*, Human skin based triboelectric nanogenerators for harvesting biomechanical energy and as self-powered active tactile sensor system, *ACS Nano*, 2013, **7**, 9213–9222.
30. Y. P. Jeon, J. H. Park and T. W. Kim, Highly-enhanced triboelectric nanogenerators based on zinc-oxide nanoripples acting as a triboelectric layer, *Appl. Surf. Sci.*, 2018, **445**, 50–55.
31. Y. Wang, Y. Yang and Z. L. Wang, Triboelectric nanogenerators as flexible power sources, *npj Flexible Electron.*, 2017, **1**, 1–9.
32. J. Shao, M. Willatzen and Z. L. Wang, Theoretical modeling of triboelectric nanogenerators (TENGs), *J. Appl. Phys.*, 2020, **128**, 111101.
33. S. Niu and Z. L. Wang, Theoretical systems of triboelectric nanogenerators, *Nano Energy*, 2014, **14**, 161–192.
34. F. Wu, *et al.*, A Flexible, Lightweight, and Wearable Triboelectric Nanogenerator for Energy Harvesting and Self-Powered Sensing, *Adv. Mater. Technol.*, 2019, **4**, 1–7.
35. B. Shi, *et al.*, Body-Integrated Self-Powered System for Wearable and Implantable Applications, *ACS Nano*, 2019, **13**, 6017–6024.
36. G. Q. Gu, *et al.*, Antibacterial Composite Film-Based Triboelectric Nanogenerator for Harvesting Walking Energy, *ACS Appl. Mater. Interfaces*, 2017, **9**, 11882–11888.
37. D. Macário, I. Domingos, N. Carvalho, P. Pinho and H. Alves, Harvesting circuits for triboelectric nanogenerators for wearable applications, *iScience*, 2022, **25**, 1–29.
38. H. Wu, *et al.*, Fully Biodegradable Water Droplet Energy Harvester Based on Leaves of Living Plants, *ACS Appl. Mater. Interfaces*, 2020, **12**, 56060–56067.
39. Y. Li, *et al.*, Low-Cost, Environmentally Friendly, and High-Performance Triboelectric Nanogenerator Based on a Common Waste Material, *ACS Appl. Mater. Interfaces*, 2021, **24**, 30776–30784.
40. H. Yang, F. R. Fan, Y. Xi and W. Wu, Bio-Derived Natural Materials Based Triboelectric Devices for Self-Powered Ubiquitous Wearable and Implantable Intelligent Devices, *Adv. Sustainable Syst.*, 2020, **4**, DOI: 10.1002/adsu.202000108.
41. X. Shi, *et al.*, A strong, biodegradable, and recyclable all-lignocellulose fabricated triboelectric nanogenerator for self-powered disposable medical monitoring, *J. Mater. Chem. A*, 2023, **11**, 11730–11739.
42. Z. Liu, *et al.*, Wearable and Implantable Triboelectric Nanogenerators, *Adv. Funct. Mater.*, 2019, **29**, DOI: 10.1002/adfm.201808820.
43. H. S. Magar, R. Y. A. Hassan and M. N. Abbas, Non-enzymatic disposable electrochemical sensors based on CuO/Co_3O_4@MWCNTs nanocomposite modified screen-printed electrode for the direct determination of urea, *Sci. Rep.*, 2023, **13**, 2034.

44. M. Jiang, Y. Lu, Z. Zhu and W. Jia, Advances in smart sensing and medical electronics by self-powered sensors based on triboelectric nanogenerators, *Micromachines*, 2021, **12**, DOI: 10.3390/mi12060698.
45. Z. Liu, *et al.*, Wearable and Implantable Triboelectric Nanogenerators, *Adv. Funct. Mater.*, 2019, **29**, DOI: 10.1002/adfm.201808820.
46. H. Zou, *et al.*, Quantifying and understanding the triboelectric series of inorganic non-metallic materials, *Nat. Commun.*, 2020, **11**, 1–7.
47. J. Luo and Z. L. Wang, Recent progress of triboelectric nanogenerators: From fundamental theory to practical applications, *EcoMat*, 2020, **2**, 1–22.
48. S. Gokhool, S. Bairagi, C. Kumar and D. M. Mulvihill, Reflections on Boosting Wearable Triboelectric Nanogenerator Performance *via* Interface Optimisation, *Results Eng.*, 2022, 100808.
49. S. N. Alam, *et al.*, An introduction to triboelectric nanogenerators, *Nano-Struct. Nano-Objects*, 2023, **34**, 100980.
50. W. G. Kim, *et al.*, Triboelectric Nanogenerator: Structure, Mechanism, and Applications, *ACS Nano*, 2021, **15**, 258–287.
51. A. K. Aliyana and G. Stylios, A Review on the Progress in Core-Spun Yarns (CSYs) Based Textile TENGs for Real-Time Energy Generation, Capture and Sensing, *Adv. Sci.*, 2023, **10**, DOI: 10.1002/advs.202304232.
52. A. K. Aliyana, *et al.*, Machine learning-assisted ammonium detection using zinc oxide/multi-walled carbon nanotube composite based impedance sensors, *Sci. Rep.*, 2021, **11**, 24321.
53. S. K. Naveen Kumar, A. Baburaj, A. K. Aliyana, H. M. Jalajamony and R. E. Fernadez, Screen Printed IDE Modified Metal Oxide Carbon Nanotube Composite Layer for Urea Fertilizer Detection, in FLEPS 2022 - IEEE International Conference on Flexible and Printable Sensors and Systems, Proceedings, 2022. DOI: 10.1109/FLEPS53764.2022.9781583.
54. S. K. N. Kumar, A. K. Aliyana, A. Baburaj, M. Adetunji and R. E. Fernandez, Impedance Sensor for Real-Time Ammonium Detection Based on MWCNT/ZnO Nanocomposites, *IEEE Trans. Nanobiosci.*, 2023, **22**, 121–127.
55. A. K. Aliyana, *et al.*, Impact of Analyte pH on the Sensitivity of Screen-Printed Flexible Ammonium Sensor, in FLEPS 2022 - IEEE International Conference on Flexible and Printable Sensors and Systems, Proceedings, 2022. DOI: 10.1109/FLEPS53764.2022.9781496.
56. A. Baburaj, *et al.*, Factors affecting the performance of flexible triboelectric nanogenerators (F-TENGs) and their sensing capabilities: A comprehensive review, *Nano Energy*, 2023, **118**, 108983.
57. J. Chen and Z. L. Wang, Reviving Vibration Energy Harvesting and Self-Powered Sensing by a Triboelectric Nanogenerator, *Joule*, 2017, **1**, 480–521.
58. H. J. Yoon, H. Ryu and S. W. Kim, Sustainable powering triboelectric nanogenerators: Approaches and the path towards efficient use, *Nano Energy*, 2018, **51**, 270–285.

59. P. Bai, *et al.*, Integrated multilayered triboelectric nanogenerator for harvesting biomechanical energy from human motions, *ACS Nano*, 2013, **7**, 3713–3719.
60. W. Yang, *et al.*, Harvesting vibration energy by a triple-cantilever based triboelectric nanogenerator, *Nano Res.*, 2013, **6**, 880–886.
61. M. L. Seol, *et al.*, Vertically stacked thin triboelectric nanogenerator for wind energy harvesting, *Nano Energy*, 2015, **14**, 201–208.
62. A. Ghosh, D. Chakraborty and A. Law, Artificial intelligence in Internet of things, *CAAI Trans. Intell. Technol.*, 2018, **3**, 208–218.
63. T. Kamilya and J. Park, Highly Sensitive Self-Powered Biomedical Applications Using Triboelectric Nanogenerator, *Micromachines*, 2022, **13**, 2065.
64. Q. Zheng, *et al.*, Biodegradable triboelectric nanogenerator as a life-time designed implantable power source, *Sci. Adv.*, 2016, **2**, e1501478.
65. Y. Mi, *et al.*, Biodegradable Polymers in Triboelectric Nanogenerators, *Polymers*, 2023, **15**, DOI: 10.3390/polym15010222.
66. W. Wang, *et al.*, Applications of nanogenerators for biomedical engineering and healthcare systems, *InfoMat*, 2022, **4**, DOI: 10.1002/inf2.12262.
67. R. Guo, X. Sun, B. Yuan, H. Wang and J. Liu, Magnetic Liquid Metal (Fe-EGaIn) Based Multifunctional Electronics for Remote Self-Healing Materials, Degradable Electronics, and Thermal Transfer Printing, *Adv. Sci.*, 2019, **6**, 1901478.
68. Z. Liu, *et al.*, Transcatheter Self-Powered Ultrasensitive Endocardial Pressure Sensor, *Adv. Funct. Mater.*, 2019, **29**, 1807560.
69. G. Jian, *et al.*, Biodegradable triboelectric nanogenerator as a implantable power source for embedded medicine devices, *NPG Asia Mater.*, 2024, **16**, 12.
70. J. Liu, *et al.*, A high-performance, biocompatible, and fully biodegradable piezo-triboelectric hybrid nanogenerator based on PVA/Glycine/PVA heterostructured piezoelectric film, *Nano Energy*, 2024, **122**, 109310.
71. A. Baburaj, *et al.*, Biodegradable based TENGs for self-sustaining implantable medical devices, *Nano Energy*, 2024, **127**, DOI: 10.1016/j.nanoen.2024.109785.
72. Z. Wang, *et al.*, Biodegradable, conductive, moisture-proof, and dielectric enhanced cellulose-based triboelectric nanogenerator for self-powered human-machine interface sensing, *Nano Energy*, 2023, **107**, 108151.
73. H. Liu, *et al.*, Fully degradable triboelectric nanogenerator using graphene composite paper to replace copper electrodes for higher output performance, *Nano Energy*, 2023, **108**, 108223.
74. B. Fatma, *et al.*, Biocompatible, breathable and degradable microbial cellulose based triboelectric nanogenerator for wearable transient electronics, *Nano Energy*, 2023, **114**, 108628.
75. L. Feng, X. Cao, Z. L. Wang and L. Zhang, A transparent and degradable bacterial cellulose-based film for triboelectric nanogenerator: Efficient

biomechanical energy harvesting and human health monitoring, *Nano Energy*, 2024, **120**, 109068.
76. C. Jiang, *et al.*, All-electrospun flexible triboelectric nanogenerator based on metallic MXene nanosheets, *Nano Energy*, 2019, **59**, 268–276.
77. R. Wang, S. Gao, Z. Yang, Y. Li, W. Chen, B. Wu and W. Wu, Engineered and Laser-Processed Chitosan Biopolymers for Sustainable and Biodegradable, *Adv. Mater.*, 2018, **30**, 1–8.
78. Y. Han, *et al.*, Fish Gelatin Based Triboelectric Nanogenerator for Harvesting Biomechanical Energy and Self-Powered Sensing of Human Physiological Signals, *ACS Appl. Mater. Interfaces*, 2020, **12**, 16442–16450.
79. G. Landi, A. Sorrentino, S. Iannace and H. C. Neitzert, Differences between graphene and graphene oxide in gelatin based systems for transient biodegradable energy storage applications, *Nanotechnology*, 2017, **28**, 054005.
80. Z. Qin, *et al.*, Flexible Janus Electrospun Nanofiber Films for Wearable Triboelectric Nanogenerator, *Adv. Mater. Technol.*, 2020, **5**, 1–9.
81. F. He, *et al.*, Stretchable, Biocompatible, and Multifunctional Silk Fibroin-Based Hydrogels toward Wearable Strain/Pressure Sensors and Triboelectric Nanogenerators, *ACS Appl. Mater. Interfaces*, 2020, **12**, 6442–6450.
82. R. Wang, *et al.*, Engineered and Laser-Processed Chitosan Biopolymers for Sustainable and Biodegradable Triboelectric Power Generation, *Adv. Mater.*, 2018, **30**, 1–8.
83. S. Chen, J. Jiang, F. Xu and S. Gong, Crepe cellulose paper and nitrocellulose membrane-based triboelectric nanogenerators for energy harvesting and self-powered human-machine interaction, *Nano Energy*, 2019, **61**, 69–77.
84. Y. Chi, *et al.*, Rice paper-based biodegradable triboelectric nanogenerator, *Microelectron. Eng.*, 2019, **216**, 111059.
85. W. B. Han, J. H. Lee, J. W. Shin and S. W. Hwang, Advanced Materials and Systems for Biodegradable, Transient Electronics, *Adv. Mater.*, 2020, **32**(51), 2002211.
86. M. Čolnik, M. K. Hrnčič, M. Škerget and Ž. Knez, Biodegradable polymers, current trends of research and their applications, a review, *Chem. Ind. Chem. Eng. Q.*, 2020, **26**, 401–418.
87. Y. Lyu and Y. Wang, Output optimization of biodegradable triboelectric nanogenerators, *Nano Energy*, 2022, **103**, DOI: 10.1016/j.nanoen.2022.107811.
88. K.-L. G. Ho, A. L. Pometto and P. N. Hinz, Effects of Temperature and Relative Humidity on Polylactic Acid Plastic Degradation 1, *J. Environ. Polym. Degrad.*, 1999, **7**, 83–92.
89. R. Solaro, A. Corti and E. Chiellini, Biodegradation of Poly(Vinyl Alcohol) with Different Molecular Weights and Degree of Hydrolysis, *Polym. Adv. Technol.*, 2000, **11**(8–12), 873–878.
90. V. R. Sinha, K. Bansal, R. Kaushik, R. Kumria and A. Trehan, Poly-ε-caprolactone microspheres and nanospheres: An overview, *Int. J. Pharm.*, 2004, **278**, 1–23.

91. Q. Zheng, *et al.*, Biodegradable triboelectric nanogenerator as a lifetime designed implantable power source, *Sci. Adv.*, 2016, **2**, e1501478.
92. S. Hu, *et al.*, Biodegradable, Super-Strong, and Conductive Cellulose Macrofibers for Fabric-Based Triboelectric Nanogenerator, *Nano-Micro Lett.*, 2022, **14**, 115.
93. Z. Li, *et al.*, Photothermally tunable biodegradation of implantable triboelectric nanogenerators for tissue repairing, *Nano Energy*, 2018, **54**, 390–399.
94. Z. Li, *et al.*, Photothermally tunable biodegradation of implantable triboelectric nanogenerators for tissue repairing, *Nano Energy*, 2018, **54**, 390–399.
95. Q. Zheng, *et al.*, Biodegradable triboelectric nanogenerator as a lifetime designed implantable power source, *Sci. Adv.*, 2016, **2**, e1501478.
96. L. Zhou, *et al.*, Study on the regularity of cellulose degradation in ionic liquids, *J. Mol. Liq.*, 2020, **308**, 113153.
97. F. Ganji, M. J. Abdekhodaie and A. Ramazani, Gelation time and degradation rate of chitosan-based injectable hydrogel, *J. Solgel. Sci. Technol.*, 2007, **42**, 47–53.
98. M. Li, M. Ogiso and N. Minoura, Enzymatic Degradation Behavior of Porous Silk Fibroin Sheets, *Biomaterials*, 2003, **24**, 357–365.
99. I. Manavitehrani, *et al.*, Biomedical applications of biodegradable polyesters, *Polymers*, 2016, **8**, DOI: 10.3390/polym8010020.
100. C. Wang, T. Chang, H. Yang and M. Cui, Surface physiological changes induced by lactic acid on pathogens in consideration of pKa and pH, *Food Control*, 2014, **46**, 525–531.
101. J. C. Middleton and A. J. Tipton, Synthetic biodegradable polymers as orthopedic devices, *Biomaterials*, 2000, **21**, 2335–2346.
102. Y. Ikada and H. Tsuji, Biodegradable Polyesters for Medical and Ecological Applications, *Macromol. Rapid Commun.*, 2000, **21**, 117–132.
103. Y. Wang, M. A. Murcia Valderrama, R.-J. van Putten, C. J. E. Davey, A. Tietema, J. R. Parsons, B. Wang and G.-J. M. Gruter, Biodegradation and Non-Enzymatic Hydrolysis of Poly(Lactic-co-Glycolic Acid) (PLGA12/88 and PLGA6/94)), *Polymers*, 2022, **14**(1), 15.
104. E. Díaz, I. Sandonis and M. B. Valle, In vitro degradation of poly (caprolactone)/nHA composites, *J. Nanomater.*, 2014, **2014**, 802435.
105. C. X. F. Lam, D. W. Hutmacher, J. T. Schantz, M. A. Woodruff and S. H. Teoh, Evaluation of polycaprolactone scaffold degradation for 6 months in vitro and in vivo, *J. Biomed. Mater. Res., Part A*, 2009, **90**, 906–919.
106. M. Ding, M. Zhang, J. Yang and J. Qiu, Study on the enzymatic degradation of PBS and its alcohol acid modified copolymer, *Biodegradation*, 2012, **23**, 127–132.
107. A. Lindström, A. C. Albertsson and M. Hakkarainen, Quantitative determination of degradation products an effective means to study early stages of degradation in linear and branched poly(butylene adipate) and poly(butylene succinate), *Polym. Degrad. Stab.*, 2004, **83**, 487–493.

CHAPTER 6

Biochemical Energy Harvesting Using Microbial Fuel Cells: Potential Applications in Implantable and Ingestible Devices

SARA AKHTAR,[a] FELIPE M. DE SOUZA[b] AND RAM K. GUPTA*[b,c]

[a] Department of Biology, Pittsburg State University, Pittsburg, KS 66762, USA; [b] National Institute for Material Advancement, Pittsburg State University, Pittsburg, KS 66762, USA; [c] Department of Chemistry, Pittsburg State University, Pittsburg, KS 66762, USA

*Email: ramguptamsu@gmail.com

6.1 Introduction

The need to find alternative sources of energy in the face of the excessive use of fossil fuels is an important approach to reduce the use of non-renewable sources. In addition, the continuous need for the proper functioning of modern society is strongly dependent on agriculture and industries such as textiles, mining and metal processing, gas and oil, food and beverages, chemical manufacturing, and power generation to name a few. Inevitably, these industries generate a considerable amount of wastewater. Because of that, when such residues are unproperly discarded in the environment they cause severe pollution of ecosystems leading to issues such as eutrophication.[1–3] Hence, finding ways to manage the waste generated by

Detection Science Series No. 33
Energy Materials for Ingestible and Implantable Healthcare Devices
Edited by Santanu Patra and Anja Boisen

Published by the Royal Society of Chemistry, www.rsc.org

agriculture and this diverse set of industries is crucial to moving towards a sustainable future. One of the technologies that is gaining attention as a candidate to address that is microbial fuel cells (MFCs) which can aid in the mitigation of large amounts of wastewater while producing electricity. Thus, MFCs can simultaneously perform waste management and generation of energy. In principle, MFCs are devices that harness chemical energy from the metabolic processes of microorganisms and convert it to electricity. These cells function by breaking down organic compounds, such as biomass or wastewater, into protons and electrons that can be used to generate electricity. An MFC functions using several key components. The anode is the electrode at which the microbes grow to form a biofilm. They oxidize the organic matter to convert it into electrons and protons as their metabolic pathway.

Usually, conducting materials are utilized as a support for the growth of microbes to facilitate the flow of electrons. Hence, graphene, graphite, carbon cloth, and other conducting materials can be utilized. The microbes are the biocatalysts that perform the actual oxidation of the organic matter at the anode's surface and perform the e^- transfer process either with a mediator or without it. The organic matter, such as wastewater, biomass, or glucose-rich mixtures, can be utilized as fuel which is metabolized by the microbes releasing H^+, e^-, and CO_2. The cathode is the electrode that combines the H^+, which passes through a proton exchange membrane (PEM) with e^-, which passes through an external circuit, in the presence of O_2, usually employed as the electron acceptor, to perform an oxygen reduction reaction (ORR). This step is necessary for the completion of the redox reaction processes as it ensures that electrons can continue to flow from the anode to the cathode, to generate electricity. The cathode can be Pt-based, metal oxides, conducting polymers, and N-doped carbon materials, among others. The PEM is another important component in MFCs as it separates both electrode compartments while allowing the passage of only H^+ generated at the anode's surface. It avoids the mixing of O_2 and organic matter and maintains the anode under anaerobic conditions. The electrolyte is typically an aqueous solution present in both electrode chambers that allows for the flow of ions. Since it is separated by the PEM, only H^+ present in the electrolyte can pass through it from the anode to the cathode. The anode and cathode chambers are the compartments in which the anode and cathode are located, respectively. Since the anode contains the microorganisms, the system is kept under anaerobic conditions to promote the microbes to donate the e^- to the anode. On the other hand, the anode chamber is aerobic, since O_2 is often utilized as the e^- acceptor to perform ORR and close the redox system. Figure 6.1 displays a typical schematic of an MFC.

This technology offers a renewable source of power while solving environmental issues, therefore it has potential for sustainable energy generation, especially when combined with waste treatment. Some of the advantages of MFCs include: (i) direct conversion of the substrate into electricity using a biofilm as an electrocatalyst, (ii) it can be operated at

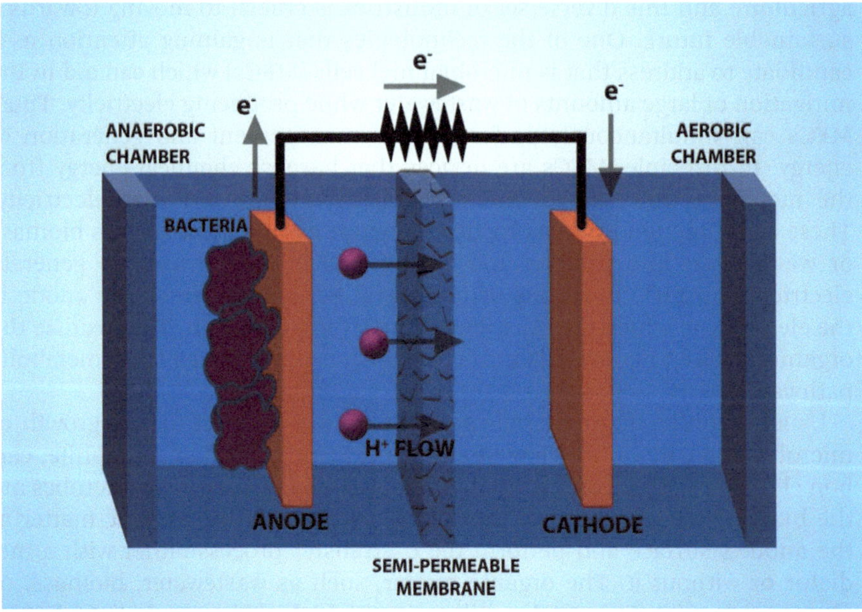

Figure 6.1 Typical schematic of an MFC. Adapted from ref. 4, https://doi.org/10.3390/mi5020171, under the terms of the CC BY 4.0 license, https://creativecommons.org/licenses/by/4.0/.

temperatures between 15 to 45 °C along with pH values close to neutrality, (iii) it generates a lower number of residues when compared to other treatments, (v) different types of wastes can be utilized as substrates, and (vi) it doesn't require input of energy for aeration.[5] However, despite these advantages, there are also some drawbacks such as (i) low power density which is attributed to inherently low metabolic rates of microbes in contrast to chemical reactions that occur in traditional fuel cells, (ii) the cost of some components such as PEMs, for instance, is relatively high, which increases the overall investment required, (iii) microbes require an inoculation time to proliferate, colonize, and stabilize in the electrode's surface to make the MFC operational, which requires time and constant surveillance to keep growth parameters in the optimal conditions, (iv) the organic substrate should be constantly supplied. However, since the concentration of organic material changes based on the waste utilized so does the microbial activity, which can lead to inconsistent energy generation, (v) the environment containing the active microbes, in most cases, should be anaerobic to avoid the diffusion of O_2 from the cathode to the anode, which can hinder microbial activity through a decrease in the anaerobic metabolic process, and (vi) lastly, there are still some challenges to translating small-scale to large-scale MFCs as it can lead to higher resistance, inconsistent energy generation due to uneven growth of microbes over the electrode, and maintaining an anaerobic system in large volumes.

Despite the recurring challenges, MFCs offer several different promising applications. From wastewater treatment to producing electricity and environmental monitoring, MFCs are useful for treating organic waste, reducing pollution, and developing the circular economy that turns residues into a resource. Their adaptability allows them to be used in a wide range of applications, such as providing energy for low-power applications, powering remote sensors, and supporting biomedical devices. The progress in research and development of MFCs focuses on improving electrode materials, microbial selection, and system design, promising enhanced efficiency, scalability, and usage. This makes them an attractive and compelling choice for communities seeking cleaner energy sources and more effective waste management systems. MFCs offer a viable path toward sustainable development, utilizing the metabolic powers of microbes to solve urgent global issues as the world strives for novel solutions to energy and environmental problems. Based on these aspects permeating MFC technology, this chapter describes the main working mechanisms of MFCs to elucidate their main aspects in terms of components, types, and main mechanisms. Then, an application section is provided containing some of the most recent examples from the literature to aid the reader in getting novel insights and understanding the main feasible approaches and bottlenecks currently faced by the scientific community. Some of the examples of applications presented in this chapter are wastewater treatment, power supply, biosensors, and implantable and ingestible MFCs. Lastly, a perspective section summarizes the main aspects, challenges, and novelties while providing an overview of the field of MFCs.

6.2 Principles of MFCs

An MFC is a device capable of generating electricity through the metabolic activity of microorganisms breaking down organic matter that leads to the release of e^- and H^+ that can be harnessed during the process. The system functions through the assembling of the components such as anode, cathode, PEM, biocatalyst, electrolyte, substrate, and an optional aeration system. The performance of an MFC is influenced by several factors such as the type of material and microorganism utilized for the fabrication of the electrode, the type of substrate, the electrolyte's ionic strength, the type of membrane, and the configuration of the MFC.

6.2.1 Components and Microorganisms

The anode is the electrode at which oxidation of the substrate takes place through the metabolic action of a microorganism leading to the release of e^- and H^+. It is often composed of carbon-based conducting materials such as graphite, carbon paper, and carbon cloth, among others. This allows the biofilm to be adhered to the electrode's surface while facilitating the e^- transfer steps. The cathode is the electrode at which reduction takes place, usually through the presence of O_2 as an e^- acceptor. It can be composed of Pt or materials such as activated carbon which present a high surface area and appreciable conductivity. The PEM separates the electrodes ionically as it allows the passage of only H^+ from the anode to the cathode while preventing the flow of e^- and other species. One of the most utilized PEMs is Nafion, which is a polymeric backbone of polytetrafluoroethylene (PTFE) with a perfluorinated-vinyl-polyether side chain that contains end groups of sulphonic acid. It has good conductivity of H^+ along with displaying high mechanical strength and chemical stability. However, it is relatively costly and can lead to fouling when utilized for wastewater applications. Ultrex is another PEM composed of polystyrene-divinylbenzene copolymer which can be more cost effective while presenting a slightly lower H^+ conductivity when compared to Nafion. Other materials utilized as PEMs include Aquivion, which, like Nafion, is a PTFE with shorter side chains containing sulphonic acid-terminated groups; polybenzimidazole (PBI); sulphonated polyether ether ketone (SPEEK); and chitosan-based membranes.

The biocatalysts can be microorganisms such as bacteria, archaea, and other microbes, or biomolecules such as enzymes which are capable of oxidizing organic matter to convert it into other components, e^- and H^+. Some examples include *Geobacter sulfurreducens* which is capable of transferring e^- directly to the electrodes without a mediator component while promoting high current densities.[6,7] *Shewanella oneidensis* is a bacterium that can perform both direct and mediated e^- transfer steps. In addition, it presents a versatile metabolism as it can degrade several organic compounds such as lactate, acetate, and butyrate. Also, during its metabolic routes, it can produce flavins which facilitate the transfer of e^-.[8,9] *Escherichia coli* is a model bacterium that is researched for genetic modification which allows for enhancements of its e^- transfer rates.[10,11] *Pseudomonas aeruginosa* has been used as a biocatalyst as it can synthesize phenazines which are soluble redox mediators that facilitate the e^- transfer steps outside the bacteria cell.[12,13] *Clostridium butyricum* is a bacterium that can biosynthesize H_2 through the consumption of bio-renewable sources such as cellulose, sugarcane bagasse, or dairy wastewater.[14,15] *Rhodopseudomonas palustris* is a bacterium that can grow in the presence of light, which allows it to be used in photo-MFCs while degrading specific types of substrates that can aid in wastewater treatment.[16,17] Other examples include *Enterobacter* sp.,[18,19] *Desulfovibrio desulfuricans*,[20,21] and *Acetobacter aceti*,[22,23] among others.

6.2.2 Types of MFCs

MFCs can be initially divided into single- and double-chambered. As implied, the single-chambered MFCs consist of one chamber that hosts both the anode and cathode at which the cathode can be exposed to air in the case of air-cathode MFCs. It presents a simpler design which can make it more affordable. However, it may display lower efficiency due to the diffusion of O_2 from the cathode to the anode which can potentially capture e^- generated from the system. The double-chambered MFCs have their electrodes separated by a PEM which often results in better efficiency of the redox reactions as it can better maintain the electrodes being separated which avoids the capturing of e^- by acceptors. However, it also displays a more complex design and higher costs. Yet, the double-chambered MFC tends to be the most common type as the addition of a PEM for ionic balance is an important step for the separation of the cathode and anode to improve redox efficiency. The e^- generated by the microorganisms on the anode's surface or in the anolyte are transferred to the cathode which contains e^- acceptor species such as O_2. For proper functioning of the MFC, it is desired that anaerobic conditions are maintained in the anode, whereas the cathode displays reducing conditions. Schematics of a double- and a single-chambered MFC are presented in Figure 6.2.

The H-shaped double-chambered MFC is a simpler design that is often utilized for lab scale, research, and educational purposes due to its simplicity, low cost, and accessibility. However, it displays a lower power output and difficulties in scaling the salt bridge. A typical example was presented by Weibel *et al.*[24] who fabricated an H-shaped double-chambered MFC to produce electricity from the metabolism of *Aspergillus niger* through the breakdown of glucose. The mediation of e^- was performed by adding 2,6-dichlorophenolindophenol, a redox dye. In addition, Pt foils were used as electrodes, KCl in a phosphate buffer was used as the electrolyte, and there was a salt bridge containing 1% agar. Other studies demonstrated different types of wastes that could be utilized as substrates for the generation of energy, such as swine wastewater[25] and goat rumen fluid,[32] for example.

The up-flow MFC is another type of design in which wastewater flows in the upper direction through the anode chamber, promoting better contact

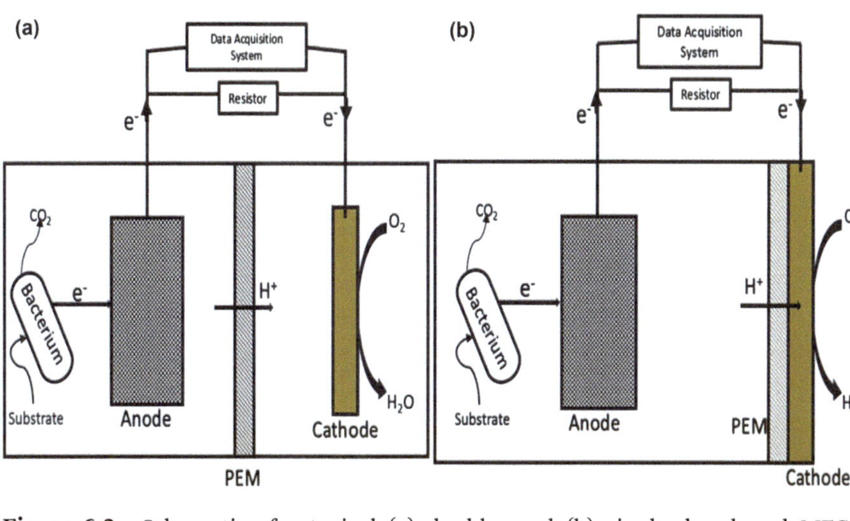

Figure 6.2 Schematics for typical (a) double- and (b) single-chambered MFCs. Adapted from ref. 26 with permission from Elsevier, Copyright 2019.

between the substrate and the active microorganisms in the biofilm. In addition, these can be continuously operated. An example was presented by Rabaey *et al.*[27] who utilized a cation exchange membrane that was folded into a sealed cylindrical structure. The anodic electrode matrix was composed of graphite granules, whereas the cathode was composed of a woven graphite mat. The anolyte (substrate) and the catholyte [$K_3Fe(CN)_6$ solution in phosphate buffer] were dripped over the anode and cathode, respectively, at a rate of 6 $L\,h^{-1}$. The schematics for the up-flow MFC are presented in Figure 6.3a and b. The system could degrade different types of substrates such as glucose, wastewater, and acetate. Among those, the latter displayed the highest power density of 52 $W\,m^{-3}$.

There are several other types of variations for MFCs which, aside from single-, double-chambered, H-shaped, and up-flow, also include stacked, membraneless, air-cathode, sediment MFCs (SMFCs), photosynthetic MFCs (photo-MFCs), and hybrid, among others. As previously mentioned, double-chambered MFCs are usually more utilized than single-chambered. Yet, both

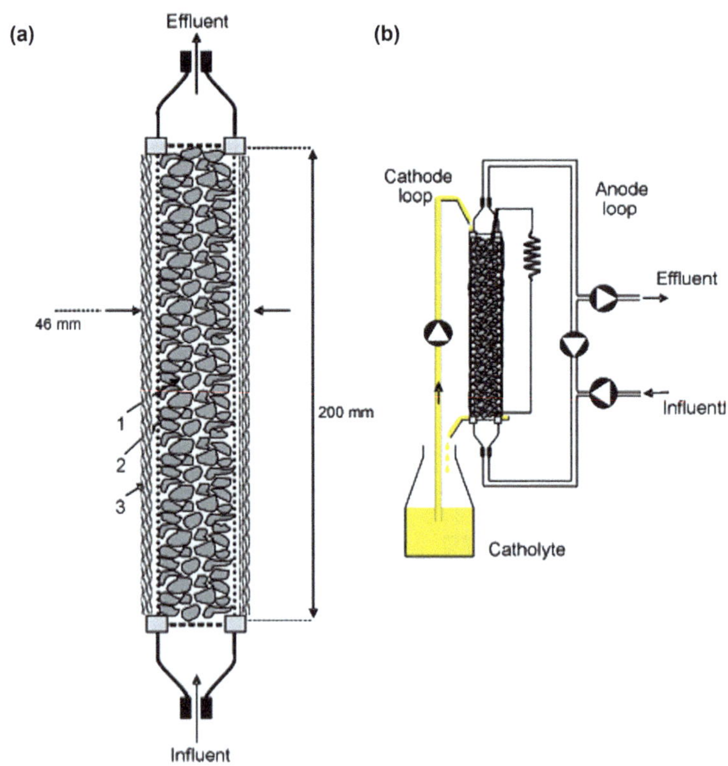

Figure 6.3 Schematic of an up-flow tubular MFC. (a) An MFC system in which 1 is the granular anode, 2 is the membrane, and 3 is the cathode. (b) Complete setup. Adapted from ref. 27 with permission from American Chemical Society, Copyright 2005.

can be used for small energy generation applications. The membraneless and SMFCs have a lower cost, and other types such as microbial electrolysis cells (MECs) and photo-MFCs can be utilized to produce H_2, a core process for sustainability soon.

6.2.2.1 *Air-cathode MFCs*

The air-cathode MFCs consist of a system in which the cathode is exposed to air to allow O_2 to react with the e^- and H^+ generated by the cell. Because of that, internal aeration of the cell is not required, which decreases the overall cost of the cell. However, reinforcement in the cathode is often required to withstand the hydraulic pressure to reduce the leakage of electrolytes. A previous study performed by Liu *et al.* [28] showed a change in performance for the generation of electricity in a single-chambered air-cathode MFC with and without a PEM. In their study, wastewater and glucose were used as substrates for a bacterium commonly present in wastewater. It was observed that with the PEM the power density was around 262 mW m^{-2}, whereas without the PEM it was 494 mW m^{-2}. The reason for that could be attributed to the reduction of internal resistance upon removal of the PEM. In addition, it could be that O_2 could diffuse more easily to the cathode, enhancing the oxygen reduction reaction (ORR), and increasing the power density. However, with a PEM, the Coulombic efficiency was around 40–55%, whereas without a PEM it dropped to 9–12%. This considerable drop occurred because O_2 present in the anode chamber can capture e^- generated by the microbes that would otherwise flow to the external circuit. This effect diminishes the amount of e^- available for the generation of electricity, hence decreasing the Coulombic efficiency. This study showed a trade-off in the design of an air-cathode MFC in which without a PEM it displayed higher power density and lower Coulombic efficiency, whereas the opposite was observed for an MFC with a PEM. Typical schematics of an air-cathode MFC with and without a PEM or ionic exchange membrane (IEM) are presented in Figure 6.4a and b, respectively. Based on these aspects, the use of a dual- or single-chambered air-cathode MFC largely depends on the desired applications. For that, some of the advantages of the single-chambered air-cathode MFC are: (i) it is not needed to aerate the catholyte as the O_2 can directly react at the cathode to perform ORR, (ii) it does not require chemical regeneration or recycling of the catholyte, and (iii) it allows for more accessible and higher volumetric power densities. However, some of the drawbacks are (i) excessive exposure to O_2 that leads to fouling of the cathode, (ii) higher diffusion of O_2 into the anodic chamber, (iii) faster evaporation rate of electrolyte, and (iv) need for a catalyst and a binder on the cathode.[29]

6.2.2.2 *Sediment MFCs*

Sediment MFCs consist of energy generation devices in which the biocatalysts utilize organic matter from soil as substrate. This is an important

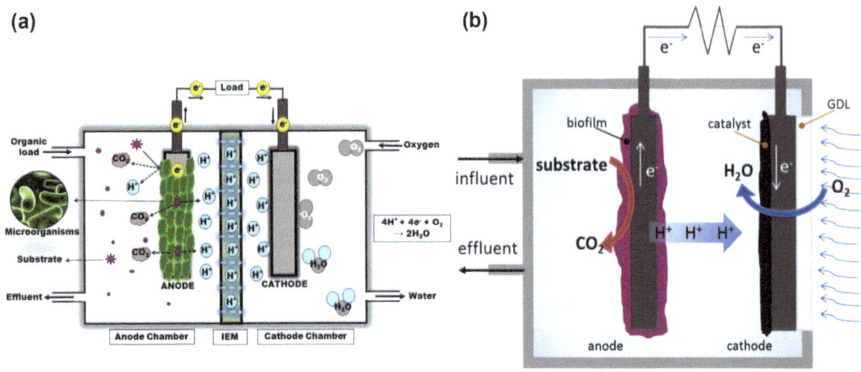

Figure 6.4 (a) Schematic of a typical dual-chambered air-cathode MFC separated by an IEM. Adapted from ref. 30 with permission from Elsevier, Copyright 2019. (b) Schematic of a typical single-chambered air-cathode MFC. Adapted from ref. 31 with permission from Elsevier, Copyright 2018.

technology as it can perform environmental remediation by removing pollutants that can serve as substrate for microorganisms, whereas it can also provide electricity in remote aquatic areas. In terms of design, one of the advantages of this type of MFC is that it does not require a membrane which facilitates the configuration. It displays better performance when soils with considerably high content of organic matter, such as those found in the bottom of aquatic regions including seas and lakes, are utilized. However, it also presents a relatively lower power density when compared to other MFC types. The components of a sediment MFC consist of an anodic and a cathodic part. The anodic part is placed at the bottom of the aquatic environment, such as a river, lake, or sea, where microorganisms can perform anaerobic oxidation of the substrates found in the sediment to produce e^-.[32] The cathode is placed in an upper water layer that contains a higher concentration of dissolved O_2, which enables reduction reactions such as ORR to take place.[33] The two parts are connected by a conducting wire to transfer the e^-. A typical schematic for a sediment MFC is presented in Figure 6.5a. It is worth noting that sediment MFCs operate in an open environment, which makes them highly influenced by their conditions based on factors such as the variety of substrates that can make contact with the biofilm in the electrode, variations in pH and temperature, among others. An example was presented by Remeirs *et al.*,[34] who positioned the cathode deep within the seabed, whereas the anode was placed in the water above the sediment which contained higher concentrations of O_2. The system provided a power density of 0.01 $W\,m^{-2}$. Despite the relatively low power density, there is a considerable variety of microorganisms that were reported as potential bioactive film components for electrodes in sediment MFCs, with some examples presented in Figure 6.5b.

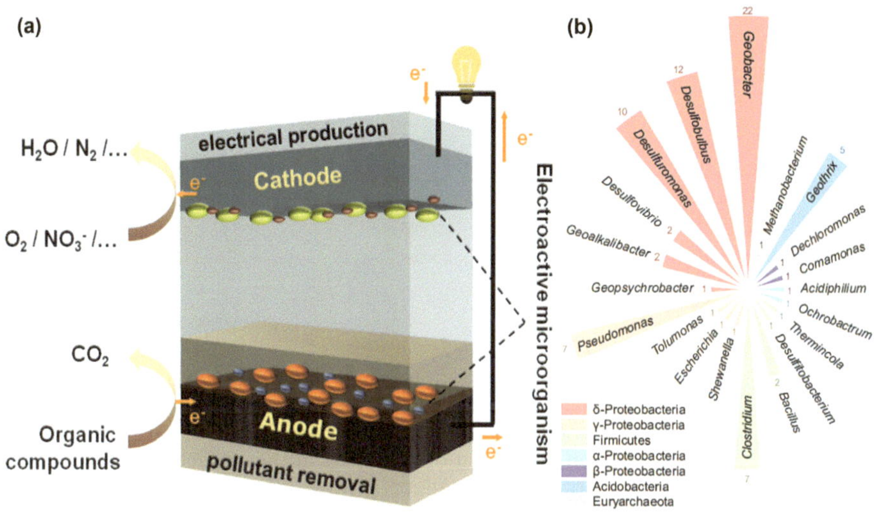

Figure 6.5 (a) Schematic of a typical sediment MFC. (b) Most cited electroactive microorganisms in sediment MFCs. A higher number represents a larger average number of publications related to that microorganism. Adapted from ref. 35 with permission from Elsevier, Copyright 2021.

6.2.2.3 *Photo-MFCs*

A photo-MFC typically features an anode containing photosynthetic microorganisms (algae or cyanobacteria) that are exposed to light which can generate e^- through photosynthetic processes. In some designs of these MFCs, a semiconductor can be used to enhance the e^- transfer processes. When irradiated by light the microorganisms perform photo biochemical processes generating e^- that are transferred to the anode, whereas reduction reactions take place at the cathode. When a semiconductor is utilized, light absorption can promote charge separation, creating e^-/H^+ pairs. The role of the microorganisms in generating the e^- is that they can improve the e^- transfer rate from the semiconductor's valence band (VB) into the conduction band (CB). Because of that, the MFC's performance is strongly influenced by the biofilm's capacity to provide e^-. Schematics for the working principles of a photo-MFC are presented in Figure 6.6a. An example of a photo-MFC can consist of an anode and cathode separated by an IEM. The electrodes are connected by wires as the photogenerated e^- flow from the anode to the cathode. Simultaneously, the H^+ generated can be used to potentially degrade a pollutant, promoting the chemical energy balance.[36] A schematic of this process is illustrated in Figure 6.6b. Based on these processes, a photocatalytic-assisted MFC can perform simultaneous photocatalysis as well as biodegradation in the anode chamber. In this sense, the photocatalytic material and the microorganism can degrade the substrate (*i.e.*, pollutant) acting as co-catalysts for the process. In addition, the

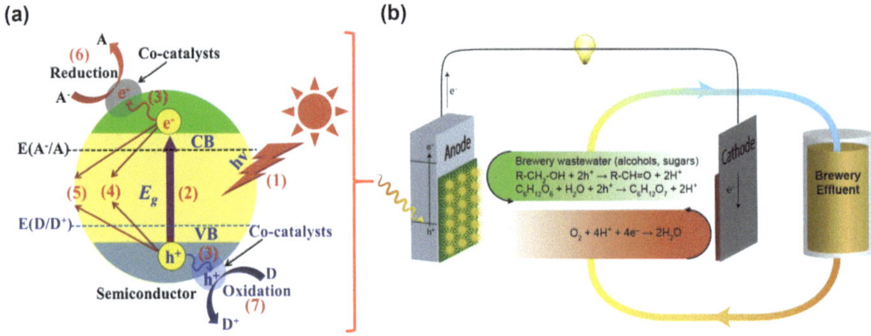

Figure 6.6 (a) Schematics for a typical photocatalytic process: (1) light absorption; (2) excitation of charges through the incident radiation; (3) separation and transfer of charges; (4) charge recombination; (5) surface charge recombination; (6) reduction reactions at the surface; and (7) oxidation reactions at the surface. Adapted from ref. 40 with permission from Elsevier, Copyright 2017. (b) Schematics for the main processes that take place in a photo-MFC with wastewater treatment. Adapted from ref. 41 with permission from Elsevier, Copyright 2019.

intermediate products formed during the degradation of the pollutant can be used as substrates for either the photocatalytic material or the microorganism. Because of that, this technology could be a valuable asset for environmental remediation through the degradation of pollutants while generating electricity. However, the complexity of the chemical structure of the pollutant that is being degraded can lead to the production of unknown components that may lead to undesired effects.[37] Despite that, there are several components obtained from renewable sources that can be utilized as fuel, such as ethanol, methanol, or glycerol. The latter is a by-product of the biodiesel industry which can potentially aid in the promotion of a circular economy.[38] In summary, the complex working process of a photo-MFC is based on a light-driven photosynthetic process performed by a microorganism at the anode to generate e^-. These e^- are then transferred to the anode and then flow through an external circuit to the photocathode. If a semiconductor is utilized at the photocathode, it can be activated by light, creating e^-/H^+ pairs. The e^- from the bio anode flows through the circuit to the photocathode, where they can perform in reduction reactions, such as ORR.[39]

6.2.2.4 *Hybrid MFCs*

A hybrid MFC system consists of an MFC conjugated with other technologies to optimize individual processes. This combination can be challenging due to the requirement for compatible components that must function properly when assembled. One example of a hybrid system is an MFC with a microbial electrolysis cell (MEC), MFC-MEC. In an MEC an external voltage is applied as a driving force for the hydrogen evolution reaction (HER) at which H^+ in the anode is reduced to H_2 in the cathode. It works under the same principle as an MFC. However, in an MFC-MEC hybrid system, the energy input required for the MEC can be supplied by an MFC, which can make the process self-sustaining. The external energy input required for the HER at the cathode is often a small amount, as the majority of the energy is provided from the chemical energy obtained from the oxidized substrates at the anode.[42] This condition allows for the production of H_2 through a bioelectrocatalytic process that is supported by an additional low-energy source.[42,43] An example was presented by Sun *et al.*[44] who proposed a MEC–MFC system in which the MFC provided a power supply for the MEC to produce H_2 from acetate. The system was based on a single-chambered MFC along with a dual-chambered MEC, both containing PEMs. The anodes were composed of carbon paper, whereas the cathodes were composed of carbon paper with Pt for both the MFC and MEC. The H_2 production rate was around 14.9 mL/L×d along with a cathodic H_2 recovery rate of 88–96%, Coulombic efficiency of 28–33%, and a yield of around 1.6 H_2/acetate per mol of each. Some of the main advantages of this approach were related to satisfactory yields of H_2 utilizing biowaste at room temperature, whereas the H_2 production could be optimized by coupling two or more MFCs in series. A schematic of a hybrid system for an MEC–MFC is presented in Figure 6.7.

6.2.3 Mechanisms of Operation

The mechanism of e^- transfer in MFCs can be classified as direct (without a mediator) also known as direct extracellular electron transfer (DEET) or indirect (with a mediator) known as mediated extracellular electron transfer (MEET).[45] DEET is a case in which microorganisms can transfer e^- to the electrode through direct contact with the electrode, which can occur through their pili or nanowires present in their outer cell structure. Some examples of microorganisms that can process the e^- without the need for a mediator include *Geobacter sulfurreducens*,[46] *Geobacter metallic reducens*,[47] *Aeromonas hydrophila*,[48] *Shewanella putrefaciens*,[48,49] and *Klebsiella pneumoniae*,[50,51] among others. Yet, to optimize the performance of mediator-less MFCs it is important to address factors that influence the generation of electricity. For that it is important to enhance the electrochemical activity of redox enzymes for efficient e^- to the anode while improving fuel oxidation at the anode, which can be performed by achieving the optimal concentration of substrate, temperature, and controlling pH, for instance. Also, optimizing the H^+ transfer through PEM to the cathode is essential since a slow H^+ permeability can greatly reduce the MFC's performance.[52] A schematic for the DEET mechanism is presented in Figure 6.8a. In the case of MEET, e^- transfer with a mediator, the microorganism lacks electroactive surface proteins or microstructures that can perform the e^- transfer process. To compensate for that, there is a need for other components that can extract the e^- from the microorganism's membrane and shuttle it to the electrode's surface. Such components are the mediators or electroactive metabolites. Depending on the case, the mediators can be secreted by the microorganism

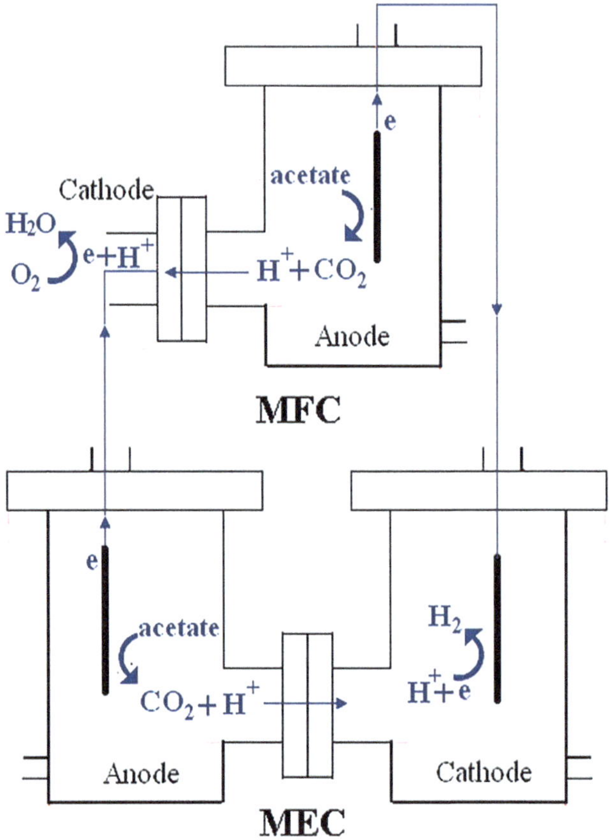

Figure 6.7 Hybrid system of an MEC–MFC for the generation of H_2 using acetate as a substrate. Adapted from ref. 44 with permission from American Chemical Society, Copyright 2008.

itself, which includes components such as pyocyanine, flavins, and phenazines, or synthetical mediators can be added for the same purpose, which include components such as neutral red, methylene blue, thionine, ferricyanide, and riboflavin, among others.[53,54] The mediators can penetrate the microorganism's outer lipidic membrane and cell wall liberating e^- that were in the transport chain, leading to the electrochemical reduction of the mediator. Then, the mediator exits the microorganism's cell and transfers the e^- to the surface of the anode making it return to its initial oxidized state. A schematic for the MEET mechanism is presented in Figure 6.8b. Even though the use of mediators improves the electricity generated by the MFC it can lead to an increase in cost and toxicity in some cases. Another important factor related to the overall performance of an MFC is the type of microbial community and their abundance in the system.[55] The microbial communities can be defined into two categories, which are the electroactive bacteria (EAB) and the non-EAB. The former is related to the microorganisms capable of transferring e^- which is a process that occurs during extracellular respiration at which the generated e^- through an anaerobic oxidation, in the absence of O_2, can be transferred to an e^- acceptor outside the cell.[56]

On the other hand, non-EAB only perform intracellular respiration processes, meaning that e^- are not transferred through the cellular membrane to a conductive media external to the cell. Because of that, EAB is majorly researched as they can directly or indirectly generate e^- through the degradation of a substrate, which can be a biowaste, to produce an electrical current.[57] Aside from that, different types of bacteria can consume specific types of pollutants. For example, Burkholderiaceae, a family of bacteria, can reduce nitrite and ammonium through nitrification and oxidation of ammonia, respectively.[58] Other examples are alphaproteobacteria, which can reduce NO_3^-, NO_2^-, and N_xO_y present in sewage, and betaproteobacteria, which can oxidize N from NH_3 into NO_2^-.[59,60] Notable examples of non-EAB are zymophytes and methanogens, among others. The zymophytes can be used to degrade larger organic molecules into smaller and easier to degrade substances, which can be further metabolized by other EABs to continue a purification process while generating electricity. Based on these processes, a consortium of bacteria can be utilized to perform a more complete degradation process of a certain type of waste at which one bacterium can act on the initial degradation steps whereas others can continue the process to generate e^-. A schematic for the mechanisms involved in a consortium of bacteria for a more complex degradation and energy generation process is presented in Figure 6.8c.

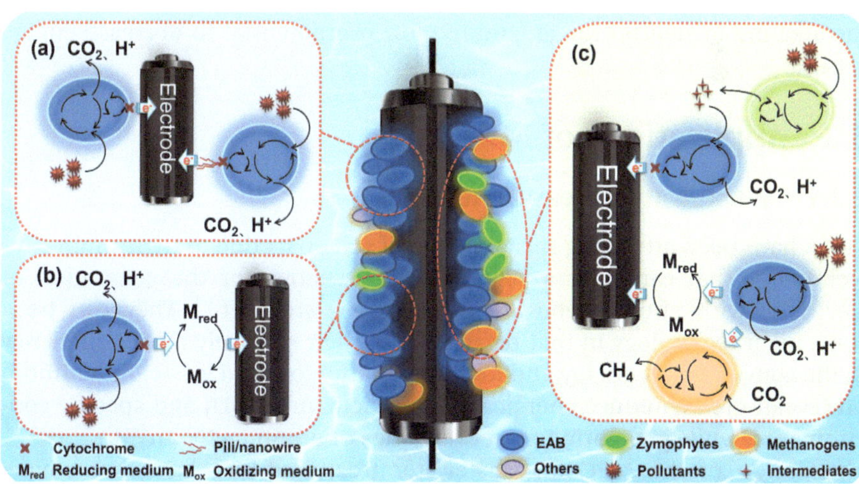

Figure 6.8 Schematic for the mechanisms of biodegradation and energy generation for different types of bacteria: (a) DEET mechanism, (b) MEET mechanism, and (c) bacteria consortium for the degradation of more complex substrates using EAB and non-EAB. Adapted from ref. 59 with permission from Elsevier, Copyright 2020.

The understanding of the e^- transfer mechanisms in an MFC provides an important insight into their functioning. In addition, it is also important to elucidate the main chemical processes that take place during its functioning. As already discussed, an MFC consists of an anode, cathode, PEM, and an external circuit for e^- flow. The anodic chamber is usually an anaerobic system in which oxidation of the substrate, *i.e.*, substrate, pollutants, or others by the microorganisms in the electrode. Such a process leads to the release of e^- and H^+. The oxidation reaction in the anodic chamber using acetate as substrate is described in eqn (6.1). Both e^- and H^+ are transported to the cathode but through different paths. The generated e^- flow is through the current collector, whereas H^+ is transferred separately through the PEM to the cathode. Upon e^- and H^+ reaching the cathode an ORR occurs which produces an electric current and H_2O. The reduction reaction in the cathodic chamber using O_2 as e^- acceptor is described in eqn (6.2).

$$\text{Oxidation reaction at the anode: } CH_3COO^- + 4H_2O \rightarrow 2HCO_3{}^- + 9H^+ + 8e^- \quad (6.1)$$

$$\text{Reduction reaction at the anode: } O_2 + 4H^+ + 4e^- \rightarrow 2H_2O \quad (6.2)$$

The improvement of the e^- transfer processes is a crucial step for the performance optimization of MFCs. In that sense, some of the factors that must be applied to the optimization of the anode are (i) high electrical conductivity, (ii) good biocompatibility and adherence with the biofilm, (iii) chemical stability at relatively low pH, and (iv) high surface area. For the cathode, the optimization approaches should be focused on: (i) high ORR activity, (ii) efficient O_2 diffusion, (iii) chemical stability and controlled pH around the alkaline or neutral range, and (iv) low overpotential to decrease energy losses.[61]

6.3 Applications of MFCs

6.3.1 Wastewater Treatment

MFCs have been attracting the attention of scientists due to their innovative technology that can offer a sustainable alternative for the generation of electricity with simultaneous treatment of wastewater.[62] They can be a convenient technology in that regard due to their relatively higher efficiency in the conversion of energy and lower generation of solid waste. In addition, MFCs can greatly reduce chemical oxygen demand (COD) and specific contaminants, while generating power densities of around 4200 mW m^{-3}.[5,63–65] The design of the MFC for wastewater treatment can be either single- (Figure 6.9a) or dual-chambered (Figure 6.9b).

The treatment of wastewater is accompanied by several factors that can influence the overall efficiency of the process, which include configuration of the MFC, origin of the wastewater, COD removal %, and power generated, among others. A complex case study for different types of wastewater

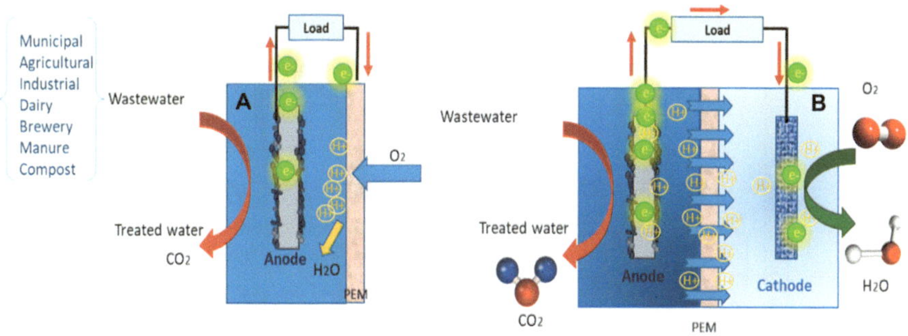

Figure 6.9 Typical designs for (a) single- and (b) dual-chambered MFCs for wastewater treatment. Adapted from ref. 62, https://doi.org/10.3390/su14020955, under the terms of the CC BY 4.0 license, https://creativecommons.org/licenses/by/4.0/.

treatment along with other effluents was analyzed by Vélez-Pérez *et al.*[66] In their study industrial acid mine drainage (I-AMD) and municipal wastewater (MWW) were co-treated using a dual-chamber MFC. For that, in the anodic chamber, MWW and sewage sludge were added as inoculum, whereas, in the cathodic chamber, I-AMD was added. Two MFCs were utilized for the analysis separately, one presented a resistor of 100 Ω to connect the electrodes (MFC-A) whereas the other was at the open circuit potential (MFC-B). Both scenarios displayed relatively similar efficiencies in the treatment of organic matter, with a removal rate of around 15% along with a decrease in the wastewater alkalinity of around half of the initial value. Also, the NO_3^- decreased by around 90% in both cells. However, some modifications of the components in the I-AMD were observed as there was an increase in the pH from 2.50 to 4.12 along with decreases in the concentration of SO_4^{2-} of 20% and 8% for the MFC-A and MFC-B, respectively. Another important factor was that metallic ions present in the wastewater were removed in appreciable percentages, such as for Cd (42%), Cu (84%), Fe (71%), Al (77%), Pb (55%), and As (42%). The MFC provided a volumetric power of 14 000 mW m^{-3}. Schematics of the dual-chambered MFC with a summary of the main findings are presented in Figure 6.10a and a photograph of the MFC is presented in Figure 6.10b.

In another study presented by Liu *et al.*,[67] a system for fecal wastewater treatment was proposed through a combined system of an anaerobic baffled reaction-MFC-MEC (ABR-MFC-MEC). The baffles of the ABR can assist in the retention of the solid components to improve the separation process during the degradation of organic matter. Through that, the microorganisms present in the MFC can optimize their contact with the organic matter and improve its efficiency for conversion into electricity. The generated electricity is used to power the MEC as it promotes the HER to generate H_2. This system could promote the separation of the processes that took place during the breaking down of the organic matter as the hydrolysis and acidification led to the formation of smaller molecules. Then, a

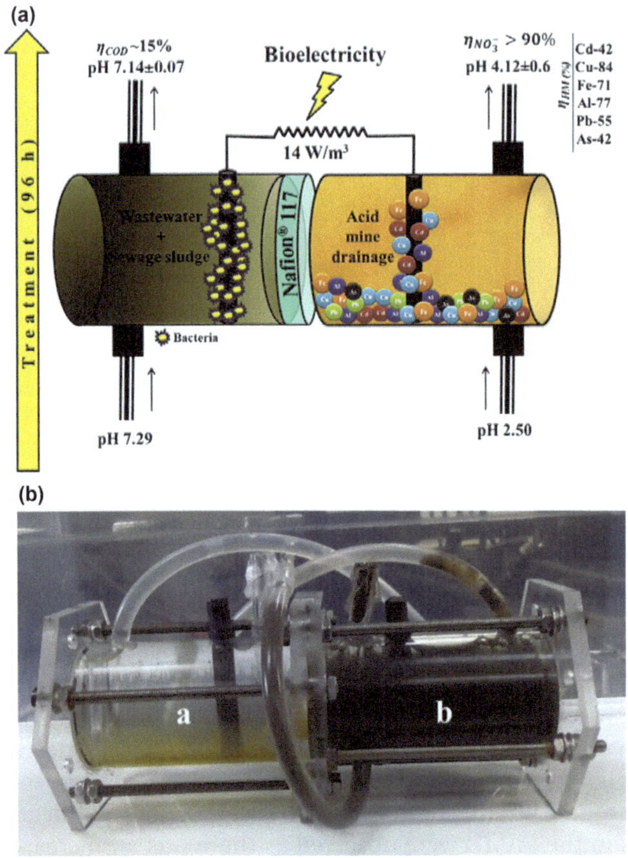

Figure 6.10 (a) General schematic for the dual-chambered MFC displaying the main aspects of its performance. (b) Photograph of the dual-chambered MFC in which "a" is the compartment filled with I-AMD and "b" is the compartment filled with wastewater-sewage sludge. Adapted from ref. 66 with permission from Elsevier, Copyright 2020.

methanogenesis process was performed to convert those small molecules into methane. The removal rates for the COD were 1.35 kg COD per m^3 per day and for NH_4^+-N were 85.0 g NH_4^+-N per m^3 per day. The average surplus voltage was around 452.5 mV, which could be obtained when the initial COD was around 1500 mg L^{-1} and the initial NH_4^+-N was 110.0 mg L^{-1}. Under such conditions, the system was then able to combine the treatment of fecal wastewater with almost zero need for energy input while enabling the production of fuel.

In another work performed by Marassi *et al.*,[68] an up-flow air-cathode MFC was built for the treatment of dairy wastewater through the metabolic activity of *Shewanella oneidensis* and *Clostridium butyricum*. The stabilization of the system was performed in a start-up step (30 days), phase 1 (75 days), and phase 2 (30 days). The start-up phase was an important step in establishing the microbial community as they adapted to the high loads of organic matter from the wastewater. This step aimed to allow the microbes to colonize the electrodes forming biofilms to achieve a consistent and constant metabolic activity, which is a requirement for continuous power generation and treatment of the wastewater. Phase 1 was a longer process as it was performed to allow the microbes to continuously treat the wastewater while finding the optimized condition of operation. Thus, this step aimed to keep the system stable and efficient in terms of chemical breakdown of organic matter with electricity generation. Removals of total COD (TCOD) and total biochemical oxygen demand (TBOD) were 94 and 96%, respectively, at a resistance of 1 kΩ. However, a relatively low Coulombic efficiency of 2.1% was obtained. The maximum power density was 3.5 $W\,m^{-3}$ with a current density of 1.1 $A\,m^{-3}$. Phase 2 was a smaller step for optimization of the system at which the concentration of substrate, and flow rates, among other operational parameters, varied to find new optimal conditions. The TCOD and TBOD were 96% and 97%, respectively, under an external resistance of 0.3 kΩ. This presented a slightly higher Coulombic efficiency of 4.4%. The maximum power densities was 2.4 $W\,m^{-3}$ and 2.4 $A\,m^{-3}$. In addition, the MFC could remove high percentages of organic components, such as organic N (47%), SO_4^{2-} (75%), P (95%), and NO_3^- (100%). Despite the low Coulombic efficiency, the MFC could treat dairy wastewater with high concentrations of organic compounds, which suggested a promising way to address industrial effluents.

The removal of metallic species from effluents is an important process to avoid pollution, which can lead to severe environmental issues. To address this, Lim *et al.*[69] fabricated an MFC that could treat organic contaminants as well as recover Zn^{2+}. To understand the MFC performance, synthetic and industrial solution samples were analyzed. It was observed that 83% and 42% of Zn were recovered from the synthetic and industrial samples, respectively. The MFC consisted of a bioanode that performed the oxidation of organic matter used as substrate which produced CO_2, H^+, and e^-. At the cathode, there was the reduction of O_2 into OH^- which could then react with the Zn^{2+} in solution to form $Zn(OH)_2$. This reaction showed that, in the industrial solution, the removal of Zn^{2+} took place through electro-precipitation rather than electrodeposition. During the removal of Zn^{2+}, the O_2 was reduced to OH^- before it could react with Zn^{2+}, which led to the formation of $Zn(OH)_2$ resulting in precipitation due to its insolubility in water. The comparatively lower recovery of Zn^{2+} in the industrial solution was attributed to the occurrence of other competitive reactions that promoted co-precipitation and co-deposition reactions which decreased the formation of $Zn(OH)_2$. Despite that, the MFC system could perform the removal of Zn^{2+} in the form of $Zn(OH)_2$ without the need for an external source of energy because the oxidation of the organic substrate generated enough e^- to promote the reduction of O_2 into OH^- followed by its precipitation reaction to obtain $Zn(OH)_2$. The schematics for this reaction are presented in Figure 6.11.

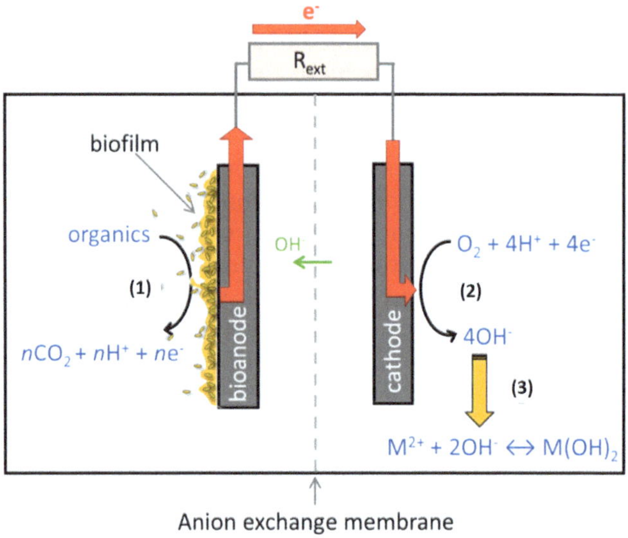

Figure 6.11 Schematic for the MFC system displaying the (1) oxidation of organic matter in the bioanode to obtain CO_2, H^+, and e^-; (2) reduction of O_2 into OH^-; and (3) precipitation of the metal ion. Adapted from ref. 69, https://doi.org/10.1016/j.scitotenv.2021.145934, under the terms of the CC BY 4.0 license, https://creativecommons.org/licenses/by/4.0/.

6.3.2 Power Supply

MFCs have been used as potential remediation for wastewater as well as renewable energy sources for smaller-scale devices. Such technology is a convenient approach in the sense that it can directly convert wastewater into electricity, which simultaneously addresses the generation of waste while providing energy to power smaller devices.[70,71] Yet, for a single MFC system, the maximum voltage provided can be in the range of 300–500 mV, which can vary based on the material and MFC design. Other factors that influence overall performance are related to diffusion limitations and high internal resistances which cause smaller MFCs to present higher power density than larger versions.[72] A common strategy to optimize the delivery of energy to a device consists of forming MFC stacks. Through that, MFC units can be connected either in series, to increase the output voltage, or in parallel, to increase the output current, with the overall performance determined by the number of MFCs stacked.[73–75] High voltages (>1.5 V) can be achieved by electrically connecting MFCs in parallel along with utilizing circuitry tools to further increase the voltage. Some studies have demonstrated the capabilities of MFCs to power electric systems through the use of a DC/DC converter.[73,76–79] To evaluate the feasibility of MFCs to power electronic devices, Walter *et al.*[80] assembled a stack of four urine-fed self-stratifying membraneless MFCs connected in series. A schematic for the single MFC is presented in Figure 6.12a. For that, the anode of the first module was the

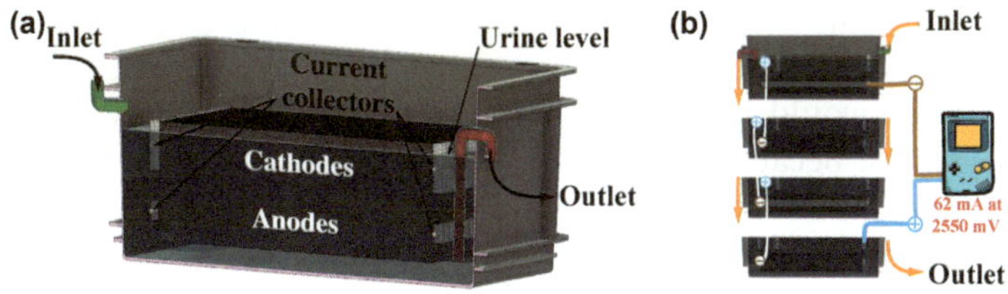

Figure 6.12 (a) Schematic for the single-MFC module. (b) Section perspective for the cascade with four modules and its conducting wiring to power the microcomputer (*Game Boy Color* ®). The orange arrows represent hydraulic connections and the white lines represent the electrical connections. Adapted from ref. 80, https://doi.org/10.1016/j.jpowsour.2019.227328, under the terms of the CC BY 4.0 license, https://creativecommons.org/licenses/by/4.0/.

negative terminal, whereas the cathode of the fourth module cascade was the positive terminal. Then, the assembled stack of MFCs was connected to the positive and negative terminals of the battery of an electronic device. Schematics for this process are displayed in Figure 6.12b. The obtained results showed that the microcomputer utilized in the study could be directly and continuously powered by the MFC assembled system as the four-module MFCs generated around 2.55 V and 61 mA, which was a total of 130 mW of potency, surpassing the requirements for proper functioning of the microcomputer. The four-module MFC system could continuously power the microcomputer for over 24 h without considerable fluctuations in the current or voltage. The challenge faced by the authors was due to the higher voltage required to switch off the microcomputer, which was around 1.832 V. Yet, such a drawback could be addressed by adding other MFC in the cascade to supply enough voltage.

In another study, Donovan *et al.*[81] proposed the use of an SMFC along with a power management system (PMS) to provide energy to a batteryless, wireless sensor. Through that, the PMS could store the microbial energy generated by the cell to transfer it to the sensor upon reaching the electric potential of 320 mV. This approach could counter two drawbacks related to the use of SMFCs. First, the low electric potential generated by MFCs is not enough for the operation of common devices. Second, the intermittent power supply of SMFCs can lead to fluctuations. Hence, attaching the SMFC with a PMS enabled storage of the energy generated by the SMFCs for later use. This system can be applied to environmental monitoring in aquatic regions, oceanographic studies, and military surveillance.[82,83] The SMFC consisted of a graphite anode buried in aquatic sediments. The microorganisms naturally present in the sediment colonize the anode's surface, forming a biofilm, while causing oxidation of the organic matter, leading to the generation of e^-. The cathode, which can be composed of graphite or stainless steel, for instance, receives the e^- to perform the ORR. In the case that the microorganisms capable of oxidizing Mn are present in the aquatic area, an increase in the potential to a more cathodic direction occurs.[84–86]

However, a fundamental challenge with the use of SMFCs is that, differently from regular MFCs, they cannot be connected in series, since, in practice, every electrode would be immersed in the same electrolyte (water from the aquatic system). Because of that, an increase in overall cell potential is negligible. Hence, the maximum potential is defined by a single SMFC, which could be around 0.8 V.[87] In another study, Khaled *et al.*[88] utilized a PMU operating at low input voltages and low power as a means to improve the efficiency of an MFC that could power a wireless sensor. The system consisted of an anode that was submerged in the sediment to allow the formation of a biofilm. The microbes then begin the breakdown of organic matter present in the sediment and transfer the generated e^- to the PMS which is then transferred to the cathode. The cathode is placed in a water column above the anode where reduction reactions take place. For that, the Mn^{2+} present in the system undergoes a series of reduction reactions forming MnOOH followed by MnO_2 through the acceptance of the e^- generated by the decomposition of the organic matter by the microorganisms at the anode. A schematic for the SMFC coupled with the PMS system to power a wireless sensor is presented in Figure 6.13. This type of system, despite generating a low amount of energy,

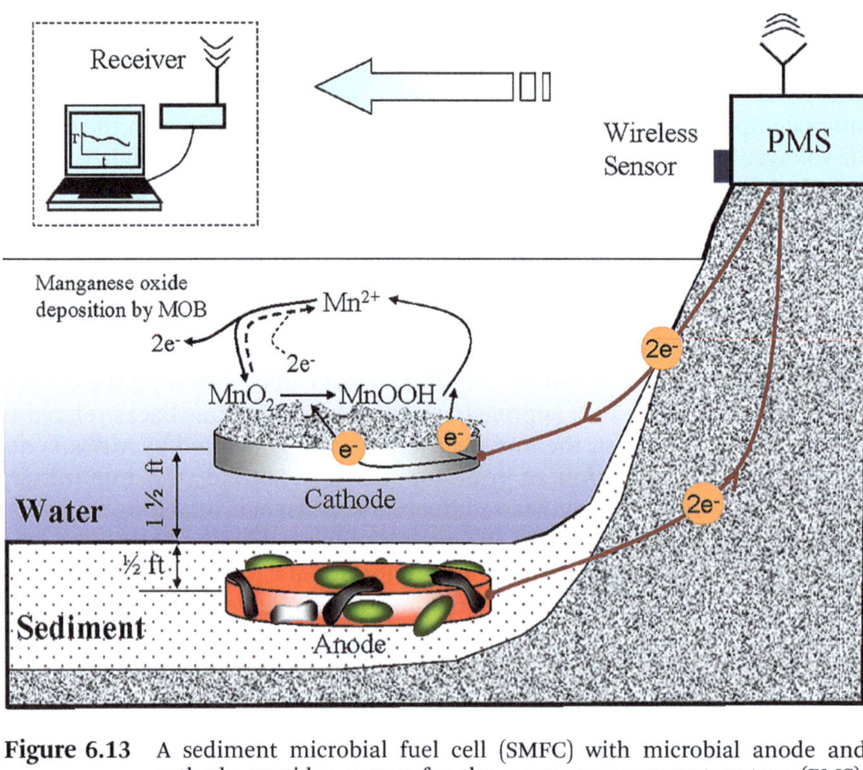

Figure 6.13 A sediment microbial fuel cell (SMFC) with microbial anode and cathode provides energy for the power management system (PMS). Adapted from ref. 81 with permission from American Chemical Society, Copyright 2008.

can be used in remote aquatic environments to monitor water quality and sediment chemistry, or be used to power a small device.

6.3.3 Biosensors

MFCs are devices capable of converting chemical energy into electricity. This process can be utilized as a signal to potentially identify a specific type of analyte that is present in the system. Because of that, developing biofilms that can break down or interact with specific types of analytes can render some degree of selectivity to identify a desired species present in the system. From that, an MFC-based biosensor system has an anodic chamber containing a biofilm attached to the anode which functions as the bioreceptor, whereas the anode functions as the transducer. MFCs can be utilized for the analysis of several analytes such as heavy metallic ions, *i.e.*, Pb^{2+}, Hg^{2+}, Cd^{2+}, other metallic ions, *i.e.*, Ni^{2+}, Cu^{2+}, and Cr^{6+}, as well as the analysis of parameters such as biochemical oxygen demand (BOD) and chemical oxygen demand (COD).[89–91] The analysis of BOD and COD are important parameters to quantify the amount of organic matter present in water, although they differ in their measurement. The BOD is the amount of dissolved O_2 required by aerobic microorganisms to digest organic matter in water, which is usually performed over five days at 20 °C (BOD_5) often reported in $mg\,L^{-1}$. Hence, a high BOD indicates a high concentration of organic matter or pollution in the system. The COD is a measurement of the total quantity of O_2 necessary to oxidize biodegradable and non-biodegradable components in a water system. For that, a strong oxidizing agent such as $K_2Cr_2O_7$ is utilized. Since BOD is a time-consuming analysis, it is important to find ways to reduce that time to allow for faster results and more data collection. With that in mind, Karube *et al.*[92] proposed the use of an MFC for the analysis of BOD through the use of *Clostridium butyricum* bacteria in the anode. By using a glucose–glutamic solution standard a linear relationship could be observed between the MFC's current output and the BOD, which demonstrated its feasibility for this type of analysis. In addition, it has been observed that mediator-less MFCs tended to provide more stable and continuous analysis for a longer period since even though the mediator could facilitate the e^- transfer between microbe and electrode it also had some degree of toxicity which led to a decrease in performance.[92,93] Furthermore, an earlier study by Kim *et al.*[94] demonstrated the use of an enriched microbial consortium in an MFC that could determine BOD for over 5 years with minor maintenance. A starch-rich wastewater from an industrial source was utilized as the substrate which presented a BOD_5 and COD_{Cr} of 520 and 1200 ppm, respectively. In addition, the wastewater contained around 25 $mg\,L^{-1}$ of total N, 10.7 $mg\,L^{-1}$ of total P, and less than 5 $mg\,L^{-1}$ of inorganic N. The pH was controlled using 50 mM phosphate buffer ($pH = 7$) along with 100 mM of NaCl. The highly concentrated media with both N and P along with the controlled pH and presence of electrolytes appeared to have generated a stable system for the continuous degradation of the wastewater while enabling a proper analysis of the BOD.

MFCs can also be utilized for the monitoring of dissolved O_2 (DO) in water, which is an important parameter for quality analysis as variations in DO provide some information about the number of organic pollutants present in the system and the biochemical reactions that might be taking place in aquatic environments.[95,96] Also, MFCs can be a more stable option for real-time monitoring of DO when compared to Clark-type O_2 electrodes. One of the reasons for that is because O_2 is commonly utilized as the terminal e^- acceptor in the cathode. Hence, the amount of O_2 dissolved in the system is directly related to the amount of current output of the cell, enabling a correlation with the DO concentration. Similarly, the power output can also be correlated with the O_2 dissolved in the cathode as a higher concentration of O_2 leads to a higher rate of oxygen reduction reaction (ORR). Another advantage is that MFCs can be more economical and sustainable for continuous DO analysis without the need for constant calibrations or maintenance. Also, it is an eco-friendly approach as it does not introduce toxic chemicals into the system. The constant monitoring of DO can also serve to signal potential cases of "dead zone" formation due to eutrophication processes.[97,98]

MFCs can offer a faster and simpler method for the analysis of bacterial activity in a given media to determine their optical growth and metabolic activity conditions, which can be analyzed through the e^- to the anode. In an earlier report, Liu *et al.*[99] fabricated a flow-cell MFC to monitor the anaerobic digestor performance which could be correlated with changes in the current based on variations of parameters such as gas flow rate, pH, and COD. In another work, Sun *et al.*[100] developed an MFC that could identify acetate in effluents through the use of a bacterial consortium. A photograph and schematic of the MFC are presented in Figure 6.14a and b. The MFC's biofilm was fed with acetate as the only source for around three months which

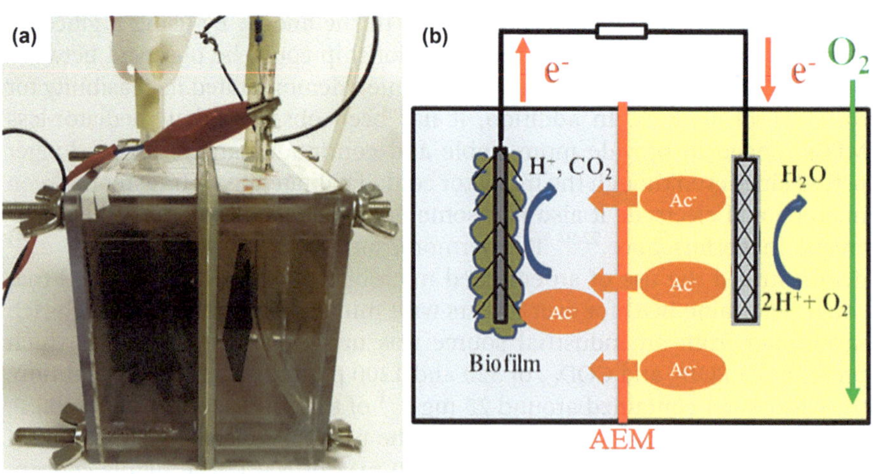

Figure 6.14 (a) Photograph of the MFC prototype and (b) schematic diagram of the MFC used for biosensing. Adapted from ref. 101 with permission from Elsevier, Copyright 2019.

could condition the bacteria consortia to only degrade that type of substrate. To test the selectivity of the biofilm towards acetate, other substrates were added to the system such as glucose, butyrate, propionate, isobutyrate, and isobutyrate propionate. After 5 hours, a negligible amount of current was generated which suggested that the biofilm displayed some degree of selectivity towards acetate. Yet, upon increasing the concentration of all the substrates the current density also increased. In addition, an increase in temperature also led to higher metabolic activity which increased the current density.

6.3.4 Potential Applications in Implantable and Ingestible Devices

A novel application that can be attained by MFCs is the development of implantable and ingestible devices. Through that, MFCs can be used to power implantable and ingestible medical devices by converting biochemical energy from the body into electrical energy. These bioelectrochemical systems utilize electroactive microorganisms that oxidize organic substrates such as glucose, urea, or other nutrients present in the environment to generate electrons. In this way, it offers a self-sustainable and biocompatible power source that can potentially reduce the need for other energy storage systems in medical devices while improving safety aspects and durability. However, its applicability strongly depends on the biocompatibility of components to avoid allergic reactions from the immune system and miniaturization to facilitate incorporation in the body and reduce invasiveness. The research into the development of implantable and ingestible electronics has made considerable progress, displaying complex ways to implant electronic devices for a plethora of functions.[102] Enzymatic biofuel cells (EBFCs) and ingestible galvanic cells (IGCs) are gaining attention as innovative technologies for powering biomedical devices. EBFCs harvest biochemical energy by catalyzing the oxidation of glucose using enzymes such as *glucose oxidase* at the anode and reducing O_2 at the cathode, defining a redox process. Through that, it generates electrons that can be harnessed to power low-energy requirement electronic devices. Moreover, wireless communication can be enabled by powering ultra-low-power wireless transmitters or sensors through the electricity produced by the EBFCs. For example, Ichi-Ribault *et al.*[103] fabricated an EBFC that was implanted in a rabbit, which could continuously generate around 16 μW/mL during each 30-min discharge per day for 16 days. The EBFC consisted of a biocathode composed of a hydraulic compressed mixture of laccase from *Trametes versicolor*, MWCNTs, and viscous chitosan-genipin. Similarly, the bioanode consisted of a pressed mixture of glucose oxidase, catalase, and naphthoquinone. A photograph of the biocathode and bioanode pellets is shown in Figure 6.15a. Based on that, the catalase in the bioanode could promote the catalysis of H_2O_2 produced by the glucose oxidase in the presence of glucose

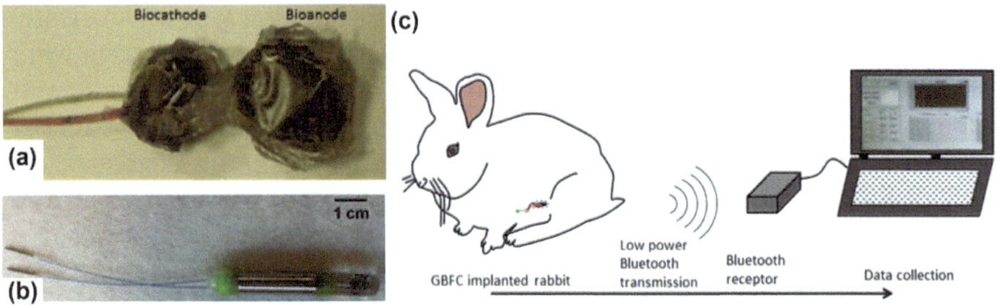

Figure 6.15 Photograph of (a) the bioelectrodes coated with dry chitosan membranes before being encapsulated in a Dacron lattice membrane and (b) the tele-transmission system. (c) Schematic of the EBFC and tele-transmission system and wireless data transfer. Adapted from ref. 103, https://doi.org/10.1016/j.electacta.2018.02.156, under the terms of the CC BY 4.0 license, https://creativecommons.org/licenses/by/4.0/.

and O_2. The tele-transmission device used to control and monitor the EBFC is shown in Figure 6.15b. Also, the monitoring system for the implanted EBFCs is shown in Figure 6.15c. It was observed that the EBFC presented an open circuit voltage (OCV) of 0.65–0.58 V, which decreased to around 0.34 V after two weeks in an *in vitro* test. Such a loss could occur due to enzymatic instability, fouling, or a decrease in glucose concentration. In addition, it was observed that the charge time after discharge took around 9 h to reach the OCV of the initial voltage value, even at an earlier stage of implantation, suggesting a slow diffusion of glucose. Another aspect is that the growth of extracellular tissue leads to an increase in impedance. Hence, preventing cell growth between the electrodes can improve the EBFC performance. However, it was shown to not fully address the substrate diffusion issue. The study also indicated that the reaction occurring at the EBFC electrodes can potentially trigger mild inflammation. In rats, a two-month implantation of an open circuit EBFC led to a moderate foreign body response, with minor degradation of chitosan and formation of fibrin. On the other hand, the inflammatory response was more pronounced in active EBFCs. The degradation could be due to unbound enzymes such as glucose oxidase that produces H_2O_2 and gluconic acid which resulted in the formation of reactive oxygen species (ROS) and a decrease in pH. Despite these issues, the EBFC could remain operational in a rabbit for two months.

In 2018, Jeerapan *et al.*[104] presented an edible biofuel cell fabricated from biocompatible, food-based materials without the use of mediators or separators. The system consisted of a combination of edible plant tissues, vegetable oils, and activated charcoal to form biocatalytic pastes for both the bioanode and biocathode. The bioanode was based on mushroom tissue mixed with charcoal, which was used to catalyze ethanol oxidation due to the presence of alcohol oxidase biocatalysts (Figure 6.16a). The cathode was based on apple tissue mixed with charcoal that contained polyphenol oxidase and phenolic compounds that could promote the O_2 reduction process.

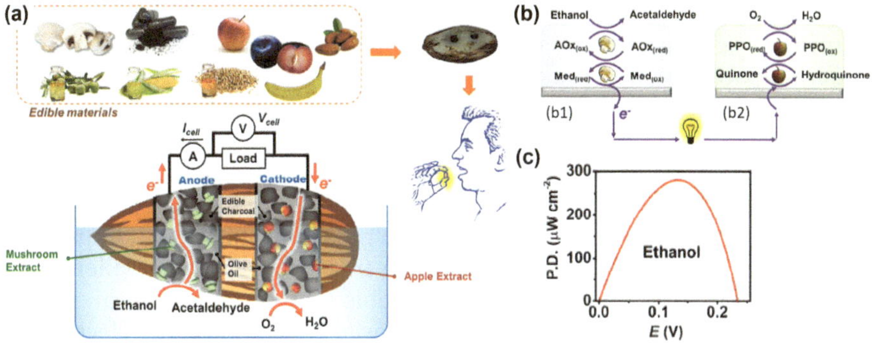

Figure 6.16 (a) Schematic for the edible membrane-free biofuel cell consisting of a bioanode and biocathode derived from mushroom and apple extracts, respectively. (b) Redox process that takes place in the (b1) bioanode and (b2) biocathode in the biofuel cell. (c) Plot of power density *vs* voltage for the ethanol-powered biofuel cell. Adapted from ref. 104 with permission from the Royal Society of Chemistry.

Hence, the system could work synergistically to close a redox system and generate electricity from ethanol and O_2 (Figure 6.16b). From that, a prototype based on mushroom, apple, and olive oil demonstrated a power density of around 282 $\mu W\,cm^{-2}$ and an OCV of 0.24 V. The system maintained a stable output of around 4 h along with a retention of 80% of the initial power. In addition, the current and voltage profiles scaled linearly in relation to the ethanol concentration (Figure 6.16c). This effect could potentially enable the system to be used as a self-powered alcohol sensor. This example highlighted one of the first fully edible biofuel cells composed of natural biocatalysts present in mushrooms, for ethanol oxidation, and apples, for O_2 reduction, instead of purified enzymes or chemical catalysts. In this way it could potentially serve as a green strategy for *in vivo* energy generation and self-powered biosensing systems.

Later, in 2022, Rezaie *et al.*[105] developed a miniaturized and ingestible MFC device which provided a practical power source for ingestible electronic devices. It presented a pH-sensitive enteric coating, a layer containing germination agents, and a microfluidic hydrogel-based anodic channel preloaded with dormant *Bacillus subtilis* spores in conducting poly(3,4-ethylenedioxythiophene)–poly(styrenesulfonate) (PEDOT:PSS) hydrogel. Upon reaching a specific pH within the gastrointestinal tract, the enteric coating dissolved, which allowed the hydrogel to absorb gut fluids. In this way, germinants can be delivered to spores, triggering their germination and initiating microbial metabolism that generates electricity. The capsule was engineered to activate in a neutral pH environment within the intestine where it reached the highest power and current densities of 64 $\mu W\,cm^{-2}$ and 435 $\mu A\,cm^{-2}$, respectively. Based on that, the design of the MFC contained the *B. subtilis* spores before and after germination due to the interaction with a neutral pH environment. The development of this MFC that can be ingested to act in the gastrointestinal tract can serve as a feasible way to provide a self-sustainable and stable system as it can perform a complete enzyme pathway and continuous regeneration of biocatalytic processes as part of the natural metabolism of the microorganisms adhered to the anodic channel.[106]

There has been continuous progress in the development of implantable and ingestible MFCs and their variations. These systems can be utilized to monitor conditions *in vivo*, such as in the gastrointestinal tract, to provide a diagnostic or potential treatment. However, it is inherently challenging to develop fully edible, miniaturized, and stable MFC systems that can perform adequately *in vivo*. The use of edible materials to compose the electrodes is an important aspect of improving their biocompatibility and decreasing potential immune responses from the body.[102,106,107]

6.4 Outlook and Challenges

The MFC is a technology that can be introduced in a myriad of core fields for the functioning of society such as the generation of bioelectricity, waste management, biosensing, remote sensing, powering of small devices, agricultural applications, medical applications, H_2 production, and water treatment for removal of heavy metals, pollutants, and desalination. This myriad of applications makes them highly desirable for integration into modern society as they can aid in the mitigation of several current issues by offering waste management, the use of bio-renewable resources, and the introduction of a sustainable and circular economy for the generation of electricity. The fabrication of MFCs also allows for plenty of versatility, as there are several types of MFCs including single-chambered, double-chambered, upflow, air-cathode, mediator-based, mediator-less, MECs, sediment, and photosynthetic. Generally, MFCs provide sustainable energy generation through their capability to degrade different types of organic wastes, potentially converting them into value-added products and/or electricity. They also emit relatively low amounts of carbon, and require low energy input and maintenance. Because of that, they may present an overall low cost while being applicable in remote areas.

On the other hand, MFCs display a low power output when compared to other energy generation devices, which are often insufficient for large-scale applications. Despite the low maintenance, there are some inherent costs associated with the manufacturing and scalability of PEMs which are core components in most MFCs. Also related to that, PEMs undergo degradation and fouling over time which require periodic cleaning or replacement of these parts, which can be relatively expensive when considering the overall cost of MFCs. Since MFCs rely on bioelectricity, their reaction rates are sluggish as they depend on microbial metabolism. Such conditions can lead to inconsistent generation of electricity until the system achieves stability. Another important factor is reaching the optimal functioning of an MFC, which often requires specific conditions for its operations related to parameters such as the concentration of substrate, pH, temperature, and removal of toxic components, among others. Thus, based on these aspects it is notable that MFCs display great potential for energy generation while filling the role of a technology that can be both sustainable and economically convenient. However, several challenges hinder its application on a larger scale as it has a low power output and high initial costs. Despite that, there has been considerable progress to improve their overall efficiency. Some of these approaches include microbial engineering, optimization of electrode materials to improve biofilm formation and electron transfer, coupling with external circuits for power management, and the use of modeling and simulation tools. In that regard, the continuous research into MFCs is paving the way for their utilization in larger-scale applications which can greatly benefit society and the environment, making it an attractive technology for research and development.

Recent advancements have demonstrated that MFCs can be effectively miniaturized and adapted for biomedical use, such as implantable and ingestible devices. These bioelectrochemical systems can function as self-powered medical electronic devices. Because of that, they can serve as a viable alternative in situations in which conventional batteries are impractical to use due to safety, size, and biocompatibility issues. The ingestible MFCs, for example, can harness energy directly from the gastrointestinal tract through the utilization of gut fluids and nutrients in the environment, which can activate dormant microbes or enzymes, allowing for on-demand power generation within specific regions of the body. This process can be achieved by developing components that are pH-sensitive, which allows them to be activated under these specific conditions. Moreover, implantable MFCs can be engineered to operate in biofluids such as blood, and interstitial fluids which can be potentially used to power biosensors, drug delivery systems, or for *in vivo* monitoring over long periods. In this way these systems can deliver power densities that are high enough to allow the functioning of low-energy demanding electronics. However, some of the main challenges lie in obtaining a consistent performance in complex *in vivo* environments while displaying good biocompatibility to mitigate potential inflammatory or biofouling formation. Despite these hindrances, the field is progressing with a focus on optimization of the biocatalyst, development of membrane-free MFC designs, and biocompatible materials for the fabrication of electrodes. These advances could allow MFCs to function as a compelling technology for the next generation of self-sustainable medical devices.

References

1. Y. Li, J. Shang, C. Zhang, W. Zhang, L. Niu, L. Wang and H. Zhang, The role of freshwater eutrophication in greenhouse gas emissions: A review, *Sci. Total Environ.*, 2021, **768**, 144582.
2. V. N. de Jonge, M. Elliott and E. Orive, in *Causes, historical development, effects and future challenges of a common environmental problem:*

eutrophication BT - Nutrients and Eutrophication in Estuaries and Coastal Waters: Proceedings of the 31st Symposium of the Estuarine and Coastal Sciences Ass, ed. E. Orive, M. Elliott and V. N. de Jonge, Springer, Netherlands, Dordrecht, 2002, pp. 1–19.
3. M. M. Dorgham, in *Effects of Eutrophication BT - Eutrophication: Causes, Consequences and Control*, ed. A. A. Ansari and S. S. Gill, Springer, Netherlands, Dordrecht, 2014, vol. 2, pp. 29–44.
4. S. Holmberg, A. Perebikovsky, L. Kulinsky and M. Madou, 3-D Micro and Nano Technologies for Improvements in Electrochemical Power Devices, *Micromachines*, 2014, **5**, 171–203.
5. L. He, P. Du, Y. Chen, H. Lu, X. Cheng, B. Chang and Z. Wang, Advances in microbial fuel cells for wastewater treatment, *Renewable Sustainable Energy Rev.*, 2017, **71**, 388–403.
6. L. Gong, M. A. Amirdehi, J. Sonawane, N. Jia, L. T. de Oliveira and J. Greener, A biomembrane grown in situ for improved microfluidic microbial fuel cell performance using a pure culture Geobacter sulfurreducens electroactive biofilm, 2022.
7. O. Guadarrama-Pérez, A. Carolina Guevara-Pérez, V. Hugo Guadarrama-Pérez, V. Bustos-Terrones, J. Hernández-Romano and R. Angélica Guillén- Garcés, G. Eleonora Moeller-Chávez, Bioelectricity production from the anodic inoculation of Geobacter sulfurreducens DL-1 bacteria in constructed wetlands-microbial fuel cells, *Bioelectrochemistry*, 2023, **154**, 108537.
8. B. Baniasadi and F. Vahabzadeh, The performance of a cyanobacterial biomass-based microbial fuel cell (MFC) inoculated with Shewanella oneidensis MR-1, *J. Environ. Chem. Eng.*, 2021, **9**, 106338.
9. H. N. Dai, T.-A. Duong Nguyen, L.-P. My LE, M. Van Tran, T.-H. Lan and C.-T. Wang, Power generation of Shewanella oneidensis MR-1 microbial fuel cells in bamboo fermentation effluent, *Int. J. Hydrogen Energy*, 2021, **46**, 16612–16621.
10. Y. Ojima, T. Kawaguchi, S. Fukui, R. Kikuchi, K. Terao, D. Koma, T. Ohmoto and M. Azuma, Promoted performance of microbial fuel cells using Escherichia coli cells with multiple-knockout of central metabolism genes, *Bioprocess Biosyst. Eng.*, 2020, **43**, 323–332.
11. M. Rethinasabapathy, A. T. E. Vilian, S. K. Hwang, S.-M. Kang, Y. Cho, Y.-K. Han, J.-K. Rhee and Y. S. Huh, Cobalt ferrite microspheres as a biocompatible anode for higher power generation in microbial fuel cells, *J. Power Sources*, 2021, **483**, 229170.
12. A. J. Slate, N. A. Hickey, J. A. Butler, D. Wilson, C. M. Liauw, C. E. Banks and K. A. Whitehead, Additive manufactured graphene-based electrodes exhibit beneficial performances in Pseudomonas aeruginosa microbial fuel cells, *J. Power Sources*, 2021, **499**, 229938.
13. R. A. Nastro, F. Flagiello, N. Silvestri, E. Gambino, G. Falcucci and K. Chandrasekhar, Use of Biochar-Based Cathodes and Increase in the Electron Flow by Pseudomonas aeruginosa to Improve Waste Treatment in Microbial Fuel Cells, *Processes*, 2021, **9**, 1941.

14. J. K. Braga, R. A. Stancari, F. Motteran, I. Malavazi and M. B. A. Varesche, Metals addition for enhanced hydrogen, acetic and butyric acids production from cellulosic substrates by Clostridium butyricum, *Biomass Bioenergy*, 2021, **150**, 105679.
15. R. J. Marassi, M. B. G. López, L. G. Queiroz, D. C. V. R. Silva, F. T. da Silva, T. C. B. de Paiva and G. C. Silva, Efficient dairy wastewater treatment and power production using graphite cylinders electrodes as a biofilter in microbial fuel cell, *Biochem. Eng. J.*, 2022, **178**, 108283.
16. M. Sogani, A. O. Pankan, A. Dongre, K. Yunus and A. C. Fisher, Augmenting the biodegradation of recalcitrant ethinylestradiol using Rhodopseudomonas palustris in a hybrid photo-assisted microbial fuel cell with enhanced bio-hydrogen production, *J. Hazard. Mater.*, 2021, **408**, 124421.
17. M. Li, P. Ning, Y. Sun, J. Luo and J. Yang, Characteristics and Application of Rhodopseudomonas palustris as a Microbial Cell Factory, *Front. Bioeng. Biotechnol.*, 2022, **10**, 897003.
18. S. S. Lim, P. S. Chong, B. C. Jong, M. H. Abu Bakar, W. R. Wan Daud, J. M. d Jahim and M. N. I. Salehmin, Microbial fuel cell-based sensor for Enterobacter sp. KBH6958 activity monitoring during hydrogen production: the effects of pH and glucose concentration, *J. Appl. Electrochem.*, 2022, **52**, 1327–1342.
19. R. Kandpal, M. Shahadat, S. W. Ali, C. Hu and S. Z. Ahammad, Material specific enrichment of electroactive microbes on polyaniline-supported anodes in a single chamber multi-anode assembly microbial fuel cell, *Mater. Res. Bull.*, 2023, **157**, 111983.
20. S. S. Kumar, V. Kumar, V. Gnaneswar Gude, S. K. Malyan and A. Pugazhendhi, Alkalinity and salinity favor bioelectricity generation potential of Clostridium, Tetrathiobacter and Desulfovibrio consortium in Microbial Fuel Cells (MFC) treating sulfate-laden wastewater, *Bioresour. Technol.*, 2020, **306**, 123110.
21. A. Almatouq, A. O. Babatunde, M. Khajah, G. Webster and M. Alfodari, Microbial community structure of anode electrodes in microbial fuel cells and microbial electrolysis cells, *J. Water Process Eng*, 2020, **34**, 101140.
22. A. Chiolerio and A. Adamatzky, Acetobacter Biofilm: Electronic Characterization and Reactive Transduction of Pressure, *ACS Biomater. Sci. Eng.*, 2021, **7**, 1651–1662.
23. J. V. Boas, L. Peixoto, V. B. Oliveira, M. Simões and A. M. F. R. Pinto, Cyclic voltammetry study of a yeast-based microbial fuel cell, *Bioresour. Technol. Rep.*, 2022, **17**, 100974.
24. M. K. Weibel and C. Dodge, Biochemical fuel cells: Demonstration of an obligatory pathway involving an external circuit for the enzymatically catalyzed aerobic oxidation of glucose, *Arch. Biochem. Biophys.*, 1975, **169**, 146–151.
25. B. Min, J. Kim, S. Oh, J. M. Regan and B. E. Logan, Electricity generation from swine wastewater using microbial fuel cells, *Water Res.*, 2005, **39**, 4961–4968.

26. E. Tamboli and J. S. Eswari, Chapter 3.2 - Microbial Fuel Cell Configurations: An Overview, ed. S. V. Mohan, S. Varjani and A. Pandey, in *Biomass, Biofuels, Biochemicals*, Elsevier, 2019, pp. 407–435.
27. K. Rabaey, P. Clauwaert, P. Aelterman and W. Verstraete, Tubular Microbial Fuel Cells for Efficient Electricity Generation, *Environ. Sci. Technol.*, 2005, **39**, 8077–8082.
28. H. Liu and B. E. Logan, Electricity Generation Using an Air-Cathode Single Chamber Microbial Fuel Cell in the Presence and Absence of a Proton Exchange Membrane, *Environ. Sci. Technol.*, 2004, **38**, 4040–4046.
29. P. Chatterjee, M. M. Ghangrekar and D. Leech, A Brief review on recent advances in air-cathode microbial fuel cells, *Environ. Eng. Manag. J.*, 2018, **17**, 152.
30. G. Palanisamy, H.-Y. Jung, T. Sadhasivam, M. D. Kurkuri, S. C. Kim and S.-H. Roh, A comprehensive review on microbial fuel cell technologies: Processes, utilization, and advanced developments in electrodes and membranes, *J. Clean. Prod.*, 2019, **221**, 598–621.
31. G. Massaglia, V. Margaria, A. Sacco, M. Castellino, A. Chiodoni, F. C. Pirri and M. Quaglio, N-doped carbon nanofibers as catalyst layer at cathode in single chamber Microbial Fuel Cells, *Int. J. Hydrogen Energy.*, 2019, **44**, 4442–4449.
32. D. R. Bond, D. E. Holmes, L. M. Tender and D. R. Lovley, Electrode-Reducing Microorganisms That Harvest Energy from Marine Sediments, *Science*, 2002, **295**, 483–485.
33. B. Erable, R. Lacroix, L. Etcheverry, D. Féron, M. L. Delia and A. Bergel, Marine floating microbial fuel cell involving aerobic biofilm on stainless steel cathodes, *Bioresour. Technol.*, 2013, **142**, 510–516.
34. C. E. Reimers, L. M. Tender, S. Fertig and W. Wang, Harvesting Energy from the Marine Sediment–Water Interface, *Environ. Sci. Technol.*, 2001, **35**, 192–195.
35. X. Yang and S. Chen, Microorganisms in sediment microbial fuel cells: Ecological niche, microbial response, and environmental function, *Sci. Total Environ.*, 2021, **756**, 144145.
36. G. Lui, G. Jiang, M. Fowler, A. Yu and Z. Chen, A high performance wastewater-fed flow-photocatalytic fuel cell, *J. Power Sources.*, 2019, **425**, 69–75.
37. Y. Tong, J. Wei, R. Mo, H. Ma and F. Ai, Photocatalytic Microbial Fuel Cells and Performance Applications: A Review, *Front. Chem.*, 2022, **10**, 953434.
38. M. Pagliaro, R. Ciriminna, H. Kimura, M. Rossi and C. Della Pina, From Glycerol to Value-Added Products, *Angew. Chem., Int. Ed.*, 2007, **46**, 4434–4440.
39. Q. Chen, J. Li, X. Li, K. Huang, B. Zhou, W. Cai and W. Shangguan, Visible-Light Responsive Photocatalytic Fuel Cell Based on WO_3/W Photoanode and Cu_2O/Cu Photocathode for Simultaneous Wastewater Treatment and Electricity Generation, *Environ. Sci. Technol.*, 2012, **46**, 11451–11458.

40. J. Wen, J. Xie, X. Chen and X. Li, A review on g-C_3N_4-based photocatalysts, *Appl. Surf. Sci.*, 2017, **391**, 72–123.
41. G. Lui, G. Jiang, M. Fowler, A. Yu and Z. Chen, A high performance wastewater-fed flow-photocatalytic fuel cell, *J. Power Sources*, 2019, **425**, 69–75.
42. M. Rezaei Talarposhti, Microbial desalination cells with efficient platinum group metal-free cathode catalysts, *ChemElectroChem*, 2017, **4**, 3322–3330.
43. A. Haghighat Mamaghani, B. Najafi, A. Casalegno and F. Rinaldi, Optimization of an HT-PEM fuel cell based residential micro combined heat and power system: A multi-objective approach, *J. Clean. Prod.*, 2018, **180**, 126–138.
44. M. Sun, G.-P. Sheng, L. Zhang, C.-R. Xia, Z.-X. Mu, X.-W. Liu, H.-L. Wang, H.-Q. Yu, R. Qi, T. Yu and M. Yang, An MEC–MFC-Coupled System for Biohydrogen Production from Acetate, *Environ. Sci. Technol.*, 2008, **42**, 8095–8100.
45. L. E. Doyle and E. Marsili, Methods for enrichment of novel electrochemically-active microorganisms, *Bioresour. Technol.*, 2015, **195**, 273–282.
46. R. Kumar, L. Singh, Z. A. Wahid and M. F. M. Din, Exoelectrogens in microbial fuel cells toward bioelectricity generation: a review, *Int. J. Energy Res.*, 2015, **39**, 1048–1067.
47. S. K. Chaudhuri and D. R. Lovley, Electricity generation by direct oxidation of glucose in mediatorless microbial fuel cells, *Nat. Biotechnol.*, 2003, **21**, 1229–1232.
48. C. A. Pham, S. J. Jung, N. T. Phung, J. Lee, I. S. Chang, B. H. Kim, H. Yi and J. Chun, A novel electrochemically active and Fe(III)-reducing bacterium phylogenetically related to Aeromonas hydrophila, isolated from a microbial fuel cell, *FEMS Microbiol. Lett.*, 2003, **223**, 129–134.
49. B. H. Kim, T. Ikeda, H. S. Park, H. J. Kim, M. S. Hyun, K. Kano, K. Takagi and H. Tatsumi, Electrochemical activity of an Fe(III)-reducing bacterium, Shewanella putrefaciens IR-1, in the presence of alternative electron acceptors, *Biotechnol. Technol.*, 1999, **13**, 475–478.
50. L. Zhang, S. Zhou, L. Zhuang, W. Li, J. Zhang, N. Lu and L. Deng, Microbial fuel cell based on Klebsiella pneumoniae biofilm, *Electrochem. Commun.*, 2008, **10**, 1641–1643.
51. L. Fan and S. Xue, Overview on electricigens for microbial fuel cell, *Open Biotechnol. J.*, 2016, **10**, 10398.
52. G.-C. Gil, I.-S. Chang, B. H. Kim, M. Kim, J.-K. Jang, H. S. Park and H. J. Kim, Operational parameters affecting the performannce of a mediator-less microbial fuel cell, *Biosens. Bioelectron.*, 2003, **18**, 327–334.
53. M. Rahimnejad, A. Adhami, S. Darvari, A. Zirepour and S.-E. Oh, Microbial fuel cell as new technology for bioelectricity generation: A review, *Alexandria Eng. J.*, 2015, **54**, 745–756.
54. H. Li, G. Li, W. Yan, G. He and L. Lin, "Synergy effect" and its application in LED-multispectral imaging for improving image quality, *Opt. Commun.*, 2019, **438**, 6–12.

55. H. Li, S. Cryer, L. Acharya and J. Raymond, Video and image classification using atomisation spray image patterns and deep learning, *Biosyst. Eng.*, 2020, **200**, 13–22.
56. A. Hirose, A. Kouzuma and K. Watanabe, Towards development of electrogenetics using electrochemically active bacteria, *Biotechnol. Adv.*, 2019, **37**, 107351.
57. C. Santoro, C. Arbizzani, B. Erable and I. Ieropoulos, Microbial fuel cells: From fundamentals to applications. A review, *J. Power Sources*, 2017, **356**, 225–244.
58. M. M. Salcher, D. Schaefle, M. Kaspar, S. M. Neuenschwander and R. Ghai, Evolution in action: habitat transition from sediment to the pelagial leads to genome streamlining in Methylophilaceae, *ISME J.*, 2019, **13**, 2764–2777.
59. W. Wang, Y. Zhang, M. Li, X. Wei, Y. Wang, L. Liu, H. Wang and S. Shen, Operation mechanism of constructed wetland-microbial fuel cells for wastewater treatment and electricity generation: A review, *Bioresour. Technol.*, 2020, **314**, 123808.
60. M. S. Painganker, K. Ahire, P. Mishra, S. Rajpathak and D. D. Deobagkar, Microbial diversity analysis in the oxygen minimum zones of the Arabian Sea using metagenomics approach, *Curr. Sci.*, 2020, **118**, 1042–1051.
61. B. E. Logan, B. Hamelers, R. Rozendal, U. Schröder, J. Keller, S. Freguia, P. Aelterman, W. Verstraete and K. Rabaey, Microbial Fuel Cells: Methodology and Technology, *Environ. Sci. Technol.*, 2006, **40**, 5181–5192.
62. S. Elhenawy, M. Khraisheh, F. AlMomani, M. Al-Ghouti and M. K. Hassan, From Waste to Watts: Updates on Key Applications of Microbial Fuel Cells in Wastewater Treatment and Energy Production, *Sustainability*, 2022, **14**, 955.
63. P. Choudhury, U. S. P. Uday, N. Mahata, O. Nath Tiwari, R. Narayan Ray, T. Kanti Bandyopadhyay and B. Bhunia, Performance improvement of microbial fuel cells for waste water treatment along with value addition: A review on past achievements and recent perspectives, *Renewable Sustainable Energy Rev.*, 2017, **79**, 372–389.
64. P. L. McCarty, J. Bae and J. Kim, Domestic Wastewater Treatment as a Net Energy Producer–Can This be Achieved?, *Environ. Sci. Technol.*, 2011, **45**, 7100–7106.
65. H. Rasouli Sadabad and G. Badalians Gholikandi, Harvesting direct electricity from municipal waste-activated sludge simultaneous with its aerobic stabilization process: Investigation and optimization, *J. Environ. Chem. Eng.*, 2017, **5**, 1174–1185.
66. L. S. Vélez-Pérez, J. Ramirez-Nava, G. Hernández-Flores, O. Talavera-Mendoza, C. Escamilla-Alvarado, H. M. Poggi-Varaldo, O. Solorza-Feria and J. A. López-Díaz, Industrial acid mine drainage and municipal wastewater co-treatment by dual-chamber microbial fuel cells, *Int. J. Hydrogen Energy*, 2020, **45**, 13757–13766.

67. H. Liu, F. Leng, Y. Guan, Y. Yao, Y. Li and S. Xu, Simultaneous Pollutant Removal and Electricity Generation in a Combined ABR-MFC-MEC System Treating Fecal Wastewater, *Water, Air, Soil Pollut.*, 2017, **228**, 179.
68. R. J. Marassi, L. G. Queiroz, D. C. V. R. Silva, F. T. da Silva, G. C. Silva and T. C. B. de Paiva, Performance and toxicity assessment of an up-flow tubular microbial fuel cell during long-term operation with high-strength dairy wastewater, *J. Clean. Prod.*, 2020, **259**, 120882.
69. S. S. Lim, J.-M. Fontmorin, H. T. Pham, E. Milner, P. M. Abdul, K. Scott, I. Head and E. H. Yu, Zinc removal and recovery from industrial wastewater with a microbial fuel cell: Experimental investigation and theoretical prediction, *Sci. Total Environ.*, 2021, **776**, 145934.
70. Y. Dong, Y. Qu, W. He, Y. Du, J. Liu, X. Han and Y. Feng, A 90-liter stackable baffled microbial fuel cell for brewery wastewater treatment based on energy self-sufficient mode, *Bioresour. Technol.*, 2015, **195**, 66–72.
71. T. Ewing, J. T. Babauta, E. Atci, N. Tang, J. Orellana, D. Heo and H. Beyenal, Self-powered wastewater treatment for the enhanced operation of a facultative lagoon, *J. Power Sources.*, 2014, **269**, 284–292.
72. Z. Ge, J. Li, L. Xiao, Y. Tong and Z. He, Recovery of Electrical Energy in Microbial Fuel Cells, *Environ. Sci. Technol. Lett.*, 2014, **1**, 137–141.
73. S. Chen, S. A. Patil, R. K. Brown and U. Schröder, Strategies for optimizing the power output of microbial fuel cells: Transitioning from fundamental studies to practical implementation, *Appl. Energy*, 2019, **233–234**, 15–28.
74. P. Aelterman, K. Rabaey, H. T. Pham, N. Boon and W. Verstraete, Continuous Electricity Generation at High Voltages and Currents Using Stacked Microbial Fuel Cells, *Environ. Sci. Technol.*, 2006, **40**, 3388–3394.
75. I. Ieropoulos, C. Melhuish and J. Greenman, in *Artificial Metabolism: Towards True Energetic Autonomy in Artificial Life BT - Advances in Artificial Life*, ed. W. Banzhaf, J. Ziegler, T. Christaller, P. Dittrich and J. T. Kim, Springer Berlin Heidelberg, Berlin, Heidelberg, 2003, pp. 792–799.
76. T. Kim, J. Yeo, Y. Yang, S. Kang, Y. Paek, J. K. Kwon and J. K. Jang, Boosting voltage without electrochemical degradation using energy-harvesting circuits and power management system-coupled multiple microbial fuel cells, *J. Power Sources*, 2019, **410–411**, 171–178.
77. X. A. Walter, I. Merino-Jiménez, J. Greenman and I. Ieropoulos, PEE POWER® urinal II – Urinal scale-up with microbial fuel cell scale-down for improved lighting, *J. Power Sources*, 2018, **392**, 150–158.
78. X. A. Walter, A. Stinchcombe, J. Greenman and I. Ieropoulos, Urine transduction to usable energy: A modular MFC approach for smartphone and remote system charging, *Appl. Energy*, 2017, **192**, 575–581.
79. C. Donovan, A. Dewan, H. Peng, D. Heo and H. Beyenal, Power management system for a 2.5W remote sensor powered by a sediment microbial fuel cell, *J. Power Sources*, 2011, **196**, 1171–1177.

80. X. A. Walter, J. Greenman and I. A. Ieropoulos, Microbial fuel cells directly powering a microcomputer, *J. Power Sources*, 2020, **446**, 227328.
81. C. Donovan, A. Dewan, D. Heo and H. Beyenal, Batteryless, Wireless Sensor Powered by a Sediment Microbial Fuel Cell, *Environ. Sci. Technol.*, 2008, **42**, 8591–8596.
82. I. F. Akyildiz, D. Pompili and T. Melodia, Underwater acoustic sensor networks: research challenges, *Ad Hoc Networks*, 2005, **3**, 257–279.
83. D. Diamond, S. Coyle, S. Scarmagnani and J. Hayes, Wireless Sensor Networks and Chemo-/Biosensing, *Chem. Rev.*, 2008, **108**, 652–679.
84. A. Rhoads, H. Beyenal and Z. Lewandowski, Microbial Fuel Cell using Anaerobic Respiration as an Anodic Reaction and Biomineralized Manganese as a Cathodic Reactant, *Environ. Sci. Technol.*, 2005, **39**, 4666–4671.
85. W. H. Dickinson, F. Caccavo and Z. Lewandowski, The ennoblement of stainless steel by manganic oxide biofouling, *Corros. Sci.*, 1996, **38**, 1407–1422.
86. A. Shantaram, H. Beyenal, R. R. A. Veluchamy and Z. Lewandowski, Wireless Sensors Powered by Microbial Fuel Cells, *Environ. Sci. Technol.*, 2005, **39**, 5037–5042.
87. C. Dumas, A. Mollica, D. Féron, R. Basséguy, L. Etcheverry and A. Bergel, Marine microbial fuel cell: Use of stainless steel electrodes as anode and cathode materials, *Electrochim. Acta*, 2007, **53**, 468–473.
88. F. Khaled, O. Ondel and B. Allard, Microbial fuel cells as power supply of a low-power temperature sensor, *J. Power Sources*, 2016, **306**, 354–360.
89. M. Kim, M. Sik Hyun, G. M. Gadd and H. J. Kim, A novel biomonitoring system using microbial fuel cells, *J. Environ. Monit.*, 2007, **9**, 1323–1328.
90. I. S. Chang, J. K. Jang, G. C. Gil, M. Kim, H. J. Kim, B. W. Cho and B. H. Kim, Continuous determination of biochemical oxygen demand using microbial fuel cell type biosensor, *Biosens. Bioelectron.*, 2004, **19**, 607–613.
91. S. V. Alferov, V. A. Arlyapov, V. A. Alferov and A. N. Reshetilov, Biofuel Cell Based on Bacteria of the Genus Gluconobacter as a Sensor for Express Analysis of Biochemical Oxygen Demand, *Appl. Biochem. Microbiol.*, 2018, **54**, 689–694.
92. I. Karube, T. Matsunaga, S. Mitsuda and S. Suzuki, Microbial electrode BOD sensors, *Biotechnol. Bioeng.*, 1977, **19**, 1535–1547.
93. I. A. Ieropoulos, J. Greenman, C. Melhuish and J. Hart, Comparative study of three types of microbial fuel cell, *Enzyme Microb. Technol.*, 2005, **37**, 238–245.
94. B. H. Kim, I. S. Chang, G. Cheol Gil, H. S. Park and H. J. Kim, Novel BOD (biological oxygen demand) sensor using mediator-less microbial fuel cell, *Biotechnol. Lett.*, 2003, **25**, 541–545.
95. O. D. Ansa-Asare, I. L. Marr and M. S. Cresser, Evaluation of modelled and measured patterns of dissolved oxygen in a freshwater lake as an indicator of the presence of biodegradable organic pollution, *Water Res.*, 2000, **34**, 1079–1088.

96. C. D. Markfort and M. Hondzo, Dissolved Oxygen Measurements in Aquatic Environments: The Effects of Changing Temperature and Pressure on Three Sensor Technologies, *J. Environ. Qual.*, 2009, **38**, 1766–1774.
97. M. Zhang, Y. Yu, Z. Yang, X. Shi and F. Kong, The distribution of phytoplankton along trophic gradients and its mediation by available light in the pelagic zone of large eutrophic lakes, *Can. J. Fish. Aquat. Sci.*, 2012, **69**, 1935–1946.
98. Q. Shen, Q. Zhou, J. Shang, S. Shao, L. Zhang and C. Fan, Beyond hypoxia: Occurrence and characteristics of black blooms due to the decomposition of the submerged plant Potamogeton crispus in a shallow lake, *J. Environ. Sci.*, 2014, **26**, 281–288.
99. Z. Liu, J. Liu, S. Zhang, X.-H. Xing and Z. Su, Microbial fuel cell based biosensor for in situ monitoring of anaerobic digestion process, *Bioresour. Technol.*, 2011, **102**, 10221–10229.
100. H. Sun, Y. Zhang, S. Wu, R. Dong and I. Angelidaki, Innovative operation of microbial fuel cell-based biosensor for selective monitoring of acetate during anaerobic digestion, *Sci. Total Environ.*, 2019, **655**, 1439–1447.
101. H. Sun, Y. Zhang, S. Wu, R. Dong and I. Angelidaki, Innovative operation of microbial fuel cell-based biosensor for selective monitoring of acetate during anaerobic digestion, *Sci. Total Environ.*, 2019, **655**, 1439–1447.
102. S.-Y. Yang, V. Sencadas, S. S. You, N. Z.-X. Jia, S. S. Srinivasan, H.-W. Huang, A. E. Ahmed, J. Y. Liang and G. Traverso, Powering Implantable and Ingestible Electronics, *Adv. Funct. Mater.*, 2021, **31**, 2009289.
103. S. El Ichi-Ribault, J.-P. Alcaraz, F. Boucher, B. Boutaud, R. Dalmolin, J. Boutonnat, P. Cinquin, A. Zebda and D. K. Martin, Remote wireless control of an enzymatic biofuel cell implanted in a rabbit for 2 months, *Electrochim. Acta*, 2018, **269**, 360–366.
104. I. Jeerapan, B. Ciui, I. Martin, C. Cristea, R. Sandulescu and J. Wang, Fully edible biofuel cells, *J. Mater. Chem. B*, 2018, **6**, 3571–3578.
105. M. Rezaie, Z. Rafiee and S. Choi, Biopower-in-gut: An ingestible bacteria-powered battery capsule, in *Technical digest SolidState Sensor Actuator and Microsystems Workshop*, 2022.
106. M. Rezaie, Z. Rafiee, S. Choi and A. Biobattery Capsule, for Ingestible Electronics in the Small Intestine: Biopower Production from Intestinal Fluids Activated Germination of Exoelectrogenic Bacterial Endospores, *Adv. Energy Mater.*, 2023, **13**, 2202581.
107. A. Yuvarajan, G. Vishnu and N. Saranraj, Multiharvested Self-Powered Systems: A Review on Harnessing Biological Energy for Implantable Medical Devices, *Biomed. Mater. & Devices*, 2025, 312–325.

CHAPTER 7

Biocompatible Batteries for Implantable and Ingestible Devices

SAGOR ROY† AND SANTANU PATRA*,†

Drug Delivery and Sensing (IDUN), Department of Health Technology, Technical University of Denmark, Ørsted Plads, 2800, Kgs. Lyngby, Denmark
*Email: sanpa@dtu.dk

7.1 Introduction

Recent decades have witnessed significant advancements in implantable and ingestible biomedical devices that can be applied to the human body for essential tests and therapies. These devices facilitate continuous biomedical research and are expected to treat countless patients. The development of implantable bioelectronics is moving toward softness, miniaturization, biodegradability, and multifunctionality. Moreover, research and advancements are taking these innovative devices toward being more patient-friendly, biodegradable, and compact sizing.[1,2] These devices are lifesaving in many aspects, as they are effective in restoring the body's normal function, such as the heart rate. They are also useful in the alteration of different body parts and pain management. Ongoing research and development are

† These authors contributed equally.

Detection Science Series No. 33
Energy Materials for Ingestible and Implantable Healthcare Devices
Edited by Santanu Patra and Anja Boisen

Published by the Royal Society of Chemistry, www.rsc.org

making these devices smaller in size, enhancing their lifetime, and powering these microdevices. Low adverse effects and patient safety make these devices more attractive for use in diagnostic and chronic disease treatment.[3]

The first implantable device, a heart pacemaker, was developed in 1958. Since that time, several implantable devices have been designed to address multiple disorders. Recent research and technologies have shown the path for making advanced pacemakers that are small, provide expedited responses, and minimize adverse effects relative to earlier models. On the other hand, cardiac defibrillators are also very popular in the context of implantable devices. They are extremely effective against fatal diseases like arrhythmia. These lifesaving, compact devices are yet powerful, making the diagnostic and treatment strategy easier.[2,4]

Ingestible devices have lately emerged as a transformative approach in the healthcare system. These devices have been extensively utilized in the diagnosis of the gastrointestinal tract. Capsule-like ingestible devices now enable the regulation of body pH and temperature while capturing video of the gastrointestinal tract. Ingestible devices can be used as sensors, for drug delivery, or for sampling. They are patient-friendly as they are orally administered and non-intrusive. Moreover, these devices have the benefits of being easy to use, enabling real-time evaluation, and monitoring physiological processes.[5–7] The significance of these devices is increasing daily inside the healthcare system. In this technological and artificial intelligence era, real-time internal evaluation and a non-invasive diagnostic technique are essential. These devices have been used in targeted drug delivery systems for a long period to deliver the drug in the exact location of the gut and to ensure the controlled release of the drug and time of stay. These microdevices are also useful in the early identification of diseases like Crohn's disease, ulcerative colitis, and even cancer, which enables the treatment to start as early as possible. Most importantly, they possess no side effects when in contact with the human body and maintain a regular environment in the gut area.[8,9]

Ongoing research and development of ingestible devices have advanced significantly. The first ingestible capsule, termed "RadioPill," was created in 1957, while the first video capsule, known as "SmartPill," was devised in 1981 and received FDA approval in 2001.[9] However, advancements in biomaterials and biochemistry, modernized drug delivery methods, and novel powering strategies have fundamentally transformed the landscape of ingestible devices. At present, a limited number of devices have gained attention for their ability to capture real-time images of gastrointestinal health and monitor drug release and pH levels. For example, PillCam SB3 (approved in 2021), CapsoCam Plus (approved in 2020), and EndoCapsule 10 (approved in 2015) are among these devices.[8,9]

Extensive study has been conducted on both ingestible and implanted devices; however, with the assistance of artificial intelligence and novel methodologies, precise diagnosis and tailored treatment are anticipated to be realized within the next decade. Among the two devices, implantable devices are more popular and somewhat more advanced; however, numerous factors must be considered in both instances, including the dimensions of the devices, the complex characteristics of the *in vivo* conditions, the duration of device residency inside the body, biocompatibility, biodegradability, and, most crucially, the power source.[2,5]

7.2 Power Sources and Supply

One of the most crucial factors for the smooth functioning of implantable and ingestible devices is power. From obtaining data to transferring the data for evaluation uninterruptedly, a steady-state energy supply is necessary. This power could be generated either *via* an energy harvesting approach inside the body/or from energy storage devices. Traditional energy sources could be harmful to the body, and their sizes are not suitable for patients. Important characteristics for the power sources for both devices should include biocompatibility, biodegradability, compactness, and the absence of adverse effects on the human body.

7.2.1 Energy Harvesting Devices

Energy harvesting devices are very efficient and possess the capability to power implanted and ingestible devices. These devices are considered the optimal substitute for conventional batteries, which exhibit hazardous effects within the body. Energy harvesting techniques within the body primarily rely on three internal stimuli: biomechanical, biochemical, and thermal. In the case of biomechanical, the primary source of energy is the biomechanical movement within the body, which can be transformed into electrical energy through approaches such as piezoelectric and triboelectric nanogenerators. In biochemical contexts, glucose or other chemicals present in body fluids serve as the source for converting chemical energy into electrical energy, a principle underlying biofuel cells. Additionally, thermal energy harvesting involves the conversion of temperature variations into electrical energy, utilizing the pyro and thermoelectric nanogenerator concept.[1–3]

However, usually, the energy output from harvesting technologies is significantly low and unsuitable for direct use. Consequently, it is a common practice to combine energy harvesting systems with energy storage components, enabling the generated energy to be stored for subsequent purposes. The harvested energy can be used to recharge the energy storage devices, like batteries or supercapacitors, for their continuous operation.

7.2.2 Energy Storage Devices

Conventional energy storage systems include batteries and supercapacitors (SCs) and have achieved remarkable success as energy storage devices and are the most frequently utilized in implantable and ingestible technologies. Batteries possess higher energy density, and they provide better cyclical stability compared to SCs. On the other side, SCs have higher power density, excellent charge/discharge rates, and better cyclic lifespans compared to batteries. Batteries are commercially available and used everywhere; however, they are often big, have little flexibility, and common batteries exhibit harmful effects on the human body. SCs also could be harmful, as their charge/discharge rate is rapid, they can create short circuits, and they are often non-degradable. However, the recent success of biocompatible systems makes these energy storage devices superior to others.[1,4]

Upon examining the uses of both energy storage systems, it is evident that each possesses distinct advantages. However, for prolonged operations, batteries are the obvious selection. In this chapter, we will concentrate on biocompatible battery systems and their applications in implantable and ingestible devices. These biocompatible batteries are safe, non-toxic, inexpensive, and biocompatible, and can remain stable for a long period.[5–8]

7.3 Batteries

Batteries are energy storage devices that generate electrical energy from chemical energy with the help of electrochemical reactions. Batteries are widely used in electronics and medical devices for their excellent characteristics; for instance, they have high energy density, excellent cyclic stability, good charge–discharge rates, *etc.* They have been instrumental in the success of implantable and ingestible devices since the first pacemaker was developed by installing a battery with it. Advanced biomaterial science and technology make batteries more feasible and attractive for medical devices. However, normal batteries are big, have little flexibility, cannot be used for a long time, and have toxic effects, and therefore often are not suitable for use in the human body. That is why new, advanced batteries have emerged to diminish these challenges.[1,5,9]

A regular battery consists of two electrode systems where the anode releases electrons and the cathode accepts electrons to complete a redox reaction. On the other hand, an electrolyte ensures that the ions can move freely in the system. The electrolyte can be obtained from inside the body or outside of the body, and the form of electrolyte can be gel, liquid, or solid. A battery also consists of a separator, which makes a barrier so that the electrode system do not collide.[9,10]

7.3.1 Common Batteries for Biomedical Devices

7.3.1.1 Lithium and Sodium Batteries

Li-ion or Li-based batteries are a common name in the battery field, and they also have been used in different biomedical devices. When the first pacemaker was designed, Li was used to power the pacemaker. Since then, Li batteries have ruled the energy system industry. These batteries have high energy density, exceptional stability and lifespan, and can be used for an extended period (5–8 years) in not only pacemakers but also other medical devices. However, they have some serious drawbacks that reduce their usage in the healthcare system nowadays. Lithium batteries degrade over time, exhibit significant toxicity to humans, are susceptible to electrolyte leakage issues, and necessitate secure packaging.[5,9]

Na-ion batteries are also very common and mainly have gained attention due to their cost-effectiveness, excellent cyclic performance, high-level stability, and super storage capacity. However, they also suffer from the biocompatibility issue, use of toxic electrolyte, compact packaging, and there is a chance of electrolyte leaking.[5,9,10] These characteristics limit their applicability in biomedical devices.

7.3.1.2 Silver and Other Metal-based Batteries

Compared to Li and Na batteries, Ag (mainly Ag-oxide) batteries have been used in these biomedical devices to a smaller extent. Nevertheless, they exhibit remarkable energy density and are clinically authorized for application in implantable and ingestible devices, having certain negative effects.[11] The high cost of silver, attributed to its scarcity and intricate production methods, makes silver batteries costly, potentially restricting their application in cost-effective scenarios. Additionally, the toxicity of silver upon disposal raises environmental concerns, requiring careful recycling procedures.

Other than these batteries, Zn–air batteries are also very popular for implantable devices. These batteries are well known for their high energy density, although their cost is considerably higher. Additionally, Zn–air batteries are not suitable for use with ingestible devices. Nickel–cadmium (Ni–Cd) batteries, once used in medical devices quite often because their charge–discharge rate was higher, had good cyclic performance and could remain stable for a long time. However, as they are extremely toxic to humans and are not sustainable, their use was banned several years ago.[9,11,12]

7.3.2 Challenges

Now, if we summarize the current status and perspectives, it is clear that batteries are extremely important and useful for medical devices and have been used for a long period of time, however their associated side effects have been a huge drawback. Batteries are particularly large, and this is one of their significant limitations. Moreover, there is a risk of potential bursting and leakage, which can lead to lower biocompatibility. The battery's lifespan is another limitation, making it unsuitable for prolonged use within the body. Degradation of battery materials can often be seen, which raises a question about the environmental friendliness of a battery. The aforementioned drawbacks are substantial, prompting the consideration of emerging biocompatible batteries for these devices, which are currently supplanting conventional batteries in various medical devices, including implantable and ingestible types, primarily due to their biocompatibility and reduced side effects.[11,13]

7.3.3 Basic Requirements of a Battery to Be Used in Implantable and Ingestible Devices

The human body is complex and composed of soft tissues, skin, muscles, *etc.* Integrating implanted or ingestible devices into the human body is complex, as these devices must meet fundamental needs and protocols. Skin, soft tissues, and internal organs are susceptible to injury, presenting a significant challenge for both devices to prevent damage while delivering optimal performance. These devices must follow some basic requirements to be applied to the body.

- Safety: the first and foremost requirement for use of these devices in the body is safety, and they should not possess any toxicity when coming into contact with the body. One of the solutions to avoid toxicity is to have a biocompatibility feature in these devices. Biocompatible materials ensure that these devices will not exhibit toxicity or irritation at the implantation or ingestion site. Furthermore, to stop thrombus development at the device's surface, the device also should have anticoagulant and antimicrobial qualities at the location of implantation.
- Stability: since implanted/ingestible energy storage devices cannot be readily replaced, they must have exceptional electrochemical, thermomechanical, and structural durability when used in bodily fluids for a prolonged period of time. Furthermore, even after extended use, the mechanical and electrochemical characteristics ought to stay consistent, making sure that tissue deformation does not jeopardize the integrity of the device's construction.
- Matching: in order to facilitate adaptation to the tissue's unevenness during dynamic deformation without applying any mechanical load, both implantable and ingestible energy storage devices should have particular mechanical properties, such as Young's modulus, that are compatible with biological tissue. Besides, corresponding mechanical qualities could lessen tissue damage and physical irritation, hence reducing negative immune responses and health risks.
- Other than these requirements, one of the most important requirements of these devices is a miniature design. These devices should be smaller in size, lightweight, and thinner.[6]

7.4 Biocompatible Batteries: A Solution

Biocompatible batteries, sometimes referred to as biobatteries, are energy storage devices designed to interact safely with living organisms, reducing adverse reactions and integrating effectively with biological tissues. They are very important in terms of powering medical implants and other bioelectronic devices that require safety and integration within the body. The essential condition for these batteries is that all battery components will be completely biocompatible and will not have any negative effects on the human body. They work like regular batteries, but the main difference is that these batteries are biocompatible and biodegradable and can be integrated with next-generation implantable and ingestible devices. Like conventional batteries, a biobattery also consists of two electrodes, an electrolyte, and a separator.[14,15]

Biobatteries are thought to be a groundbreaking approach for powering next-generation ingestible and implantable devices. Previously, Li-based batteries were the preferred choice because of their high energy density and efficiency. They have been widely employed in implantable devices and are still used in electronics and appliances. However, their toxicity and lack of biocompatibility make them unsuitable for use in these biomedical devices and emerging biobatteries being considered as the replacement. New-generation biobatteries are safe, biocompatible, inexpensive, environmentally friendly, and have longer lifespans than ordinary lithium-ion batteries.[5,12] Their integration with these batteries in the biomedical devices has yet to be extensively explored, but these batteries will undoubtedly revolutionize the landscape of diagnostic and treatment approaches in the near future. In terms of these biocompatible batteries, Zn- and Mg-based batteries are leading due to their high energy production and biocompatibility phenomenon. In this chapter, we will discuss their progress, challenges, and perspectives in terms of their applications in implantable and ingestible devices.

7.4.1 New-generation Biobatteries

7.4.1.1 Zn-based Batteries

Zn-based batteries are one of the most effective and most used biobatteries in medical devices. They are safe, inexpensive, have high energy density and energy efficiency, and have excellent stability and biocompatibility. Zn as an anode has attained significant popularity and is frequently employed in electronics and medical devices owing to its biocompatibility, making it suitable for both implantable and ingestible devices. Furthermore, they pose no toxicity risk upon bodily contact and are suitable for use with these devices. Additionally, the high energy density enhances the stability and efficiency of the charging–discharging cycle of the zinc battery.[12,16] Some of the most commonly used Zn batteries are Zn ion, Zn metal, Zn air, *etc.* They typically have moderate to high energy density and acceptable stability.[17,18] The structure and inherent mechanism of Zn batteries are shown in Figure 7.1. However, the use of Zn ions as a cathode adversely affects the reversibility and decreases the output voltage of the battery, leading to their replacement by more efficient metal ions. Table 7.1 describes some potential Zn-based biobatteries that can be integrated with ingestible and implantable devices.

Table 7.1 indicates that Zn–Cu batteries provide the best energy density and voltage capacity, while Zn–$MgMn_2O_4$ exhibits superior cycle performance. Zn–MnO_2 excels in retention capacity and stability.

The overall reaction in a Zn–Cu battery involves the oxidation of zinc and the reduction of copper ions, resulting in electron transfer and the formation

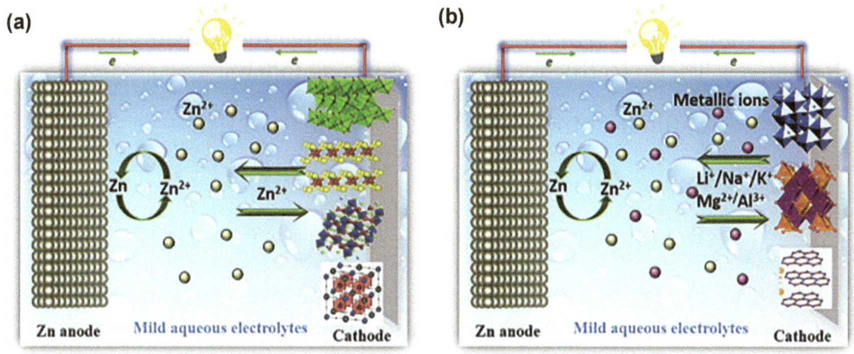

Figure 7.1 (a) Regular Zn ion batteries and (b) different metal ions replacing Zn ions as the cathode. Reproduced from ref. 17 with permission from Elsevier, Copyright 2019.

Table 7.1 Potential Zn-based biobatteries for implantable and ingestible devices.

Battery composition	Zn–MnO_2[23,24]	Zn–Mo[5,12,25]	Zn–Cu[5,13,19,26]	Zn–$MgMn_2O_4$[27]
Anode	Zn	Zn	Zn	Zn
Cathode	MnO_2	MoO_3	Cu	$MgMn_2O_4$
Electrolyte	Zn–Alg	$ZnSO_4/ZnCl_2$	$ZnSO_4/CuCl_2$	$ZnSO_4/MnSO_4/MgSO_4$
Energy density W h Kg^{-1}	158.5	39	380	370
Voltage capacity	1.38	0.7	1.7	1.5
Cyclic performance (after 100 cycles)	225 mA g^{-1}	273 mA g^{-1}	200 mA g^{-1}	500 mA g^{-1} (after 500 cycles)
Stability	>500 h	>450 h	>150 h	—
Retention capacity (after 100 cycles)	99.9%	73%	83%	94% (after 500 cycles)

of zinc and copper ions. The retention capacity and stability of these batteries are moderate. The electrochemical reaction can be written as follows:[19]

$$\text{Anode (Oxidation): Zn (s)} \rightarrow \text{Zn}^{2+}\text{ (aq)} + 2\text{e}^-$$

$$\text{Cathode (Reduction): Cu}^{2+}\text{(aq)}+2\text{e}^- \rightarrow \text{Cu(s)}$$

$$\text{Overall Reaction: Zn(s)} + \text{Cu}^{2+}\text{(aq)} \rightarrow \text{Zn}^{2+}\text{(aq)} + \text{Cu(s)}$$

A battery where Zn was an anode and a combination of Mg and Mn acted as a cathode, and their energy density was excellent, and this battery's retention capacity was more than 90%. Their reaction is as follows:[20]

$$\text{Anode: Zn} \rightarrow \text{Zn}^{2+} + 2\text{e}^-$$

$$\text{Cathode: MgMn}_2\text{O}_4 + \text{Zn}^{2+} + 2\text{e}^- \leftrightarrow \text{ZnMgMn}_2\text{O}_4$$

$$\text{Overall: Zn} + \text{MgMn}_2\text{O}_4 \leftrightarrow \text{ZnMgMn}_2\text{O}_4$$

Zn–MnO_2 exhibits moderate energy density and voltage capacity; nevertheless, its stability and retention capacity are unparalleled compared to the alternatives. Their electrochemical process is as follows:[12]

$$\text{Cathode: } MnO_2 + 4H^+ + 2e^- \leftrightarrow Mn^{2+} + 2H_2O$$

$$\text{Anode: } Zn \rightarrow Zn^{2+} + 2e^-$$

$$\text{Overall: } MnO_2 + Zn + 4H^+ \leftrightarrow Mn^{2+} + Zn^{2+} + 2H_2O$$

The Zn–Mo battery is also very stable and can give moderate cyclic performance. The electrochemical reaction of this battery is as follows:[21,22]

$$\text{Anode: } Zn \rightarrow Zn^{2+} + 2e^-$$

$$\text{Cathode: } MoO_3 + Zn^{2+} + 2e^- \rightarrow ZnMoO_3$$

$$\text{Overall: } Zn + MoO_3 \leftrightarrow ZnMoO_3$$

Nadeau *et al.* reported a self-sufficient ingestible device in capsule form utilizing a biobattery with zinc as the anode and copper as the cathode.[19] The device provided an average power output of 0.23 $\mu W\,mm^{-2}$ for an average duration of 6.1 days in the GI tract of a swine model, along with wireless transmission. Even though the cell voltage and power fluctuated during trials, the system was still able to power these more advanced electronics by combining the cell with a boost converter in the energy harvesting integrated circuit.

The ingestible battery can function in the stomach, intestines, and colon, among other areas of the GI system. However, because of the distinct physiological conditions in the colon and intestine, the power output was significantly lower. The device demonstrated enhanced biocompatibility, as evidenced by biocompatibility tests and animal studies, and can remain in the body for an extended duration.

In a different study, Xie *et al.* developed a zinc and alginate green battery using the superionic interaction of alginate carboxylate groups with electrooxidized free-Zn^{2+}.[23] As can be seen in Figure 7.2, the battery exhibits superior electrochemical performance. The two usual redox peaks of approximately 1.25 and 1.38 V (*vs.* Zn^{2+}/Zn) in the cyclic voltammetry (CV) (Figure 7.2a) are in good agreement with the conventional liquid electrolyte (2 M $ZnSO_4$ + 0.2 M $MnSO_4$). The cell exhibits excellent electrochemical stability at a current density of 1 $A\,g^{-1}$, retaining 95% of its initial capacity and achieving a specific capacity of 206 $mA\,h\,g^{-1}$ after 100 cycles (Figure 7.2b). This stability is also seen in batteries with lower current densities of 0.5 and 0.2 $A\,g^{-1}$. Figure 7.2c displays the voltage profiles for the 1st, 50th, and 100th cycles, respectively. The execution of the rate capacity is depicted in Figure 7.2d. Figure 7.2e shows the stability and flexibility of the twisted wire-shaped ZBs, and Figure 7.2f shows that they can be easily rolled and able to power an LED.

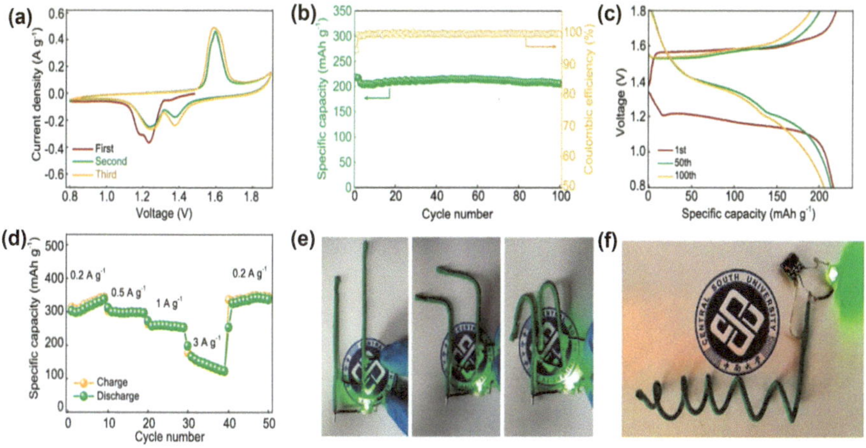

Figure 7.2 (a) Full wire-shaped battery cyclic voltammetry curves at 0.1 $mV\,s^{-1}$. (b) Full wire-shaped battery's cycling performance at 1000 $mA\,g^{-1}$ and (c) the associated charge–discharge curves. (d) Performance of rate capacity. (e) Photographs of wire-shaped zinc batteries following bending at angles of 90°, 0°, and 180°. (f) Zinc battery in the form of rolling wires. Reproduced from ref. 23, https://doi.org/10.1093/nsr/nwac281, under the terms of the CC BY 4.0 license, https://creativecommons.org/licenses/by/4.0/.

Using rabbit models, researchers examined the battery's biocompatibility and safety with the Zn–Alg-5 polymer (Figure 7.3a and b). Figure 7.3c illustrates that the batteries cycled with liquid and Zn–Alg-5 polymer electrolytes exhibited structural integrity, mucosal secretions, and an absence of injury in the stomach mucosa. The perforated batteries were positioned on the freshly removed duodenum for 6 hours to assess biocompatibility. Figure 7.3d illustrates that the duodenum treated with the Zn–Alg-5 electrolyte-based battery exhibits little intestinal mucosa damage, likely attributable to the biocompatibility of the alginate components. The stained segment of the duodenal mucosa (Figure 7.3e) further supports the structural integrity and absence of necrosis in the intestinal villus following treatment with the Zn–Alg-5 electrolyte-based battery.

7.4.1.2 Mg-based Batteries

Similar to Zn, Mg has garnered significant interest for its role as an important biobattery in medical devices and microdevices. They serve as direct alternatives to lithium-based batteries and possess advantages similar to zinc batteries. Mg batteries are safe, non-toxic, biocompatible, and have higher energy density and are much cheaper than Li batteries. They have been utilized in implantable devices and also possess the potential for integration into ingestible devices.[20] Different types of Mg-based batteries and their performances have been discussed in Table 7.2.

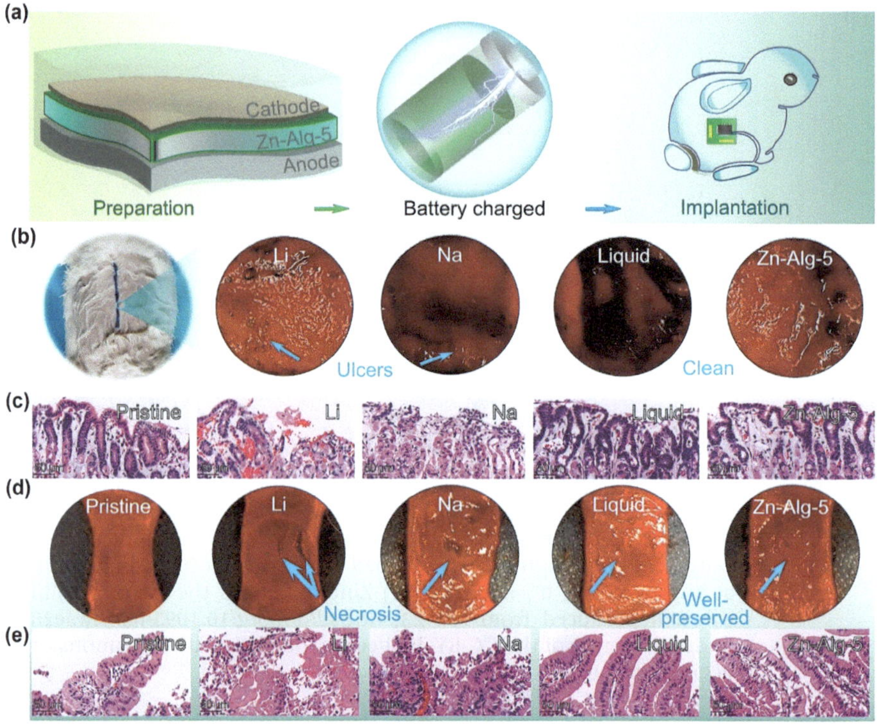

Figure 7.3 Biocompatibility and safety studies of batteries implanted in the rabbit. (a) Schematic diagram of biocompatible applications using rabbit models. (b) Photographic picture of gastric walls and (c) the HE-stained section of its mucosa after the undamaged batteries were implanted in the stomach for 6 hours. (d) Pictures of freshly extracted duodenum tissue and (e) the HE-stained section of its mucosa after the punched batteries were implanted in the duodenum organ for 6 hours. Reproduced from ref. 23, https://doi.org/10.1093/nsr/nwac281, under the terms of the CC BY 4.0 license, https://creativecommons.org/licenses/by/4.0/.

In comparison to other magnesium batteries, magnesium-iron has a worse energy capacity and cycle performance; however, its retention capacity exceeds 85%. In the case of Mg–Mo, stability is significantly superior to that of the others, while Mg–MnO_2 offers an enhanced voltage capacity. Mg–S has superior cycle performance and energy capacity compared to its counterparts. The electrochemical reactions for all of these batteries have been mentioned below.[20,28]

The electrochemical reaction for the Mg–Fe battery is as follows:

$$\text{Anode: } Mg \rightarrow Mg^{2+} + 2e^-$$

$$\text{Cathode: } Fe^{3+} + e^- \rightarrow Fe^{2+}$$

$$\text{Overall: } Mg + Fe^{3+} \rightarrow Mg^{2+} + Fe^{2+}$$

Table 7.2 Potential Mg-based biobatteries for biomedical devices.

Battery composition	Mg–Fe[12,20,29]	Mg–Mo[5,20,29]	Mg–MnO_2[29]	Mg–S[28,30]
Anode	Mg	Mg	Mg	Mg
Cathode	Fe	MoO_3	MnO_2	S
Electrolyte (with Mg-salt)	PCL	Alginate hydrogel	THF/C_4H_8O	Mg $[B(hfip)_4]_2$
Energy capacity $mA\,h\,g^{-1}$	125	220	231	400
Voltage capacity	2.3	1.8	2.8	1.5
Cyclic performance (after 100 cycles)	20 $mA\,g^{-1}$	210 $mA\,g^{-1}$	164 $mA\,g^{-1}$ (after 500 cycles)	545 $mA\,g^{-1}$ (after 300 cycles)
Stability	6 h	>300 h	—	48 h
Retention capacity	87% after 50 cycles	80.9% after 10 cycles	78% after 500 cycles	53% after 50 cycles

The electrochemical reaction for the Mg–Mo battery is as follows:

$$\text{Anode: } Mg \rightarrow Mg^{2+} + 2e^-$$

$$\text{Cathode: } MoO_3 + Mg^{2+} + 2e^- \rightarrow MgMoO_3$$

$$\text{Overall: } Mg + MoO_3 \leftrightarrow MgMoO_3$$

The electrochemical reaction for the Mg–MnO_2 battery is as follows:

$$\text{Anode: } Mg \rightarrow Mg^{2+} + 2e^-$$

$$\text{Cathode: } MnO_2 + Mg^{2+} + 2e^- \rightarrow MgMnO_2$$

$$\text{Overall: } Mg + MnO_2 \leftrightarrow MgMnO_2$$

The electrochemical reaction for the Mg–S battery is as follows:

$$\text{Anode: } Mg \rightarrow Mg^{2+} + 2e^-$$

$$\text{Cathode: } S_8 + 2Mg^{2+} + 4e^- \rightarrow MgS_8$$

$$\text{Overall: } Mg + S \leftrightarrow MgS$$

Magnesium biobatteries can integrate with ingestible devices due to their superior characteristics, similar to those of zinc batteries. They have demonstrated outstanding performance in implanted devices and other equipment over the years. In addition to pure Mg batteries, Mg alloys may also serve as an optimal choice for biomedical devices. For instance, MZZ is a magnesium alloy including 3% Zn and 0.8% Zr, AZ61 consists of 6% Al and 1% Zn, and AZ31 contains 3% Al and 1% Zn. These have all demonstrated superior properties in comparison to conventional Mg batteries.[5]

The voltage capacity was comparable in both standard Mg batteries and Mg alloys. Furthermore, regarding corrosion resistance, magnesium alloys

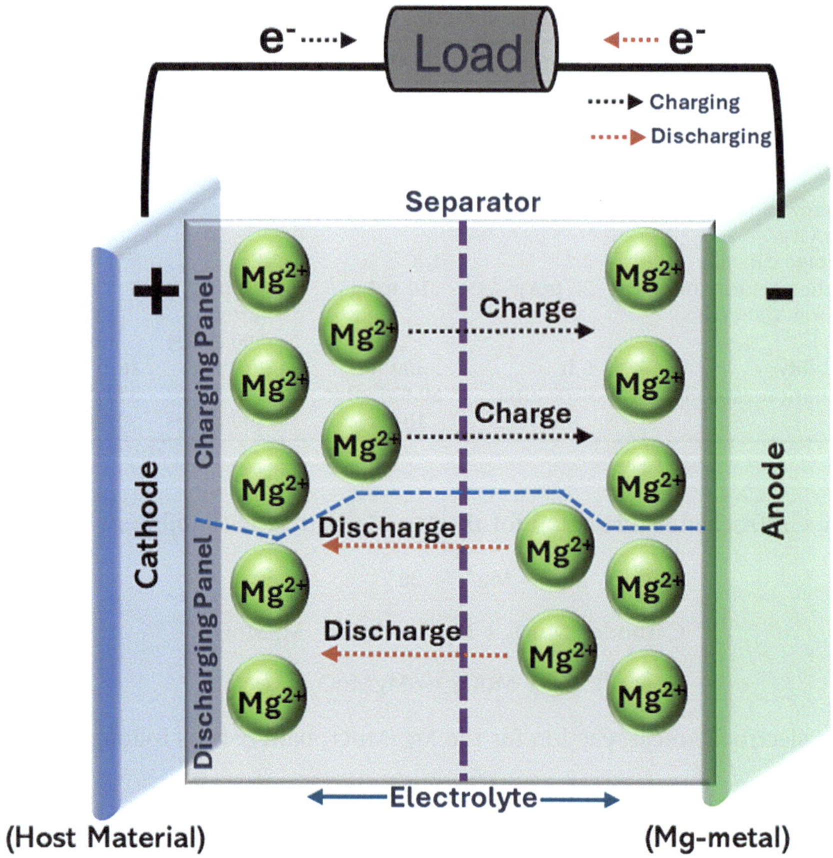

Figure 7.4 Schematic illustration of Mg batteries.

exhibited superior performance and possess a longer lifespan in comparison to their individual counterparts.[5,31] Therefore, along with Zn and pure Mg, Mg alloys could also be a potential candidate for being incorporated into biomedical devices. Figure 7.4 presents a schematic illustration of Mg-based batteries with their degradation mechanism in a bodily fluid medium, and their charge and discharge mechanism.

Tsang *et al.* presented the design and testing of biodegradable magnesium/iron batteries utilizing polycaprolactone (PCL) as both packaging and a functional material.[32] Compared to state-of-the-art biodegradable batteries, after the PCL encapsulation, the reported battery had smaller electrochemical cells, longer discharge lifetimes, and greater discharge rates. At these miniature scales, PCL-coated Mg/Fe batteries performed better than medical-grade non-degradable lithium-ion batteries and biodegradable batteries. These batteries had an energy density of 694 $W\,h\,kg^{-1}$ and a total volume of 0.02 cm^3. A solution containing 0.1 M $MgCl_2$, 0.9 wt% NaCl, and 1× PBS was used to discharge fully developed PCL dip-coated Mg/Fe batteries (Figure 7.5a).

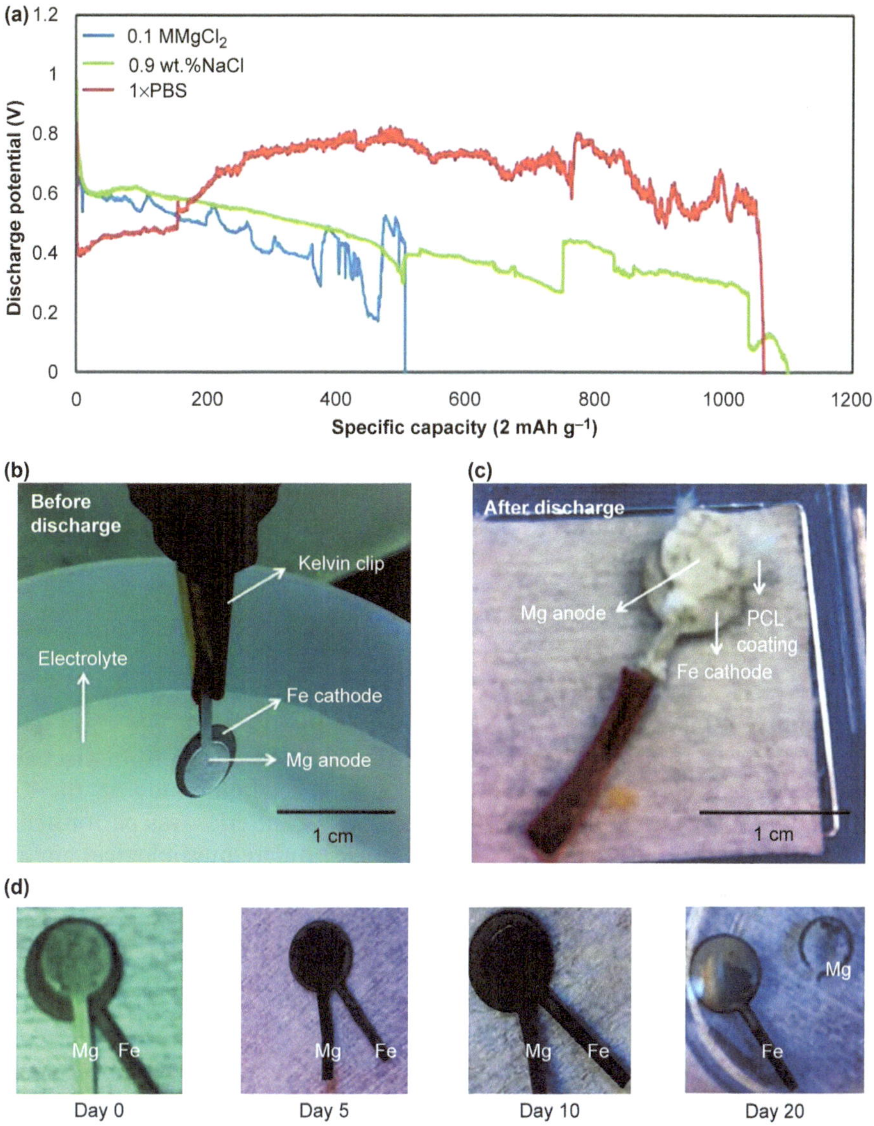

Figure 7.5 Evaluation of PCL dip-coated Mg/Fe batteries. (a) Discharge profiles of PCL dip-coated batteries under galvanostatic discharge at 45 μA in 0.1 M $MgCl_2$, 0.9 wt% NaCl, and 1× PBS. Optical images of the assembled PCL dip-coated Mg/Fe battery (b) before and (c) after discharge at 45 μA. (d) Degradation tests were performed in 1× PBS at 37 °C, and samples were not discharged prior to testing to demonstrate the complete dissolution of the biodegradable battery within physiological solution. Reproduced from ref. 32, https://doi.org/10.1038/micronano.2015.24, under the terms of the CC BY 4.0 license, http://creativecommons.org/licenses/by/4.0/.

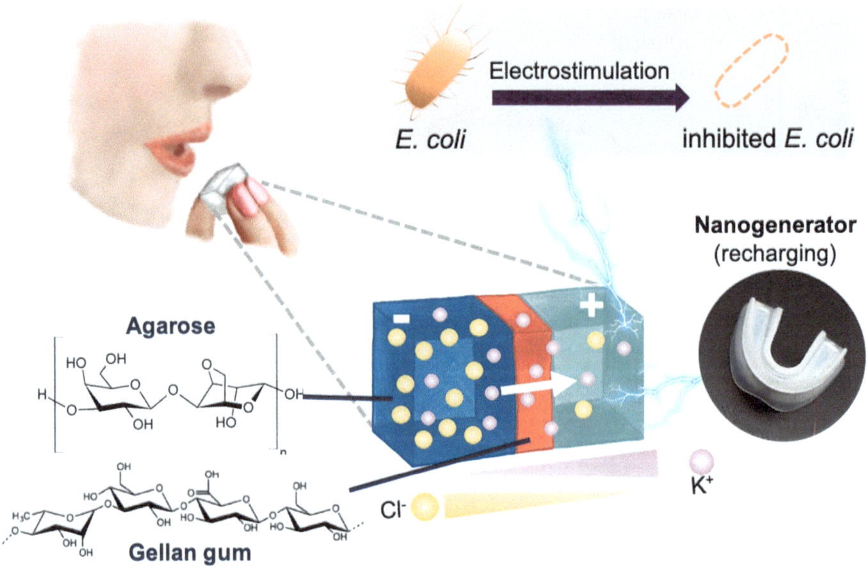

Figure 7.6 Ingestible polysaccharide gel battery using the Donnan dialysis effect and self-charging phenomena. Reproduced from ref. 36 with permission from Elsevier, Copyright 2020.

Optical photos of the PCL dip-coated batteries, both pre- and post-discharge, illustrate the degradation of the battery resulting from discharge and inherent corrosion (Figure 7.5b and c). A separate study on the degradation of the embedded PCL dip-coated Mg/Fe batteries was performed in 1× PBS at 37 °C with agitation at 50 rpm to simulate physiological circumstances; the results are depicted in Figure 7.6d. The favorable discharge performance and confirmed biodegradation of PCL-coated Mg/Fe batteries endorse their potential as energy sources for transient biomedical devices.

7.4.2 Edible Batteries

Thus, research into ingestible systems is being directed toward the development of biodegradable and biocompatible batteries that provide excellent performance while remaining compatible with the body's natural functions.[33–35] These also eliminate the need for any surgeries for device recoveries and open up new possibilities for ingestible systems. In this regard, edible batteries are one of the important areas to be considered.

In one of the earlier works, Lin *et al.* proposed a proof-of-concept ingestible polysaccharide gel battery using the Donnan dialysis effect.[36] The main components of the proposed battery were gellan gum, a naturally occurring linear anionic heteropolysaccharide, and agarose, a hydrogel made from red seaweed (Figure 7.6). The battery can be charged further using a self-charging TENG, which produces a voltage of 300 mV, which is sufficient to recharge the battery without the need for any additional phenomenon.

The developed ingestible battery was applied for the killing of pathogenic Gram-negative bacteria like *Escherichia coli* (*E. coli*), commonly present in the human intestine and which may cause food poisoning. Although, as per the current size, the direct swallowing of this battery is not possible, there is always the possibility of reducing the size so that it can be used as a harmless energy source in medical devices.

When the requirement is for it to be ingestible, edible electrode materials developed from foods or food ingredients that are completely digestible appear to be a more promising advancement. Recently, Ilic *et al.* proposed an edible rechargeable battery based on organic redox-active food ingredients.[37] All the materials used for the fabrication of the battery were typical food ingredients and additives that humans can consume in significant quantities (more than 100 mg per day) without any harmful effects. The battery was composed of riboflavin and quercetin as the cathode and anode, respectively. Both compounds were immobilized on activated charcoal (a food ingredient) and then encased in beeswax. The choice of materials makes the battery fully edible, and it operates at 0.65 V, sustaining a current of 48 μA for 12 min. The proposed proof-of-concept could pave the way for a more sustainable approach to ingestible battery systems.

7.4.3 Gastric Battery

The concept of a gastric battery involves using the body's gastric juice or gastric fluid (GF) as an electrolyte.[38] It is considered a plentiful source of electrolytes, replacing the harmful electrolyte constituents found in traditional batteries. These batteries are generally electrochemical galvanic cells where stomach juice serves as both an electrolyte and a source of ions. The operational principle of a gastric battery resembles that of a conventional battery, utilizing two electrodes; the sole distinction lies in the substitution of stomach acid for the hazardous electrolyte used in standard batteries. It is generally accepted that the battery is housed within a capsule, which, once reaching the stomach, is activated by the stomach acid. The battery is expected to maintain power supply as long as it gets continuous access to electrolytes.

Mostafalu and Sonkusale reported a biobattery containing a combination of Zn as the anode and Pd as the cathode, with gastric fluid as the electrolyte.[39] It was developed for performing endoscopy using the body's electrolytes so that it could remain stable in the stomach's acidic environment, and they concluded that it could be an ideal choice for endoscopy (Figure 7.7). Though Pd has demerits as a material, this study provides an initial overview of the gastric battery concept.

In another work, Nadeau *et al.* investigated the design and operation of an energy harvesting galvanic cell intended for wireless communication and continuous *in vivo* temperature monitoring.[19] The galvanic cell was constructed utilizing a Zn and Cu redox pair, achieving a power output of 0.23 $\mu W\,mm^{-2}$. This energy harvesting method can power ingestible electrical devices for extended durations within the gastrointestinal tract.

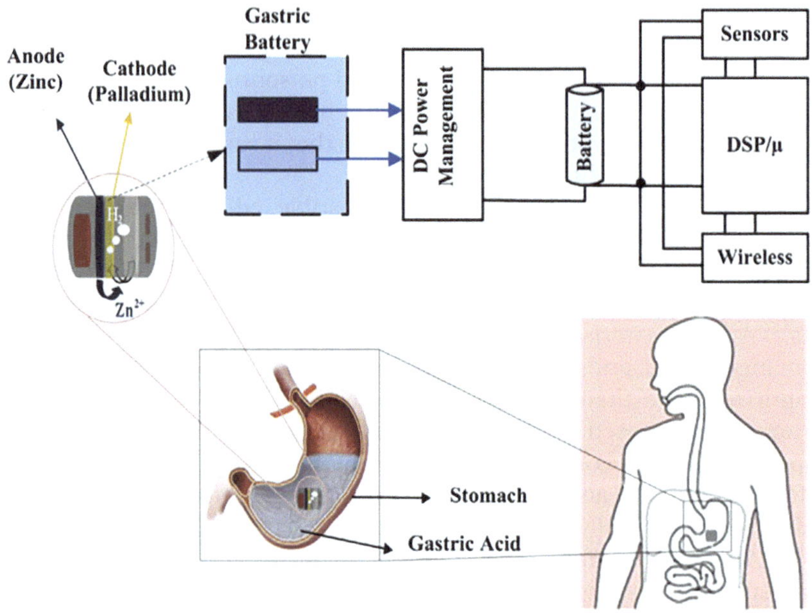

Figure 7.7 A proposed gastric battery using Zn and Pd. Reproduced from ref. 39 with permission from Elsevier, Copyright 2014.

7.5 Challenges and Future Outlook

Ingestible and implantable devices are becoming an integral component of medical technology and the healthcare system, having already transformed diagnostic and therapeutic approaches.[33,38,40,41] In contrast to implantable devices, ingestible technologies have not been thoroughly investigated. Their potential application in drug delivery is highly popular, as these devices are predominantly safe and exhibit no adverse effects on the human body. Batteries have been used in these devices since the first implantable device was developed. Regular batteries are not currently used in the human body, as they possess severe side effects, and they are generally not flexible. If we consider the challenges, the size of the batteries remains the biggest challenge. Furthermore, batteries must be biocompatible, biodegradable, and flexible to be acceptable for use in implantable and ingestible devices.[34] Selecting the appropriate electrode is crucial for the development of a bio-battery. When selecting electrode materials for ingestible and implantable batteries, the safety of the materials is the most important factor. Additionally, the solid electrolyte enhances battery safety and can be fabricated into a compact structure that is easily implantable in the human body. Furthermore, the improvement of the battery preparation process is crucial when integrating it into ingestible and implantable devices. Batteries produced by the traditional slurry casting method are bulky and susceptible to electrolyte leakage problems. Advanced techniques such as electrodeposition and dry or wet etching can be employed in battery preparation. Additionally, 3D printing technology provides advantages of high complexity, low cost, scalability, and manufacturing efficiency, facilitating the reduction of battery size.[42] Along with the advancements in this field, further research and preclinical testing are needed before utilizing them in humans. Moreover, their comparable features (high energy density, excellent cyclic performance, compact size) make them more convenient than normal batteries. On the other hand, there is a possibility to develop a gastric battery using gastric acid entirely as the electrolyte, but more research and investigations should be conducted to successfully integrate them into an ingestible battery. To make these biobatteries more compatible with ingestible and implantable devices, we need to understand how battery performance affects human organs. We are confident that these problems will soon be solved thanks to ongoing progress, leading to small, biocompatible, and very efficient batteries for healthcare devices.

References

1. X. Huang, L. Wang, H. Wang, B. Zhang, X. Wang and R. Y. Z. Stening, *et al.*, Materials Strategies and Device Architectures of Emerging Power Supply Devices for Implantable Bioelectronics, *Small*, 2020, **16**, 1902827.
2. A. Esmaeilian, F. Yazdian, A. Saberi, A. Hosseini, A. Rahdar and S. Fathi-karkan, *et al.*, Innovations in Self-Powered glucose Biosensors: From energy harvesting to Continuous, Non-Invasive monitoring, *Inorg. Chem. Commun.*, 2025, **174**, 114076.
3. A. Khan, R. Joshi, M. K. Sharma, A. Ganguly, P. Parashar and T. W. Wang, *et al.*, Piezoelectric and triboelectric nanogenerators: Promising technologies for self-powered implantable biomedical devices, *Nano Energy*, 2024, **119**, 109051.
4. A. Ghani Olabi, Q. Abbas, A. Al Makky and M. Ali Abdelkareem, Supercapacitors as Next Generation Energy Storage Devices: Properties and Applications, *Energy*, 2022, **248**, 123617.
5. L. Zhai, J. Duan, T. Lin and H. Shao, Recent advances in implantable batteries: Development and challenge, *J. Alloys Compd.*, 2024, **979**, 173551.
6. M. Yu, Y. Peng, X. Wang and F. Ran, Emerging Design Strategies Toward Developing Next-Generation Implantable Batteries and Supercapacitors, *Adv. Funct. Mater.*, 2023, **33**, 2301877.
7. K. Kalantar-Zadeh, N. Ha, J. Z. Ou and K. J. Berean, Ingestible Sensors, *ACS Sens.*, 2017, **2**, 468–483.
8. P. A. Thwaites, C. K. Yao, E. P. Halmos, J. G. Muir, R. E. Burgell and K. J. Berean, *et al.*, Review article: Current status and future directions of ingestible electronic devices in gastroenterology, *Aliment. Pharmacol. Ther.*, 2024, **59**, 459–474.

9. T. R. Abraham, K. Sunil, M. Shah, N. Ashok and S. Thomas, Energy storage devices: Batteries and supercapacitors, in *Nanobiohybrids for Advanced Wastewater Treatment and Energy Recovery*, IWA Publishing, 2023, pp. 61–84.
10. Y. Lv, X. Liu, J. Liu, S. Wu, S. Sun and P. Wu, *et al.*, Implantable and biocompatible Na–O_2 battery, *Chem*, 2024, **10**(6), 1885–1896.
11. S. Y. Yang, V. Sencadas, S. S. You, N. Z. X. Jia, S. S. Srinivasan and H. W. Huang, *et al.*, Powering Implantable and Ingestible Electronics, *Adv. Funct. Mater.*, 2021, **31**, 2009289.
12. S. Lei, Z. Liu, C. Liu, J. Li, B. Lu and S. Liang, *et al.*, Opportunities for biocompatible and safe zinc-based batteries, *Energy Environ. Sci.*, 2022, **48**, 4911–4927.
13. S. Stauss and I. Honma, Biocompatible batteries-materials and chemistry, fabrication, applications, and future prospects, *Bull. Chem. Soc. Jpn.*, 2018, **91**(3), 492–505.
14. X. Yan, K. Zu, J. Tang, Y. Ding and X. Xiao, A membrane-less Zn-air biobattery/supercapacitor hybrid device, *J. Energy Storage*, 2024, **88**, 111469.
15. M. M. Biswas, M. S. Azim, T. K. Saha, U. Zobayer and M. C. Urmi, Towards Implementation of Smart Grid: An Updated Review on Electrical Energy Storage Systems, *Smart Grid Renewable Energy*, 2013, **4**(1), 122–132.
16. G. He and I. P. Parkin, Sustainable and biocompatible Zn-based batteries, *Natl. Sci. Rev.*, 2023, **10**, nwad055.
17. X. Zeng, J. Hao, Z. Wang, J. Mao and Z. Guo, Recent progress and perspectives on aqueous Zn-based rechargeable batteries with mild aqueous electrolytes, *Energy Storage Mater.*, 2019, **20**, 410–437.
18. L. Tang, H. Peng, J. Kang, H. Chen, M. Zhang and Y. Liu, *et al.*, Zn-based batteries for sustainable energy storage: strategies and mechanisms, *Chem. Soc. Rev.*, 2024, **53**, 4877–4925.
19. P. Nadeau, D. El-Damak, D. Glettig, Y. L. Kong, S. Mo and C. Cleveland, *et al.*, Prolonged energy harvesting for ingestible devices, *Nat. Biomed. Eng.*, 2017, **1**(3), 0022.
20. R. Xu, X. Gao, Y. Chen, C. Peng, Z. Zhang and C. Wang, *et al.*, Research advances of the electrolytes for rechargeable magnesium ion batteries, *Mater. Today Phys.*, 2023, **36**, 101186.
21. Y. He, W. Xue, Y. Huang, H. Tang, G. Wang and D. Zheng, *et al.*, Boosting the capacity and stability of a MoO_3 cathode via valence regulation and polypyrrole coating for a rechargeable Zn ion battery, *RSC Adv*, 2023, **13**(22), 15295–15301.
22. M. R. Dunkin, J. Kuang, S. Yan, S. T. King, L. M. Housel and L. Ma, *et al.*, Unveiling Charge Transport and Degradation Mechanisms of Aqueous Zn|α-MoO_3 Batteries in Conventional Concentration and Water-in-Salt Electrolytes: A Multi-Modal In Situ and Operando Study, *Adv. Mater. Interfaces*, 2022, **9**(29), 2201125.
23. X. Xie, J. Li, Z. Xing, B. Lu, S. Liang and J. Zhou, Biocompatible zinc battery with programmable electro-cross-linked electrolyte, *Natl. Sci. Rev.*, 2023, **10**(3), nwac281.

24. N. Zhang, F. Cheng, J. Liu, L. Wang, X. Long and X. Liu, *et al.*, Rechargeable aqueous zinc-manganese dioxide batteries with high energy and power densities, *Nat. Commun.*, 2017, **8**(1), 1–9.
25. L. Wang, Achieving Stable Molybdenum Oxide Cathodes for Aqueous Zinc-Ion Batteries in Water-in-Salt Electrolyte, *Adv. Mater. Interfaces*, 2021, **8**(1), 2002080.
26. C. Xu, C. Lei, J. Li, X. He, P. Jiang and H. Wang, *et al.*, Unravelling rechargeable zinc-copper batteries by a chloride shuttle in a biphasic electrolyte, *Nat. Commun.*, 2023, **14**(1), 2349.
27. V. Soundharrajan, B. Sambandam, S. Kim, V. Mathew, J. Jo and S. Kim, *et al.*, Aqueous Magnesium Zinc Hybrid Battery: An Advanced High-Voltage and High-Energy $MgMn_2O_4$ Cathode, *ACS Energy Lett.*, 2018, **3**(8), 1998–2004.
28. L. Wang, S. Riedel, J. Drews and Z. Zhao-Karger, Recent developments and future prospects of magnesium–sulfur batteries, *Front. Batteries Electrochem.*, 2024, **3**, 1358199.
29. R. Xu, X. Gao, Y. Chen, X. Chen and L. Cui, Research status and prospect of rechargeable magnesium ion batteries cathode materials, *Chin. Chem. Lett.*, 2024, **35**, 109852.
30. L. Wang, S. Riedel, A. Welle, S. Vincent, S. Dinda and B. Dasari, *et al.*, Advancing Reversible Magnesium−Sulfur Batteries with a Self-Standing Gel Polymer Electrolyte, *ACS Appl. Energy Mater.*, 2024, **7**(14), 5857–5868.
31. F. Bella, S. De Luca, L. Fagiolari, D. Versaci, J. Amici and C. Francia, *et al.*, An Overview on Anodes for Magnesium Batteries: Challenges towards a Promising Storage Solution for Renewables, *Nanomaterials*, 2021, **11**(3), 810.
32. M. Tsang, A. Armutlulu, A. W. Martinez, S. A. B. Allen and M. G. Allen, Biodegradable magnesium/iron batteries with polycaprolactone encapsulation: A microfabricated power source for transient implantable devices, *Microsyst. Nanoeng.*, 2015, **1**, 15024.
33. S. Y. Yang, V. Sencadas, S. S. You, N. Z. X. Jia, S. S. Srinivasan and H. W. Huang, *et al.*, Powering Implantable and Ingestible Electronics, *Adv. Funct. Mater.*, 2021, **31**(44), 2009289.
34. L. Zhai, J. Duan, T. Lin and H. Shao, Recent advances in implantable batteries: Development and challenge, *J. Alloys Compd.*, 2024, **979**, 173551.
35. X. Huang, H. Hou, B. Yu, J. Bai, Y. Guan, L. Wang, K. Chen, X. Wang, P. Sun, Y. Deng, S. Liu, X. Cai, Y. Wang, J. Peng, X. Sheng, W. Xiong and L. Yin, Fully Biodegradable and Long-Term Operational Primary Zinc Batteries as Power Sources for Electronic Medicine, *ACS Nano*, 2023, **17**(6), 5727–5739.
36. Z. H. Lin, W. S. Hsu, A. Preet, L. H. Yeh, Y. H. Chen and Y. P. Pao, *et al.*, Ingestible polysaccharide battery coupled with a self-charging nanogenerator for controllable disinfection system, *Nano Energy*, 2021, **79**, 105440.
37. I. K. Ilic, V. Galli, L. Lamanna, P. Cataldi, L. Pasquale and V. F. Annese, *et al.*, An Edible Rechargeable Battery, *Adv. Mater.*, 2023, **35**(20), 2211400.

38. N. Golafshan, S. Patra, N. D. Huynh, N. Gadegaard and A. Boisen, Self-Powered Implantable and Ingestible Devices: Harvesting Energy Within the Body, *Mater. Today*, 2025, **87**, 403–421.
39. P. Mostafalu and S. Sonkusale, Flexible and transparent gastric battery: Energy harvesting from gastric acid for endoscopy application, *Biosens. Bioelectron.*, 2014, **54**, 292–296.
40. P. A. Thwaites, C. K. Yao, E. P. Halmos, J. G. Muir, R. E. Burgell and K. J. Berean, *et al.*, Review article: Current status and future directions of ingestible electronic devices in gastroenterology, *Aliment. Pharmacol. Ther.*, 2024, **59**(4), 459–474.
41. S. Mukherjee, S. Suleman, R. Pilloton, J. Narang and K. Rani, State of the Art in Smart Portable, Wearable, Ingestible and Implantable Devices for Health Status Monitoring and Disease Management, *Sensors*, 2022, **22**(11), 4228.
42. Z. Lyu, G. J. H. Lim, J. J. Koh, Y. Li, Y. Ma and J. Ding, *et al.*, Design and Manufacture of 3D-Printed Batteries, *Joule*, 2021, **5**(1), 89–114.

CHAPTER 8

Supercapacitors in Healthcare Devices

METE BATUHAN DURUKAN,[a,b] SİMGE ÇINAR AYGÜN[c] AND HUSNU EMRAH UNALAN*[c]

[a] Helmholtz Institute Ulm (HIU), Ulm 89081, Germany; [b] Karlsruhe Institute of Technology (KIT), Karlsruhe 76021, Germany; [c] Department of Metallurgical and Materials Engineering, Middle East Technical University, Ankara 06800, Türkiye
*Email: unalan@metu.edu.tr

8.1 Introduction

8.1.1 Overview of Supercapacitors

Supercapacitors (SCs) have existed for six decades.[1] However, their origins date back to Volta's Pile, as researchers in the 18th century used the Leyden Jar for energy storage. This early energy storage device consisted of silver foils positioned inside and outside a glass jar, utilizing the glass jar itself as the dielectric layer. This marked the introduction of the Leyden Jar as the world's first energy storage device functioning as a capacitor.[2,3] Early and traditional SCs primarily relied on carbonaceous materials with high surface areas.[4] More recently, electrochemical capacitors have gained significant attention for their long cycling life, excellent capacity retention, and superior power delivery and uptake compared to conventional Li-ion batteries.[5,6]

As of 2025, numerous companies are manufacturing a variety of SCs. Maxwell Technologies,[7,8] CAP-XX,[9,10] Skeleton Technologies,[11] Capacitech,[12] and Kyocera-AVX[13] are among the companies producing supercapacitors in various configurations. These devices come in different shapes and

Detection Science Series No. 33
Energy Materials for Ingestible and Implantable Healthcare Devices
Edited by Santanu Patra and Anja Boisen

Published by the Royal Society of Chemistry, www.rsc.org

capacities, tailored for specific applications. Small cylindrical SCs are utilized in various electronic circuits, while pseudocapacitors, distinguished by their higher energy densities, are employed explicitly in start–stop systems for heavy-duty vehicles. Wire-shaped flexible capacitors are specifically used in situations requiring flexibility. SCs can also be designed as pouch-type. Large modules find applications in transportation and grid/power correction.[10]

SCs offer advantages over other energy storage solutions with their ultra-high-power density (10–100 $kW\,kg^{-1}$) and wide operating temperature range (-40 to 80 °C).[14] The ability to deliver this level of power density within seconds is a notable advantage, particularly for the rapidly evolving bioelectronics.

8.1.2 Relevance to Healthcare Devices and Scope of This Chapter

As mentioned earlier, SCs are currently available in various sizes and shapes within the industry. However, a notable demand remains in a specific sector, particularly within healthcare devices. From wearable patches to implantable and ingestible devices, SCs show great potential for powering various health-monitoring devices and bioelectronics. However, the requirement of being biocompatible, bioresorbable (also widely called bioabsorbable in the literature), and even ingestible means that more than the conventional solutions are needed, as they are generally bulky, not flexible, and mostly contain toxic materials to the human body. Compared to batteries, SCs offer more flexibility in the structural components and materials. In particular, the use of carbon-based materials with solid-state or gel electrolytes provides SCs with the potential to be safer and more biocompatible than traditional batteries. Moreover, SCs can offer significant mechanical flexibility, enabling improved integration into various parts of the human body where bioelectronics are needed. The biggest disadvantage of SCs, on the other hand, is their low energy density and non-steady discharge voltage. This makes them suitable only for short-term applications or for delivering quick-burst signals.

Quick progression and advancement of the Internet of Things (IoT) also pave the way for more effective health biomonitoring with advanced bioelectronics. Implantable and ingestible bioelectronics will allow decentralized home-care systems with fewer readmissions and shorter stays at the hospital by enabling long-distance monitoring.[15,16] A key factor in realizing this potential is the development of advanced energy storage devices. The development of cutting-edge biocompatible, bioresorbable, and ingestible supercapacitors can be a solution to this demand.

This chapter reviews the overall research on SCs for healthcare devices and their applications in bioelectronics. We briefly introduce the SCs, their working mechanisms, types, and everyday materials that fit the vision of future bioelectronics. Following the introduction, the research on SCs in various healthcare devices, mainly wearable, ingestible, and implantable

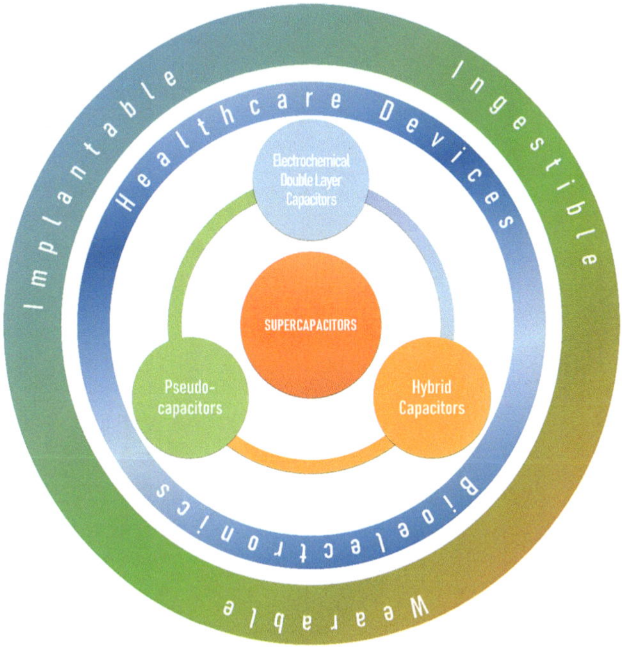

Figure 8.1 Schematic diagram of the overview of this chapter.

devices, is outlined (Figure 8.1). The challenges of involving SCs in healthcare devices are examined in light of recent developments.

8.2 Fundamentals of Supercapacitors

Both batteries and SCs rely on electrochemical processes to store energy. The main difference between these two devices lies in their relative energy and power densities. Lithium-ion batteries (LIBs) store energy through the intercalation of Li^+ ions into the bulk of the electrode, allowing redox reactions to occur in a diffusion-controlled manner, resulting in slow charge and discharge rates. On the other hand, energy storage in SCs is confined to the surface of the electrodes. Therefore, SCs have much faster reaction rates than batteries, but have much lower energy density.[17] As a result, SCs are a practical solution for bridging the gap between batteries and capacitors. This is best represented in the Ragone Plot, which positions the different electrochemical energy storage devices according to their specific power ($W\,kg^{-1}$) and specific energy densities ($W\,h\,kg^{-1}$) (see Figure 8.2). Research on ECs is primarily focused on enhancing energy densities while maintaining specific power output. The plot highlights the evolving trends in both batteries and ECs, indicating that the boundary between the two is progressively diminishing with ongoing technological developments.[18]

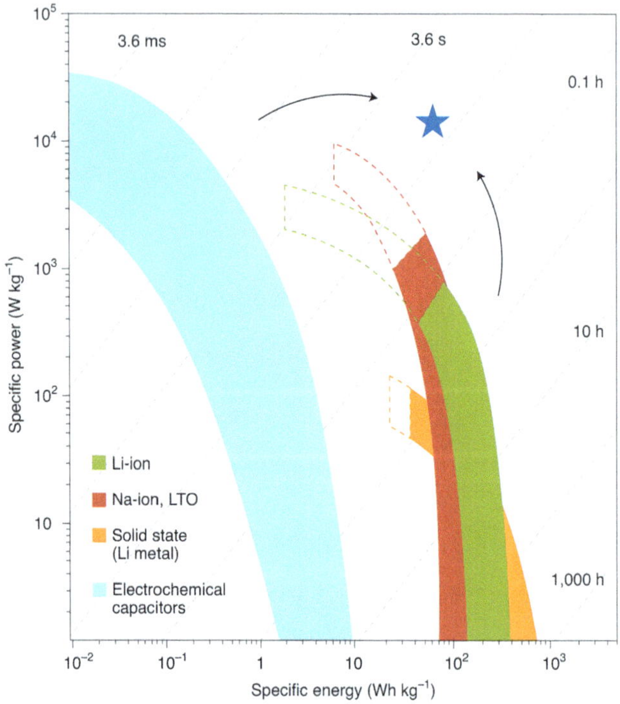

Figure 8.2 Electrochemical capacitors on the Ragone Plot. Arrows show the general trend/aim for electrochemical capacitors and batteries. LTO = Lithium titanium oxide ($Li_4Ti_5O_{12}$). Reproduced from ref. 18 with permission from Springer Nature, Copyright 2020.

8.2.1 Working Principles of Supercapacitors

Non-electrolytic capacitors are designed with two parallel plate electrodes and a dielectric material between them. An example illustration is given in Figure 8.3a. When a potential difference is applied, the positive and negative charges move towards the surface of the electrodes with opposite polarity, storing the charge electrostatically. The capacitance of a capacitor is measured in farads (F) using eqn (8.1), where ε_0 and ε_r are the permittivity of free space and dielectric constant, respectively. A is the area of the electrodes, and D is the distance between the electrodes.[3]

$$C = \frac{\varepsilon_0 \varepsilon_r A}{D} \tag{8.1}$$

These standard capacitors can store only a small amount of electrical charge, resulting in a very low energy density. This type of charge storage is known as non-faradaic, meaning it relies on electrostatic storage.[5]

A standard SC setup includes two electrodes with a separator placed in between, as depicted in Figure 8.3b. This separator might be made of materials like glass fiber, polyimide, polytetrafluoroethylene (PTFE), or any

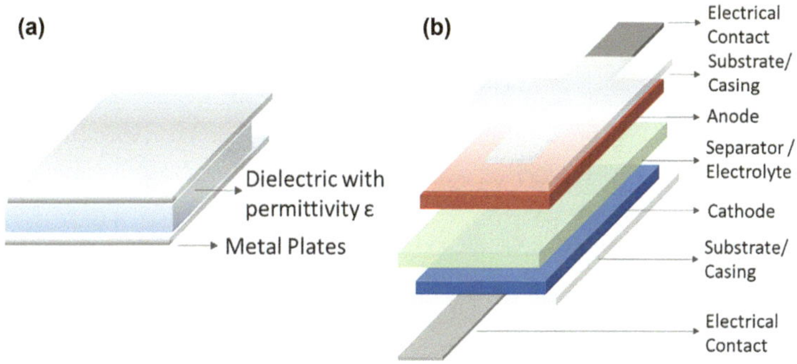

Figure 8.3 Schematic illustrations of (a) non-electrolytic capacitors and (b) supercapacitors.[19]

other type of porous membrane that permits the passage of ions. The separator is soaked with either aqueous or organic electrolytes, whereas solid-state electrolytes serve the dual purpose of separator and ionic conductor.[19]

Although SCs share a typical structure, the charging process varies according to the type of active material used in the electrodes. Based on the charging process, there are two primary categories of SCs based on this criterion: electrical double-layer capacitors (EDLCs) and pseudocapacitors, a.k.a. redox capacitors.

8.2.1.1 Electrical Double-layer Capacitors

EDLCs make use of carbonaceous materials with very high specific surface areas (SSAs). This allows improved charge accumulation at the surface of the carbonaceous materials *via* adsorption. Hence, the overall capacitance of the system depends primarily on the electrode–electrolyte interface, with SSA and ion permeation as the main determining factors.[20] Controlling pore sizes - from micropores to meso- and macropores - and controlling the surface chemistry through functional groups can also enhance the overall capacitance of the system.

One of the biggest differences between EDLCs and batteries is the charge storage mechanism. Unlike batteries, EDLCs have no chemical reactions, namely redox or faradaic reactions, to store charge. Instead, this mechanism utilizes the double-layer charge storage at the interface of high SSA carbonaceous electrode/electrolyte interfaces.[5] Since this type of energy storage does not utilize relatively slow faradaic reactions, the power delivery of the EDLCs can exceed 10 $kW\,kg^{-1}$; offering long charge/discharge durability of over 10^6 cycles.[21]

The first conceptualization of the electrical double layer was proposed by von Helmholtz,[22] utilizing a quasi-two-dimensional system in which opposite charges are separated by an atomic-scale distance (see Figure 8.4a). Initially applied to colloidal interfaces, this concept was extended to electrode interfaces, acknowledging that ions are prone to thermal movements

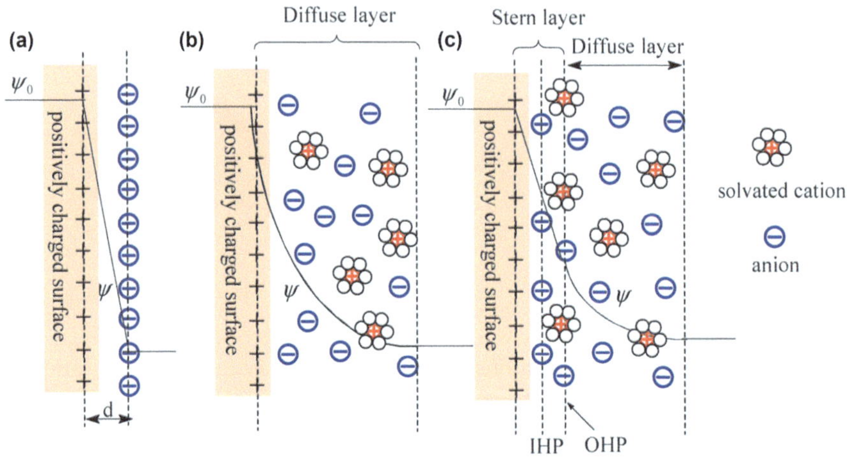

Figure 8.4 Different models for an electrical double-layer with respect to their chronological evolution: (a) Helmholtz model, (b) Gouy–Chapman model, and (c) Stern model. In the Stern model, IHP (inner Helmholtz plane) refers to the surface-adsorbed ions, while OHP (outer Helmholtz plane) consists of non-specifically adsorbed ions. ψ_0 and ψ represent the potentials at the surface of the electrode and the interface between the electrode and the electrolyte, respectively. Reproduced from ref. 20 with permission from the Royal Society of Chemistry.

and would constantly move on the solution side of the double layer.[23] This insight led to the development of the "diffuse" double-layer capacitance model by Gouy. However, this model had its flaws, notably the assumption that ions are point charges, which resulted in inaccurate predictions of the potential profile at the electrode surface and an overestimation of the potential difference. This discrepancy arises from defining the potential difference as the change in net ionic charge on the solution side relative to the metal solution across the interface. Chapman[24] refined this model by integrating the diffuse layer model with Boltzmann's energy distribution equation and Poisson's equation. This integration helped to better understand how the ionic space charge density in the interface region relates to the distance from the electrode surface and the second derivative of electric potential, known as the Poisson–Boltzmann equation (see Figure 8.4b). Stern[25] further refined the model to address the overestimation of double-layer capacitance by Gouy–Chapman. He incorporated Langmuir's adsorption isotherm to describe the ion distribution within the inner region (inner Helmholtz layer) while treating the area beyond the adsorption layer as a diffuse region of distributed ionic charge (outer Helmholtz layer). The schematic illustration of the Gouy–Chapman–Stern interface is given in Figure 8.4c. Stern also considered the finite size of ions, allowing for an estimation of the geometric limit of the ion adsorption region. By combining a Helmholtz-type double layer with capacitance C_H and a diffuse double layer with capacitance C_{diff} in series, Stern was able to calculate the overall double-layer capacitance, C_{dl}, as an equivalent circuit, using the equation given below;[5]

$$\frac{1}{C_{dl}} = \frac{1}{C_H} + \frac{1}{C_{diff}} \tag{8.2}$$

Carbon derivatives are the primary active materials employed in commercial SCs for energy storage, with activated carbon (AC)[26] being the most widely used form. However, various other forms and carbon structures have been extensively researched, including graphenes,[27,28] carbon fibers,[29–31] carbon nanotubes (CNT),[32,33] nano-onions,[34,35] aerogels,[36,37] and glassy carbons.[38] These carbon materials are valued for their high conductivity, controllable surface area and pore structure, corrosion resistance, temperature stability, and processability.

8.2.2 Pseudocapacitors

As mentioned earlier, double-layer capacitance is relatively low because the charge is only physically stored on the surface of the carbonaceous materials, and there are limitations in manipulating ion accessibility and porosity.[39] Instead of EDLC-type materials, pseudocapacitive materials can store much more charge through electrochemical reduction–oxidation reactions, also known as redox or faradaic reactions. In fact, the term "pseudo" originates from Greek and means "false", "apparent", or "resembling". Conway and Gileadi[40] first used this term to investigate the dependence of capacitance value on the electrochemical adsorption of ionic species onto electrode surfaces. Therefore, a pseudocapacitor refers to an electrode with similar electrochemical behavior to an EDLC but with a different energy storage mechanism.[41]

Pseudocapacitive materials efficiently store energy through fast and reversible redox reactions occurring at the surface of active materials. Current improvements in electrochemical energy storage based on pseudocapacitance are largely due to the transition metal oxides.[42] The reactions begin with the transfer of reactant mass toward the electrode interface to facilitate electron transfer at the electrode/electrolyte interface. Subsequently, chemical reactions occur at the electrode surface through adsorption or desorption mechanisms. Examples of such materials include but are not limited to ruthenium oxide (RuO_2),[43] iron oxide (Fe_3O_4),[44] zinc oxide (ZnO),[45] manganese oxide (MnO_2),[46,47] cobalt oxide (Co_3O_4),[48,49] and nickel oxide (NiO),[50] their sulfides, phosphides, and nitrides, while polyaniline[51] and polypyrrole[52] are examples of conductive polymers utilized in SCs.[53]

The first pseudocapacitive material discovered for pseudocapacitive charge storage was RuO_2. In these devices, charge storage involves the storage of protons from the electrolytes through faradaic reactions. Trasatti and Buzzanca[54] (1971) reported capacitor-like electrochemical behavior of RuO_2 films in perchloric acid, distinct from bulk single-crystal RuO_2. Their

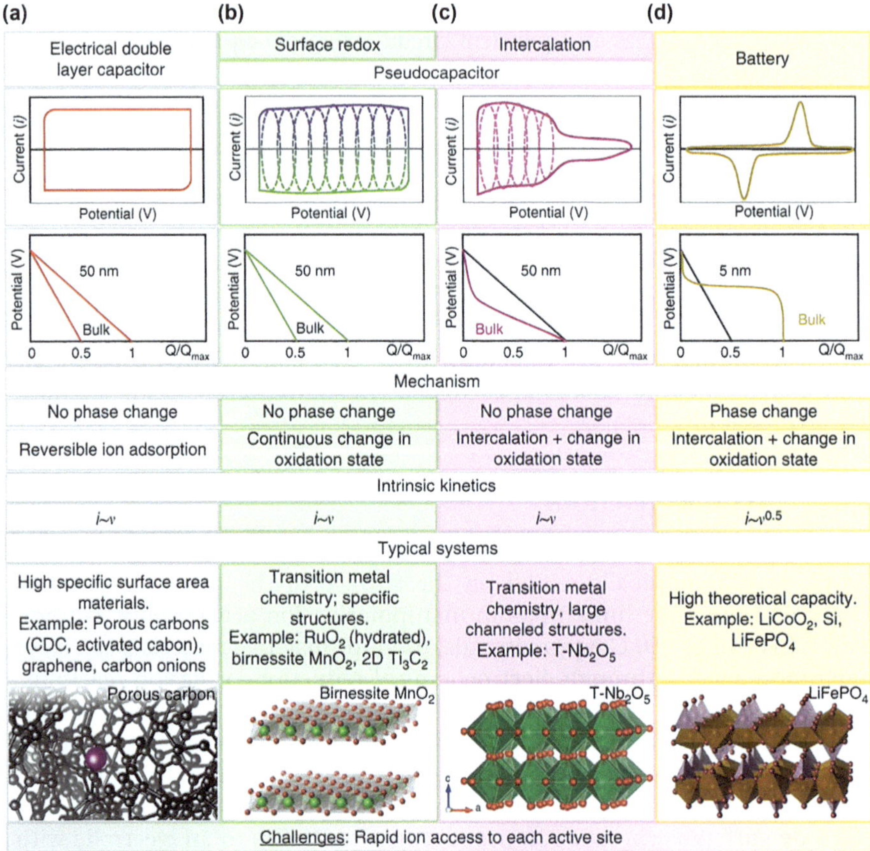

Figure 8.5 Typical behavior of CV and discharge curves for (a) EDLCs, (b) surface redox pseudocapacitors, (c) intercalation supercapacitors, and (d) batteries. Reproduced from ref. 56, https://doi.org/10.1038/ncomms12647, under the terms of the CC BY 4.0 license, http://creativecommons.org/licenses/by/4.0/.

research indicated that it is possible to achieve pseudocapacitance by transforming the material from bulk to a film structure. The storage occurred *via* faradaic reactions, resulting in a much higher energy density than carbonaceous materials.[55]

In Figure 8.5, an analysis and examples of capacitive materials, pseudocapacitive materials, and faradaic materials (exhibiting battery-like behavior) are presented in relation to their cyclic voltammetry (CV) and discharge profiles. As seen in Figure 8.5a, EDLCs have a rectangular CV shape with little/no dependence of current on potential while having linear discharge profiles. Pseudocapacitive materials can exhibit electrochemical behavior of more than one type. Figure 8.5b depicts pseudocapacitive materials with only surface redox reactions, such as MnO_2 or 2D materials such as titanium carbides or MXenes (Ti_3C_2), molybdenum disulfide

(MoS_2), or titanium disulfide (TiS_2); intercalation-type materials, such as niobium oxide (Nb_2O_5); or materials with both intercalation mechanism and partial redox, such as MXenes and transition metal dichalcogenides (TMDs), respectively.[56] It is imperative to differentiate batteries from the common EDLCs and pseudocapacitive supercapacitors. One of the common mistakes in the literature is to use battery-type materials or energy storage mechanisms, as in Figure 8.5d, and publish them as high-energy-density SCs. This has been thoroughly discussed by Brousse[57] and Simon,[18] who showed that the presence of a discharge plateau and well-defined redox peaks in CV correlates with battery-type behavior governed by semi-infinite diffusion.[56] According to Fleischmann,[58] to call a material pseudocapacitive, it should exhibit the following behaviors:

i. The relationship between the applied potential and state of charge is either linear or pseudolinear.
ii. The electrochemical reversibility is nearly ideal.
iii. The kinetics are controlled by the surface.

Pseudocapacitors generally have a much higher energy density compared to conventional EDLCs. However, this comes at the expense of lower power densities in general, as the electrochemical charge–discharge relies on the overall kinetics of faradaic redox reactions. The presence of these reactions can also lead to irreversible damage to the pseudocapacitive material caused by lattice distortions or partial dissolution/degradation of the active material. As a result, this can shorten the lifetime of pseudocapacitors compared to EDLCs. Despite these drawbacks, the enhanced energy density is essential for expanding the use of SCs.

8.2.3 Hybrid Capacitors

Hybrid capacitors are simply a combination of EDLCs and pseudocapacitors. They represent the combination of such materials in a single electrode or the combination of different electrodes in a cell. There are several reasons for obtaining hybrid structures for SCs. The primary motivation is to improve overall energy density, as EDLC type carbonaceous materials exhibit significantly lower values than batteries. In this context, pseudocapacitive materials – particularly metal oxides, sulfides, and phosphides – generally exhibit much lower electrical conductivity. Hybridized structures, whether for a single electrode or a device, are promising to enhance SCs' energy density and kinetics.[59]

8.2.4 Key Performance Parameters

Up to now, SCs have been introduced, and their working mechanisms and types have been discussed briefly. Depending on the discussed mechanisms, defining key performance parameters and their metrics is important. Factors such as cell capacitance, energy density, power density, and operating voltage are essential for effectively powering bioelectronic devices, whether they operate continuously or intermittently. Simply put, for EDLCs, the porous structure and surface functionalities of the carbonaceous materials are important to maximize the capacitance of SCs. For pseudocapacitors, limiting the faradaic processes to the surface of the materials by controlling the sizes and kinetics of the reactions becomes important. Yet, it is essential to acknowledge that these performance parameters represent only the initial stages of a more significant design aspect.

A useful summary was provided by Zhang[60] and the important parameters of SCs in a simplified network are given in Figure 8.6. Overall, three core parameters (depicted with yellow colors) can be defined as the most important ones that govern and/or comprise all others: specific capacitance (C_{sp}), operating voltage (V_o), and equivalent series resistance (R_{ES}).

C_{sp} can be controlled by tuning the material properties, as well as the electrodes' preparation formulation/procedure. Units that are generally used in the literature are F g^{-1} for gravimetric capacitance, F cm^{-2} for areal capacitance, and F cm^{-3} for volumetric capacitance. Parameters like electrode thickness, electrode density, and mass loading become critical for these metrics. The electrode preparation process can also affect the electrode's

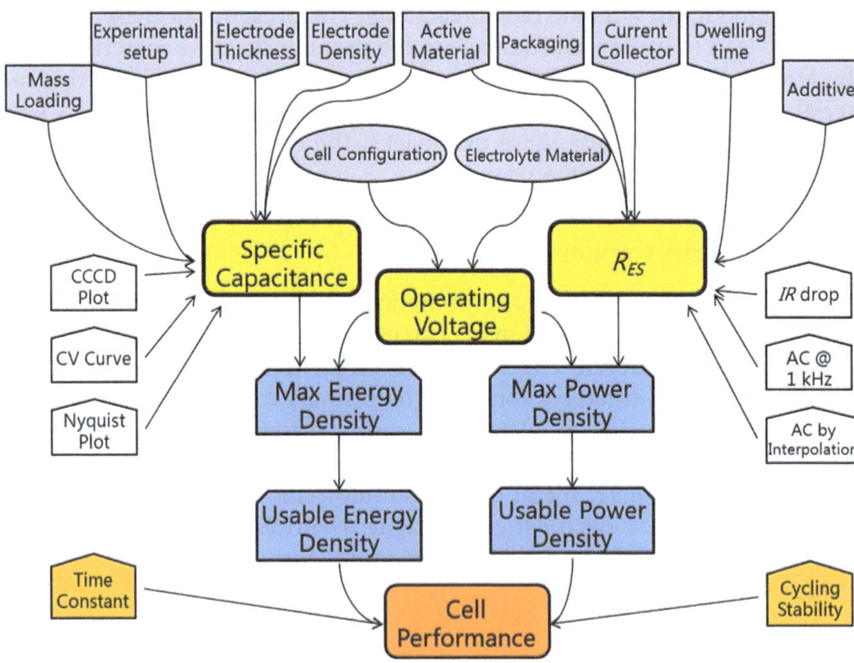

Figure 8.6 Schematic representation of the parameters and their relevant metrics for supercapacitors. Reproduced from ref. 60 with permission from John Wiley & Sons, Copyright 2014 WILEY-VCH Verlag GmbH & Co. KGaA, Weinheim.

conductivity, contact material, additives, and so on. V_o is another parameter immensely contributing to the design as its combination with C_{sp} and R_{ES} governs the maximum energy density and maximum power density. The design parameters for V_o are comprised of the electrolyte selection as well as the cell configuration.[60] Overall, SCs are generally compared using the C_{sp} at slow scan rates (~1–10 mV s^{-1}) or slow discharge currents (~0.1 A g^{-1}), and energy (W h kg^{-1}) and power density (W kg^{-1}). In these terms, SCs are able to supply power densities ranging from 1 to 10 000 W kg^{-1} and energy densities of 1 to 10 W h kg^{-1}, whereas batteries typically provide power densities of 1 to 1000 W kg^{-1} and energy densities of 10 to 1000 W h kg^{-1}.[61] Typical bioelectronic applications and their energy/power requirements are provided in Table 8.1.

Regarding these power requirements, some bioelectronic devices are listed in Table 8.1, along with the typical power requirements for devices such as deep brain stimulators, spinal cord stimulators, retinal prostheses, cochlear implants, pacemakers, wireless capsule endoscopes, glucose monitoring,

Table 8.1 Typical power requirements for implantable, ingestible, and wearable devices.

Applications	Power requirement	Working time	Ref.
Deep brain stimulator	10–1000 μW	2–5 years (non-rechargeable)	62
Spinal cord stimulator	2 μW–5 mW	2–5 years (non-rechargeable) 10–25 years (rechargeable)	63–65
Retinal prosthesis	2–10 mW cm^{-2}	5–10 years (non-rechargeable)	66 and 67
Cochlear implant	0.5 μW–100 mW	Wearable: 12–48 hours (rechargeable) Wearable: 5–10 days (non-rechargeable) Implants: 5–10 years (both rechargeable and non-rechargeable)	68 and 69
Pacemaker	5–30 μW	Continuous, from 5 to 10 years	70 and 71
Wireless capsule endoscope	5–500 mW	8–12 hours	72 and 73
Continuous glucose monitoring	~100 μW–300 μW (approximate values based on used batteries)	Continuous, from 15 to 30 days	74 and 75
Smart watches (oxygen saturation, ECG, heart rate monitoring, blood pressure)	~2–50 mW (based on average power consumption for everyday usage)	Continuous or used on demand, from 1 to 20 days (rechargeable)	76–78

implantable drug delivery systems, and smart watches. Although most of the power requirements are rather low in the range of μW, many systems must operate for prolonged periods, often for several years, especially implantable ones. For non-rechargeable cases, implantable devices must be surgically removed to replace the batteries.

8.2.5 Design Aspects of Supercapacitors for Bioelectronics

It is imperative to realize energy storage devices that can safely operate in the required application while satisfying the energy requirements. Although the requirements can change regarding the application, generalizations can be made as below:

- Biocompatibility/biodegradability: first and foremost, all materials used in the architecture of SCs—including electrodes, electrolytes, and even packaging—must be biocompatible or biodegradable, depending on the specific application. It is worth noting that conventional batteries are still employed in implantable and ingestible devices, despite the risk of toxicity in the possible event of failure. Therefore, developing fully biocompatible SCs is essential and should be a key objective for advancing healthcare technologies.
- Energy and power densities: SCs must provide sufficient energy and power densities to ensure the reliable operation of the intended healthcare device.
- Stability: SCs should exhibit both electrochemical and chemical stability. Electrochemical stability is important so that the SCs can be used without performance loss. Chemical stability is particularly critical for implantable and ingestible applications, where exposure to physiological environments can compromise performance and safety.
- Mechanical properties: for wearable and implantable healthcare monitoring devices, SCs must possess mechanical durability to withstand continuous flexing, stretching, and compression without performance loss or failure throughout their service life.

Despite the low energy densities, SCs offer unique advantages compared to batteries. These are tabulated in Table 8.2. In essence, supercapacitors facilitate the development of biocompatible and biodegradable energy storage units with mechanical flexibility and reduced form factors. For example, typical pacemakers today still rely on Li-based batteries, which are still bulky and use toxic materials and electrolytes. Although encapsulated, an unexpected failure still carries the risk of toxicity to the human body.[70,71] Although SCs are not yet fully capable of addressing this problem because of their low energy densities, self-rechargeable SCs, sometimes coupled with energy harvesters, have been proposed.[79] This is enabled by SCs' capability to deliver high power at low energy densities, thus requiring relatively low power for charging.

Table 8.2 Key design aspects of SCs compared to batteries for bioelectronics.

Key properties	Supercapacitors	Batteries	Ref.
Biocompatibility and biodegradability	More flexibility in design, allowing both biocompatibility and biodegradability	Performance is often limited with toxic electrolytes or non-degradable electrodes	80 and 81
Safety	Possible use of biocompatible and biodegradable electrolyte gels—which leads to safer device architecture with a low risk of leakage or toxicity	Traditional counterparts have a risk of explosions and leakage, or the electrodes can lead to inflammation due to reactive components	19 and 82–84
Mechanical flexibility	High—can be designed as flexible/stretchable	Low—generally rigid	85 and 86
Size	Can be miniaturized easily to fit various applications	Generally bulky due to material requirements	87 and 88
Power density	Very high—ideal for bursts of power	Moderate	18 and 57
Energy density	Low	High—better for long-term applications	18 and 57
Cycle life	Excellent, can reach up to 1 000 000 cycles	Limited—typically a few thousand cycles	89
Self-discharge	High—loses stored charge quickly—not suitable for long-term use	Low—can retain charge over long periods	90
Suitable applications	Best for transient or short-term applications such as signaling and wireless transmission	Best for continuous, long-term applications such as pacemakers, continuous monitoring, long-term sensors	

8.3 Supercapacitor Applications in Healthcare Devices

8.3.1 Biocompatibility Considerations

In the context of this book, it is essential to address the biocompatibility of the materials that will be used in the fabrication of the SCs. The intended application and the desired (target) lifespan govern a device's specific requirements. For instance, wearable electronics intended to conform to the skin must be non-toxic to ensure safety for the human body. Conversely, implantable devices may need to be resistant to dissolution while still being biocompatible, or they may need to undergo bioresorption (also called bioabsorption).

Ingestible devices, in general, must be safe to be ingested and then excreted without complications. Yet, edible versions of ingestible electronics are also expected to be digested in gastric fluid. Thus, *biocompatibility* has become a broad concept encompassing many aspects of bioelectronics.

Introducing different classes of materials in the context of biocompatibility is a crucial step in understanding how various materials can be used in bioelectronics. Referring to the traditional structure of the SCs given in Figure 8.3b, it is possible to divide the materials into two parts: electrode materials and electrolytes. A schematic representation of the classifications for achieving biocompatible SCs is given in Figure 8.7. The scope is rather big, considering all the different materials that can be used in bioelectronics. Rather, the following sections will focus on the applications while taking into account this variety of materials.

This chapter focuses on SCs for ingestible and implantable devices while briefly introducing wearable applications. Considering the healthcare devices that may utilize SCs, wearable applications are also quite important and should be mentioned—especially skin-conformal and minimally invasive devices. Figure 8.8 summarizes some of the uses of bioelectronics. These include but are not limited to brain–computer interfaces, deep brain stimulators, cochlear implants, pacemakers, nerve stimulators, physiological monitoring sensors, gastrointestinal tract sensors, and imaging devices. In all cases, developing biocompatible energy storage devices—particularly SCs in this chapter—is essential for advancing innovative bioelectronics.

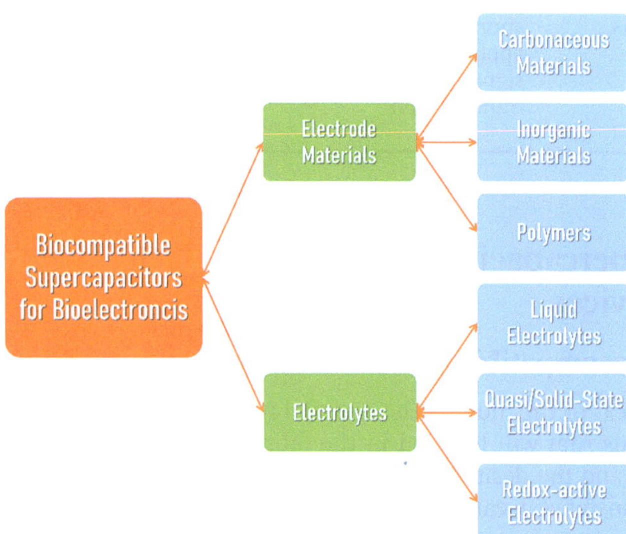

Figure 8.7 Schematic representation of the classification of different parts for biocompatible SCs.

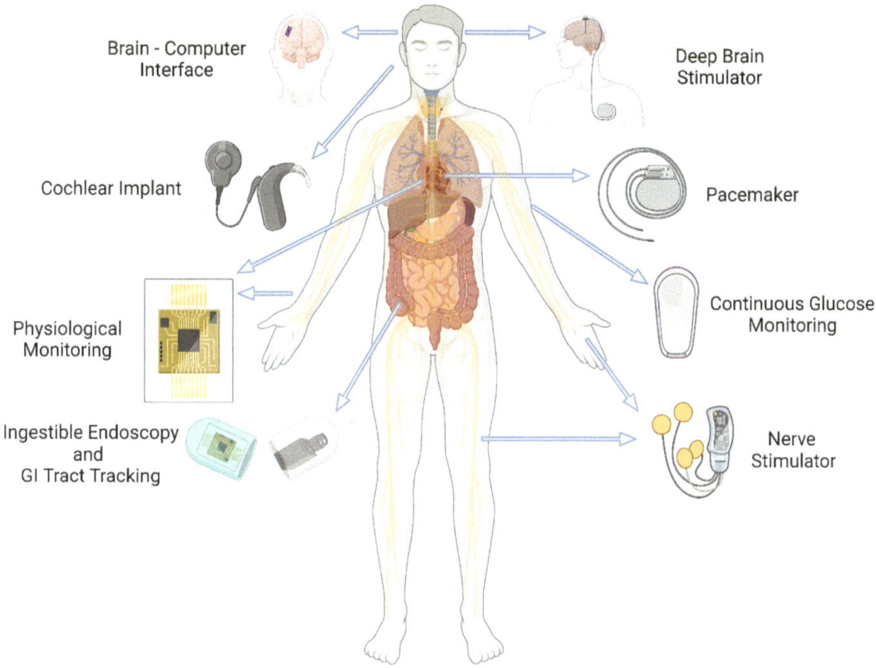

Figure 8.8 Examples of wearable, implantable, and ingestible electronics. Created in BioRender. Durukan, M. (2024) https://BioRender.com/m34u448.

8.3.2 Supercapacitors for Wearable Healthcare Devices

Although they are beyond the scope of this book, wearable healthcare devices are worth discussing within the field of bioelectronics. In addition to ingestible and implantable SCs, wearable ones also present a promising solution for bioelectronics. They offer seamless integration with wearable bioelectronics and can be designed to be flexible, stretchable, and lightweight. The development of wearable and conformal SCs is vital for the future of bioelectronics, especially in terms of health monitoring, disease-assisted therapy, biomarker analysis, and brain–computer interaction.[91] In the context of bioelectronics, it is reasonable to assume that wearable SCs must be biocompatible. Developing wearable SCs that can be integrated with skin-interfaced wearable sensors is crucial to satisfy remote monitoring of important metabolic processes.[92]

A notable example is provided by Lee and coworkers[93] using only metal-based electrodes. These researchers fabricated biodegradable micro-supercapacitors (MSCs) with magnetron sputtering of water-soluble tungsten (W), iron (Fe), and molybdenum (Mo). The idea is essentially developed in the aforementioned publication dealing with transient electronics.[94] "*Transiency*" in electronics is a relatively new concept that refers to the ability of electronic devices to disappear physically.[95–98] Unlike traditional

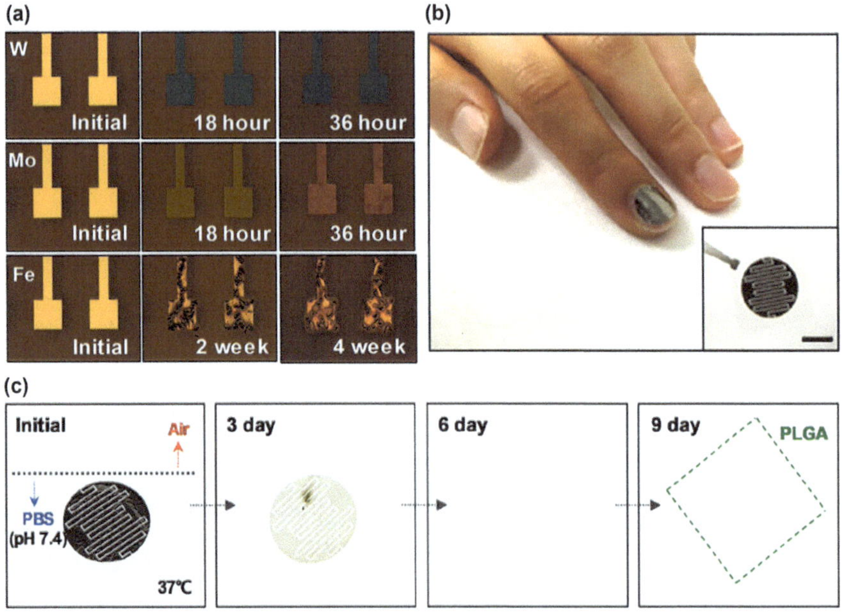

Figure 8.9 Fully biodegradable metal-based MSCs for transient and wearable electronics. (a) Dissolution of metal-based electrodes. (b) A conceptual image of a wearable MSC. (c) Dissolution of the MSC in PBS. Adapted from ref. 93 with permission from John Wiley & Sons, Copyright 2017 WILEY-VCH Verlag GmbH & Co. KGaA, Weinheim.

electronics, which are designed for long-term use, transient electronics are intended to degrade partially or completely once their function is fulfilled. These devices are typically designed for short-term, predefined applications and can be activated on-demand, triggered by specific conditions, or degraded at a controlled rate. The primary objective is their complete and untraceable disappearance after use. In this case, researchers show the degradation of the electrodes in phosphate buffer solution (PBS), as shown in Figure 8.9a. As a proof-of-concept, an MSC was fabricated on poly(lactic-*co*-glycolic) acid (PLGA) with sodium chloride (NaCl)/agarose gel electrolyte and put on a fingernail (Figure 8.9b). The fabricated MSC was completely degraded after 9 days, demonstrating the potential of transient and wearable SCs for bioelectronics.[93]

Fiber- or yarn-based SCs stand out due to their flexibility and possible integration into casual clothing.[99] An example was provided by Xiao and coworkers[100] who investigated a sweat-activated SC for sweat biosensing. The researchers coated polypyrrole (PPy) and poly(3-4-ethylenedioxythiophene)/poly(styrenesulfonate) (PEDOT:PSS) on stainless-steel yarns, and separated them with cotton fibers to fabricate yarn-based flexible SCs. These yarns were also proven to be wearable and washable up to 17 times without any loss in performance. These SCs were activated with artificial sweat. Their performance was found to depend on the ionic strength of the sweat and they were used to power up a sweat pH sensor.[100] A similar study was conducted by Luo and coworkers[101] utilizing self-standing polyaniline (PANI)/carbon nanotube (CNT) electrodes and biocompatible sodium chloride (NaCl) electrolyte. The innovative part is the addition of a zinc (Zn) foil with silver paste on one side and filter paper on the other. This primitive Zn–air battery was attached to the supercapacitor to obtain a "sweat-chargeable" device. Zn foil interacted with the artificial sweat of 0.5 wt% NaCl, 0.1 wt% potassium chloride (KCl), 0.1 wt% lactic acid, and 0.1 wt% urea and had spontaneous redox reactions with the oxygen in the air to provide an energy source for the SCs. The device was proven to charge up a wearable watch, indicating its possibility to be used in wearable devices.[101]

Fabric-based SCs are also the subject of extensive research to offer an energy storage solution to wearable electronics. They provide the same advantages as the fiber/yarn-type SCs—being wearable and integrable into clothing. Sun and coworkers[102] offered a unique application of wearable fabric-based SCs. These researchers utilized a two-step electrodeposition of nickel (Ni)–cobalt (Co) precursor and PPy, followed by an annealing step to obtain nickel–cobalt oxide nanosheets ($NiCoO_2$) coated with nitrogen (N)-doped carbon. The overall structure of the supercapacitor is presented in Figure 8.10a. To charge them, researchers integrated commercial flexible solar cells on top of the wearable SC and connected the fabricated power pack to a portable workstation. The same electrode, $NiCoO_2$@N–C on carbon textile, was utilized as a glucose sensor. The overall schematic and performance of the integrated device are given in Figure 8.10b–e. Overall, a wearable fabric electrode capable of dual functionality as both a SC electrode and a glucose sensor showed significant promise for wearable SC technology.[102]

Apart from the electrode materials, electrolytes for wearable SCs are also being actively investigated.[103] For wearable applications intended for skin contact, flexibility and biocompatibility are key requirements. Given the requirements of flexibility and biocompatibility, novel electrolytes must also be developed. In such a work, Jiang and coworkers[104] developed hydrogels to be used in SCs for wearable electronics. Their motivation for this study was to eliminate the mechanical mismatch between the electrodes and electrolytes, resulting in a more resilient SC device that can be stretched and compressed. They developed a self-adhesive, compressible, flexible, and stretchable all-hydrogel-based SC utilizing polyacrylamide (PAM). PAM-based hydrogels containing polydopamine (PDA)–reduced graphene oxide (rGO), cobalt–nickel oxide, and cobalt nitride are used as electrodes, while those containing rGO and potassium hydroxide (KOH) are used as electrolytes. The electrodes could be stretched up to 36 times their original length. An all-hydrogel SC was then used to provide myoelectric stimulation for electromyography (EMG), showing the promise for such hydrogel-based SCs for wearable bioelectronics.[104] Another example of an all-hydrogel SC is provided by Xu and coworkers,[105] who combined polyvinyl alcohol (PVA) with polyaniline (PANI) with a 3D honeycombed structure. The intricate

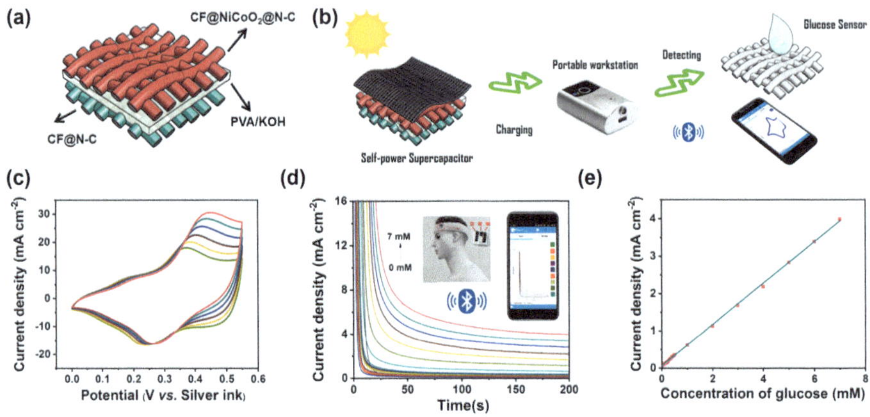

Figure 8.10 Wearable textile SCs for enzyme-free smart sensors. (a) Schematic structure of the textile-based SCs. (b) Schematic representation of the self-powered SCs for glucose detection. (c) CV curves of the $NiCoO_2$@N–C-based glucose sensor in glucose concentrations of 0–5 mM. (d) Amperometric curves of wearable self-powered devices in different glucose concentrations. (e) The relationship between the concentration of glucose and current density. Adapted from ref. 102 with permission from American Chemical Society, Copyright 2020.

structure is provided both schematically and through scanning electron microscopy (SEM) images in Figure 8.11a. When combined with a PVA-sulfuric acid (H_2SO_4) gel electrolyte, wearable and flexible SCs (Figure 8.11b) were obtained with capacitance values reaching 295 F g^{-1}. Fabricated SCs could be stretched, bent, twisted, and compressed without performance loss and could even self-heal (Figure 8.11c). The same electrodes could also be used for pressure sensing, allowing the tracking of physiological movements (Figure 8.11d).[105] This study highlighted the multifunctionality of SCs, demonstrating their utility as both an energy source and in sensing applications.

Last but not least, MSCs can be assembled for wearable bioelectronics, especially to meet the need for miniaturized energy sources.[106,107] Planar and 2D architectures of MSCs are particularly promising for bioelectronics, especially for electronic skin (e-skin). Their advantage comes from their interdigitated structure with a narrow distance between the electrodes, removing the need for a separator while allowing fast kinetics. Moreover, their integration into various devices is straightforward, particularly because the structure can be fabricated in 2D.[108] In one such example, Say and coworkers[109] fabricated interdigitated planar MSCs with PEDOT:PSS/cellulose nanofibril (CNF) inks on previously prepared parylene-C/gold (Au) electrodes with sequential spray coating of the electrode and 1-ethyl-3-methylimidazolium ethyl sulfate (EMIM-ES)/hydroxyethyl cellulose (HEC) electrolyte. A maximum volumetric capacitance of 7 F cm^{-3} is achieved with a capacitance retention of 90% after 10 000 cycles. The fabricated MSCs are mechanically robust and can be attached to flexible surfaces. Two MSCs

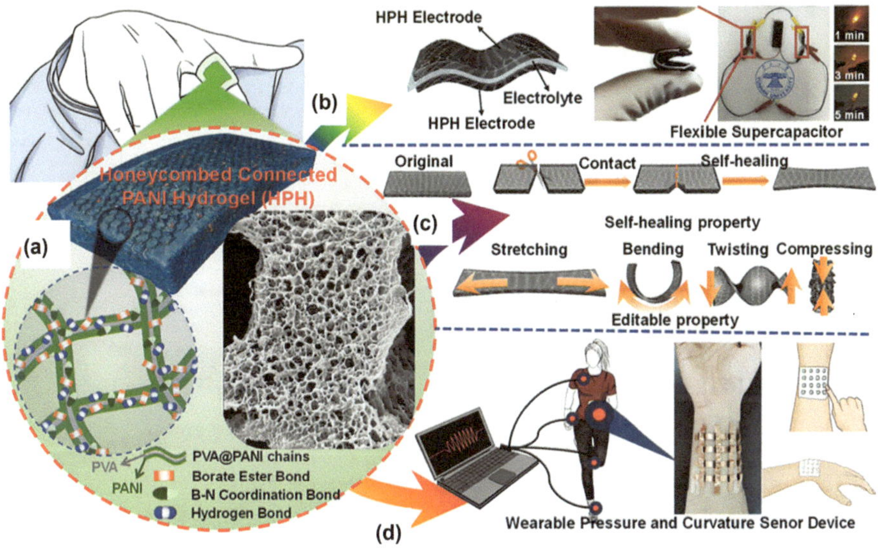

Figure 8.11 An integrated flexible supercapacitor-sensor device for wearable health monitoring. (a) Schematic illustration of the PVA@PANI hydrogels. The inset shows the SEM image of the porous structure. (b) Schematic structure and photographs of the flexible SC. (c) Schematic illustration of the hydrogel's stretchability, flexibility, and self-healing properties. (d) Schematic representation and photograph of the pressure sensor. Reproduced from ref. 105 with permission from John Wiley & Sons, Copyright 2023 Wiley-VCH GmbH.

connected in series successfully powered electrochromic devices, demonstrating their potential for electronic skin applications. In another application, Wang and coworkers[110] fabricated MXene/TiS_2 interdigitated MSCs for both energy storage and tactile sensing. 2D-MXenes are fabricated through a hydrofluoric acid (HF) etching route. Bulk TiS_2 powders were exfoliated *via* ball milling in a high-energy planetary ball mill. Their suspensions in distilled (DI) water are mixed and filtrated using a Celgard separator to obtain MXene/TiS_2 films. Later, the films are laser-marked to form interdigitated electrodes and assembled to obtain planar MSCs (Figure 8.12a and b). Cross-sectional SEM images provided in Figure 8.12c–e show the layered arrangement of MXene/TiS_2 electrodes at different magnifications. A maximum specific capacitance of 403.6 F g^{-1} is reported. What is intriguing is that the assembled SC can also function as a pressure sensor. When flexed or compressed, each interdigitated electrode is polarized, altering the current response based on different physiological movements.[110]

8.3.3 Supercapacitors for Ingestible/Edible Bioelectronics

The overall health of the gastrointestinal (GI) tract, from the start of the esophagus to the colon, is vital for the whole body. Although the main

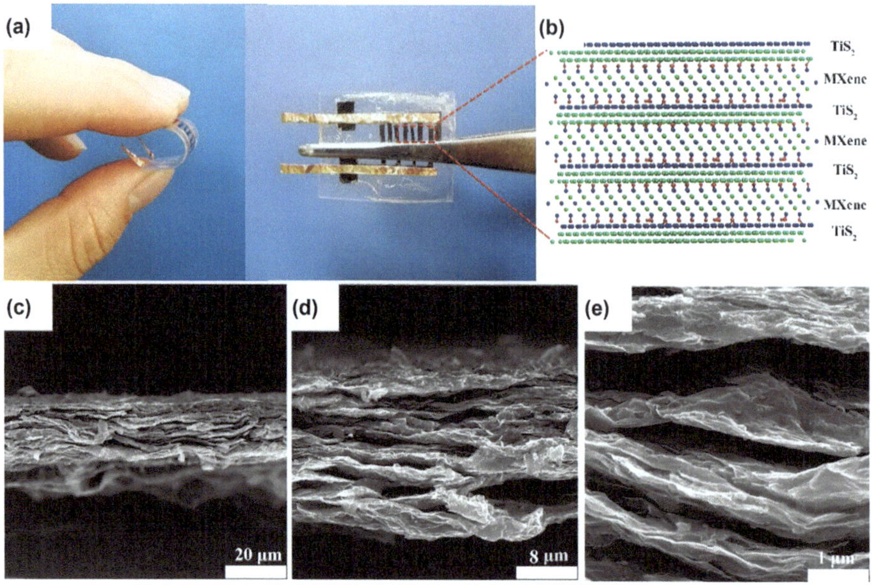

Figure 8.12 Wearable MSCs. (a) Photographs of the MXene/TiS_2-based MSCs. (b) Schematic representation of the electrode layers. (c)–(e) SEM images of MXene/TiS_2 at different magnifications. Adapted from ref. 110 with permission from American Chemical Society, Copyright 2022.

function of the GI tract is digestion, it influences nearly the entire body.[111] An overall depiction of the GI tract, its functions, and possible diseases and affected organs is provided in Figure 8.13. In order to check on the GI tract and identify illnesses, patients are generally admitted for endoscopy or colonoscopy, which are quite invasive and require sedation.

To reduce the invasiveness of the procedures and health monitoring, ingestible electronics are proposed. Ingestible electronics are a niche field that has been under development for more than 50 years. Mackay and Jacobson[112] fabricated a radio transmitter that can be swallowed and transmit information on internal temperature, pressure, and chemical composition. In the same year, Farrar and coworkers[113] fabricated a pill with telemetry and pressure-sensitive units focused on GI motility and recorded pressure changes in the gastrointestinal tract. Much later, under the sponsorship of NASA and Goddard Space Flight Center, an ingestible pill was devised to monitor body core temperature with a telemetry link.[114] Developments of such devices brought the innovation to a complete tracking of the GI tract. Devices such as PillCam SB and CapsoCam Plus fabricated pills with cameras, allowing complete endoscopy and colonoscopy imaging without the need for sedation or radiation exposure.[115] The overall design of the common ingestible electronics is given in Figure 8.14. As can be seen in the figure and the overall literature, conventional battery counterparts, such as primary batteries or LIBs, are used to power the ingestible pills rather than

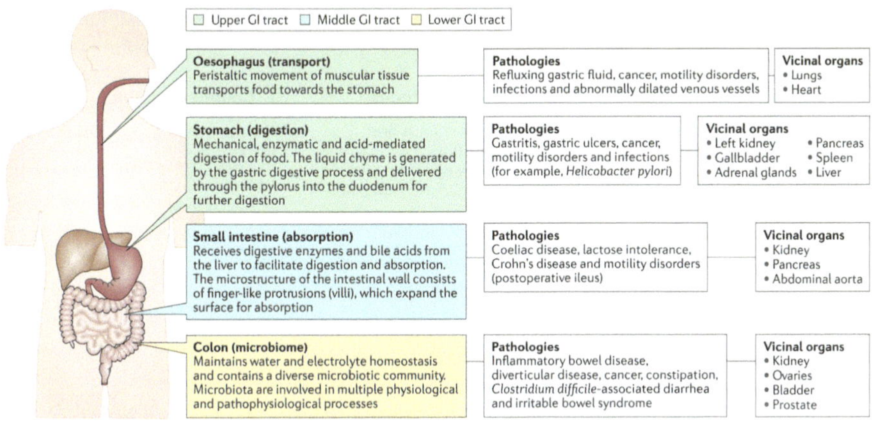

Figure 8.13 Representation of the GI tract, its parts, possible diseases, and organs that may be affected due to the GI tract. Reproduced from ref. 111 with permission from Springer Nature, Copyright 2018.

SCs. However, these ingestible devices carry the risk of toxicity in case of failure, leading to serious complications for patients. Moreover, the pills are still required to be excreted from the body, which may result in further complications.[16,116] In light of this, a more convenient and visionary field has emerged: edible electronics.[117,118]

Edible electronics can be viewed as an emerging branch of ingestible electronics, although it remains in its early stages of development. Being "edible" offers a more decentralized healthcare solution by eliminating the need for post-procedure removal of the electronic device.[16] The idea of digesting the electronics completely, rather than being only "swallowable," opens up a whole new area for healthcare monitoring. It is important to note that edible electronics can help reduce electronic waste generated by commercial devices, as they can be consumed after use. However, this chapter will focus exclusively on edible electronics for health monitoring.

The power source is essential for the functionality of ingestible electronics, typically provided by conventional batteries, primarily LIBs or silver oxide batteries. Batteries have been used since the first inventions of ingestible devices.[111,119] Still, the same problems with ingestible electronics are valid for traditional batteries. They mostly contain materials toxic for the human body, or encapsulated with a non-digestible casing, and should be excreted after monitoring. Therefore, the same solution for ingestible devices can be explored for batteries: edible energy storage devices. Among the edible energy storage devices, edible SCs offer higher power densities and can be flexible and stretchable. Yet there are only a few examples of edible SCs due to their low energy densities, limiting their application areas.

Achieving food-based SCs is feasible owing to their biocompatible, biodegradable, ingestible, and edible nature. Therefore, fabricating SCs based on already-consumed foods and products is valuable.[118] One of the first

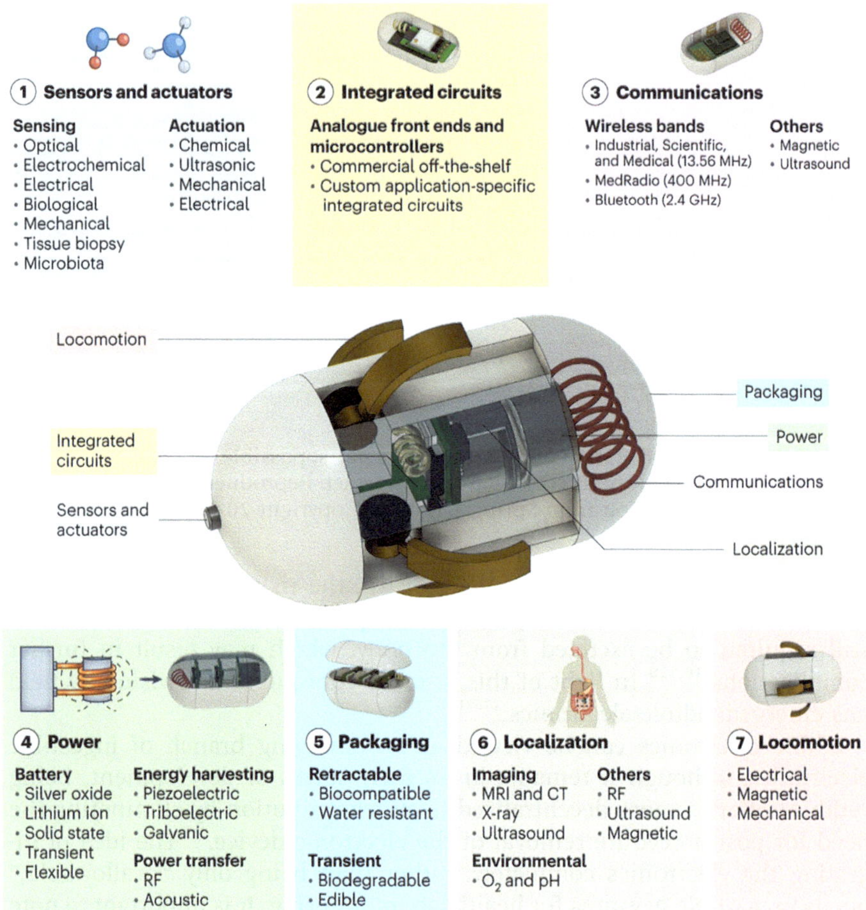

Figure 8.14 Design of a conceptual ingestible pill. Reproduced from ref. 115 with permission from Springer Nature, Copyright 2024.

conceptualizations of edible capacitors comes from this food-based approach. Wang and coworkers[120] replaced the traditional noninert or non-biocompatible electrode, electrolyte, and separator materials with food-based products. The main components of edible SCs were formed with AC on top of gold (Au) leaves for electrodes, seaweed as the separator, and cheese as a segregation layer. Carbons are electrochemically stable and non-toxic (if less than 100 grams per day are consumed) to the human body, which is appropriate for edible SCs. Au leaves were chosen to be current collectors because of their high conductivity and edible nature, with a purity of 23 karats or higher. Seaweed is an already popular food ingredient and acted perfectly as a separator when roasted. Gatorade, being a polyelectrolyte drink consisting of high amounts of sodium, potassium, citrate, and stabilizing agent, was utilized as an electrolyte due to its high ionic conductivity. Cheese layers were employed to separate the gelatin packaging from the electrolyte. The authors also tried different food products such as V8 vegetable and Monster Energy drinks as the electrolytes, monosodium glutamate (MSG), BBQ sauce, and Jello as electrolyte additives, carboxymethyl cellulose (CMC) as food additives, silver leaves as current collectors, and more. They achieved a maximum capacitance of 129 $F g^{-1}$ with the addition of MSG to the Gatorade. The authors further demonstrated edibility, underscoring the potential of edible SCs for applications in edible electronics.

Wang and coworkers[121] fabricated edible SCs based on dough-based gel electrolytes. The gluten network in the dough allows superior mechanical ability and ionic conductivity. The electrolytes were prepared in a common way, using flour, water, NaCl, and potassium carbonate (K_2CO_3), followed by mixing and kneading. Excess starch was removed by immersing the dough into the water. Then, researchers directly used the dough without any further modifications, as the 3D gluten network in the dough allows ionic conductivity. K_2CO_3 was added to increase the pH and improve the dough's hydration. NaCl was used to enhance the ionic conductivity. An ionic conductivity of 2.5×10^{-3} $S cm^{-1}$ was reported. The researchers utilized multi-walled carbon nanotube (MWCNT) films as the electrodes and achieved a maximum gravimetric capacitance of 53.7 $F g^{-1}$. Moreover, they fabricated edible supercapacitors with Au foils as current collectors, medicinal AC as the electrodes, and dough as the electrolyte and showcased its edible behavior.[121]

Seeing the promise in food-based materials for edible SCs, Durukan and coworkers[82] fabricated shoyu-based (soy sauce) zwitterionic gel electrolyte for edible SCs. The overall schematic for the fabrication of edible SCs is given in Figure 8.15a. The researchers again used carbonaceous materials for the electrode. The electrodes were prepared in ink form and doctor-bladed, which resulted in flexible carbon films that removed the need for a current collector. As a polymeric matrix, HEC and glycerol, both food additives, were utilized for the electrode and electrolyte. Gelatin was added to the electrolyte to improve water retention. The ionic conductivity was provided by shoyu (Kikkoman), which is rich in NaCl, soy proteins, calcium (Ca), sulfates (SO_4^{-2}), and phosphates (PO_4^{-3}). But most importantly, the soy protein content enhanced the ionic conductivity. The key point was the adjustment of the gel pH to benefit from its zwitterionic properties. Zwitterionic molecules carry both positively and negatively charged groups and, hence, are electrically neutral. When the molecule, here the soy protein, is at its isoelectric point (electrically neutral point), it is in the zwitterionic state. Zwitterionic molecules tend to be more ionically conducting and mechanically stable.[122,123] Herein, the isoelectric point, and hence the most ionically conductive state of the gel electrolyte, was achieved at pH 4.5. A conductivity of 1.3×10^{-1} $S cm^{-1}$ and a maximum specific capacitance of 3.75 $F g^{-1}$ were reported (Figure 8.15b). This specific capacitance was approximately three times that of the simple HEC/NaCl electrolyte. Moreover, the electrolytes demonstrated resistance to freezing and drying. To prove its biocompatibility,

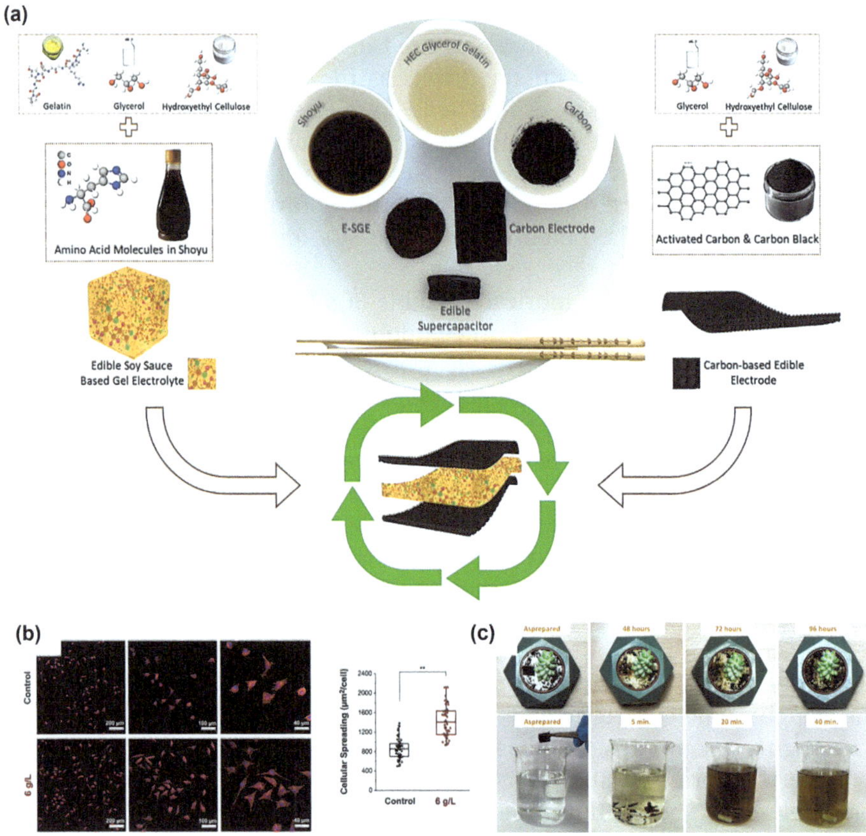

Figure 8.15 Edible supercapacitors based on zwitterionic shoyu-based gel electrolyte. (a) Overall schematic for the fabrication of edible supercapacitors. (b) L929 proliferation *via* confocal microscopy images of a standard culture sample and media containing shoyu-based gel electrolyte and their cellular spreading results. (c) Biodegradation of the edible supercapacitor by watering on soil and dissolving in the simulated gastric fluid. Adapted from ref. 82 with permission from John Wiley & Sons, Copyright 2023 The Authors. Advanced Functional Materials published by Wiley-VCH GmbH.

the researchers conducted cytotoxicity tests using L929 fibroblasts (ISO 10993). These tests showed that the electrolytes were not only biocompatible but also support cell proliferation and enhance cell spreading, thereby facilitating cellular growth (Figure 8.15). Last but not least, biodegradability and possible digestion were checked on a simple soil and in simulated gastric fluid, respectively. Edible supercapacitors were disintegrated in both cases, proving their edible nature.[82]

Caironi's group is recognized for their work on green and edible electronics.[116,124,125] In one of their studies, they assembled an edible SC using AC/ethyl cellulose (EC) composites as electrodes, gold leaves as the current

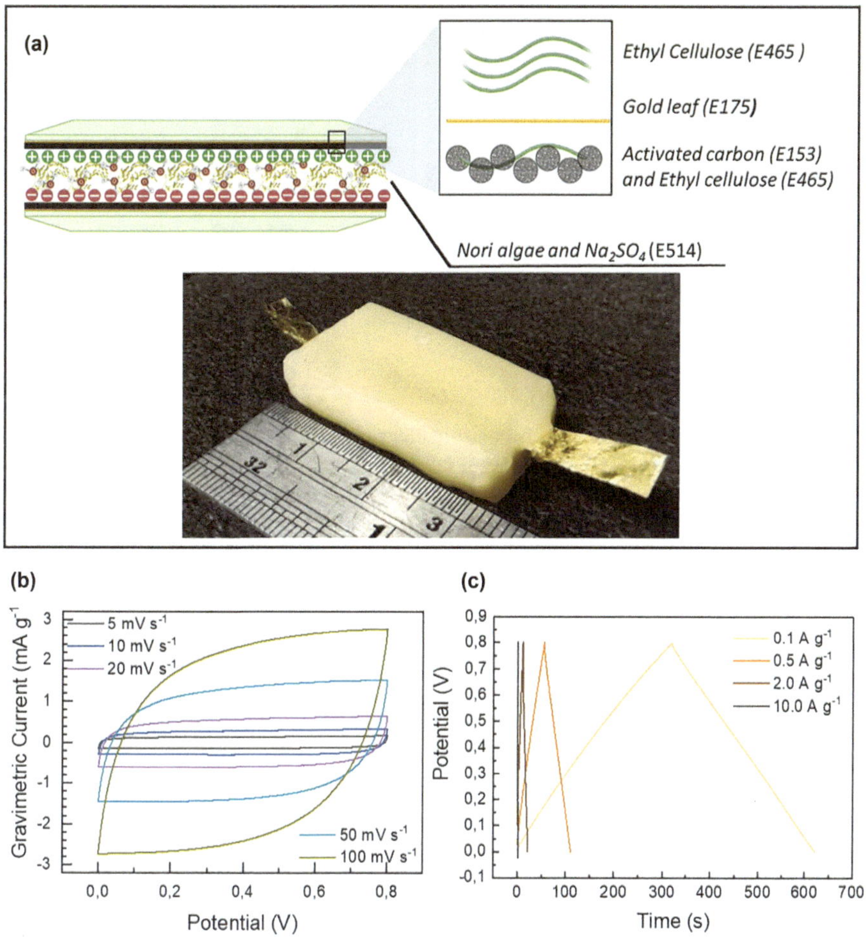

Figure 8.16 Cellulose-based edible supercapacitors. (a) Overall schematic of the edible supercapacitor and optical image of the edible supercapacitor encapsulated with beeswax. (b) CV and (c) GCD graphs of cellulose-based edible supercapacitors. Adapted from ref. 126, https://doi.org/10.1016/j.nanoen.2023.108168, under the terms of the CC BY 4.0 license, https://creativecommons.org/licenses/by/4.0/.

collector, and a nori algae separator soaked in sodium sulfate (Na_2SO_4) as the electrolyte. The edible supercapacitor was then encapsulated with beeswax to prevent drying. The overall schematic and optical image can be seen in Figure 8.16a. CV and GCD measurements showed an almost ideal EDLC behavior (Figure 8.16b and c). A maximum gravimetric capacitance of 40 F g^{-1} was reported. Although the structure was relatively simple, the novelty lies in fabricating a triboelectric nanogenerator with similar materials, demonstrating that these materials hold promise for use in edible SCs for energy storage and harvesting.[126]

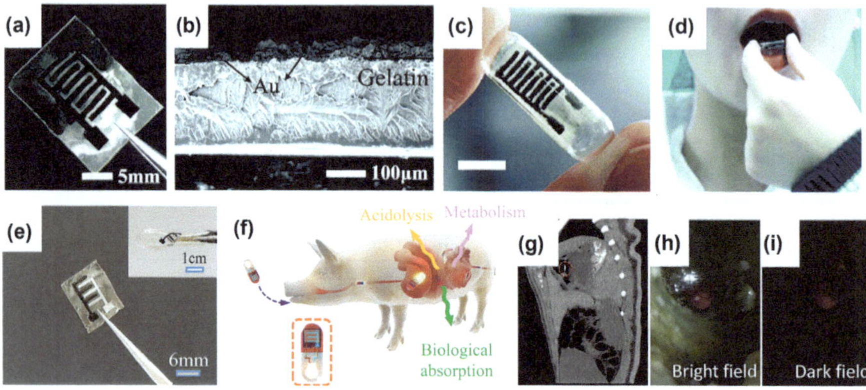

Figure 8.17 Edible MSCs for ingestible electronics. (a) Optical image and (b) cross-sectional SEM image of an AC-based microsupercapacitor. (c) and (d) The MSC can be encapsulated in a gelatin capsule and swallowed. Adapted from ref. 127 with permission from the Royal Society of Chemistry. (e) Optical image of Zn-based MSC. (f) Schematic diagram of the *in vivo* experimental procedure, from administration to excretion in pigs. (g) Computed tomography and (h) and (i) gastroscope images of a red LED powered by ZMSC. Adapted from ref. 128 with permission from American Chemical Society, Copyright 2022.

MSCs can also be fabricated for edible electronics to obtain planar energy storage devices requiring flexibility and a small volume. Gao and co-workers[127] fabricated MSCs based on gelatin substrates *via* the simple transfer of Au leaves using a laser-scribed aluminum (Al) template. AC was mixed with gelatin and sugar to obtain highly viscous ink, which was poured inside the Al template to form the electrode on top of Au. The optical image and cross-sectional SEM image of the edible MSC are given in Figure 8.17a and b, respectively. A volumetric capacitance of 11.17 F cm^{-3} was achieved. The MSCs were shown to be inserted inside gelatin capsules (Figure 8.17c), and the edible nature of the fabricated MSCs was demonstrated through swallowing of the devices (Figure 8.17d).[127] In another work, Chen and co-workers[128] fabricated Zn-ion MSCs (ZMSCs) to improve the overall energy density of edible MSCs. Furthermore, the Zn microanode also served as a nutritional supplement within the daily recommended dosage range. Researchers followed a similar route in their previous paper. Using masks and layer-by-layer stacking, Au electrodes were obtained on a gelatin substrate, which was then coated with AC to obtain the cathode. For the anode, a Zn foil is laser patterned and assembled directly with the cathode. Coupled with zinc sulfate ($ZnSO_4$) gel electrolyte, the ZMSC is finalized. An optical image of the final ZMSC structure and its flexed condition is provided in Figure 8.17e. ZMSCs showed a maximum areal capacitance of 605 mF cm^{-2} and volumetric capacitance of 39 F cm^{-3}, which was reported to exceed miniaturized commercial button batteries. Cytotoxicity tests were conducted using intestinal adenocarcinoma cells (Caco-2), in which ZMSCs showed similar cell viability with the culture medium and even supported cell growth due to the Zn supplement from both the anode and electrolyte. To demonstrate the possible application of ZMSCs, a microdevice containing an LED and ZMSC was fabricated and stored in a capsule, and then tested *in vivo*. A schematic representation is provided in Figure 8.17f. The illumination of the LED was followed through ingestion to excretion by gastroscopy, and LED could be easily observed throughout the procedure (Figure 8.17g–i).[128]

In summary, edible SCs hold significant potential for enhancing both ingestible and edible bioelectronics, offering a new route for advancing health-monitoring devices. While traditional batteries are commonly used in ingestible electronics, research into edible batteries is still ongoing. However, SCs have yet to be widely translated, clinically adopted, or commercially applied in ingestible devices. Nonetheless, SCs could represent the next significant advancement in edible bioelectronics, although research in this area remains somewhat limited.

8.3.4 Supercapacitors for Implantable Bioelectronics

Implantable electronics, such as pacemakers, neurostimulators, and drug delivery systems, have revolutionized modern medicine by enabling continuous monitoring and therapeutic interventions within the body. These devices are crucial in managing chronic diseases, enhancing patient outcomes, and restoring lost functions. The first notable milestone was the implantation of a pacemaker by surgeon Åke Senning in 1958 in a 40-year-old patient. The patient, Arne Larsson, suffered from extremely low blood pressure and needed to be resuscitated 20–30 times a day.[129] Along with his engineer associate, Rune Elmquist, they developed the pacemaker based on myocardial electrodes and a pulse generator using a nickel–cadmium battery as a power source. Despite the initial failure within hours, the patient received 26 successive improved versions and ultimately lived to the age of 86.[129,130]

Following the pacemaker, various implantable devices have been developed over time to combat a range of diseases. Most make use of power sources for both health monitoring/treatment and communications with external systems.[131] The energy requirements vary for each component in the integrated circuits (see Table 8.1). Biosensors generally require below 35 μW. Radiofrequency (RF) transmissions can require around 10 μW to 1.5 mW. Stimulators such as pacemakers may need around 5–30 μW, while muscle stimulators may require 4–21 mW.[70,71,131,132] To ensure the functionality of the entire device, researchers are focusing on energy storage solutions. Energy storage is crucial for the performance and longevity of these devices. As implantable bioelectronics advance, there is an increasing emphasis on developing reliable, miniaturized power sources. Although traditional batteries like LIBs are commonly used, recent research has shifted toward biocompatible solutions to minimize the risk of toxicity and reduce the need for surgical replacement of devices or batteries. Therefore, the overall research focuses on innovative biocompatible energy storage devices with comparable or higher energy density to ensure the seamless operation of implantable bioelectronics. Moreover, researchers have been focusing on a unique aspect, bioresorbability. Bioresorbable energy storage devices and the rest of bioelectronics offer bioresorption of the device in a predetermined time, eliminating the need for additional surgery to remove the device.[133,134] Among the solutions for implantable bioelectronics, SCs have been researched extensively (compared to edible solutions) for the last decade.

Robert A. Huggins introduced the use of SCs as the energy source for implantable devices for the first time in 1995 while reviewing the definition and working mechanism of SCs.[135] In the meantime, several other studies were delving into optimizing conventional capacitors for implantable defibrillators.[136–139] Yet the research shifted its direction toward biocompatible energy storage devices. For implantable SCs, biocompatibility can be related to safety and stability. In the case of safety, SCs should have no cytotoxicity, cause no coagulation or inflammation, and be antibacterial. In the case of stability, electrochemical and thermomechanical stability is a must. Apart from safety and stability, a new generation of implantable SCs is expected to be flexible and smaller for miniaturized bioelectronics.[131] The overall requirements are schematically shown in Figure 8.18.

One of the first implantable SCs was proposed by Hur and coworkers[140] in 2013, where they developed a porous DNA hydrogel (Dgel) combined with

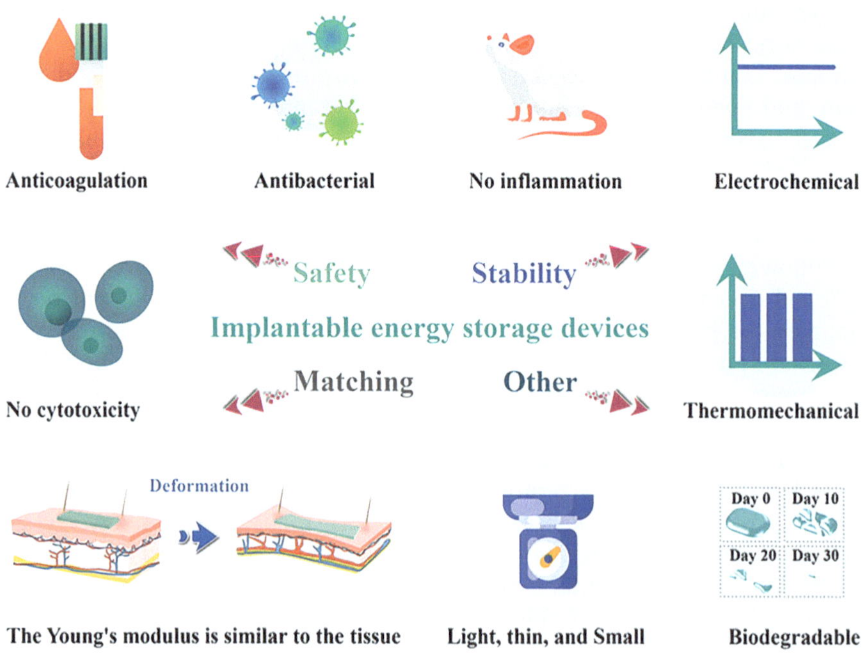

Figure 8.18 Overall requirements for implantable supercapacitors. Reproduced from ref. 131 with permission from John Wiley & Sons, Copyright 2023 Wiley-VCH GmbH.

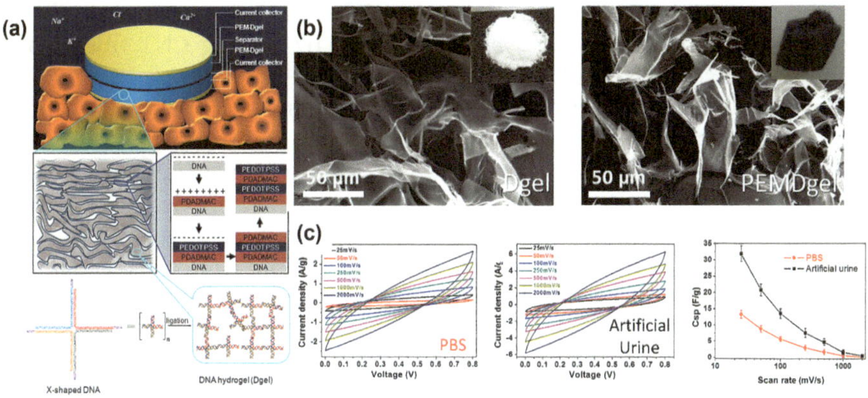

Figure 8.19 PEM-Dgel hydrogels for implantable supercapacitors that utilize biofluids. (a) Overall structure of PEM-Dgel supercapacitors. (b) SEM images of Dgel and PEM-Dgels. (c) CV measurements of PEM-Dgel supercapacitors in PBS and artificial urine and their corresponding gravimetric capacitance values. Adapted from ref. 141 with permission from Springer Nature, Copyright 2013.

CNTs and PANI to obtain biocompatible SCs. Poly-(allylamine hydrochloride) (PAH) was used as a cationic polyelectrolyte. They assembled PANI/CNT/PAH/Dgel electrodes symmetrically with 1 M H_2SO_4 electrolyte and reported a maximum gravimetric capacitance of 141.6 F g^{-1}. The researchers also employed *in vitro* cytotoxicity tests by immersing the device into mouse embryonic fibroblast cells (NIH3T3) and human dermal fibroblast cells (HDF). The cells sustained 1000 charge–discharge cycles without any drastic performance loss, and they remained intact during the charge–discharge cycles, proving the structural stability of the fabricated SCs.[140] In the same year, Hur and coworkers[141] fabricated again Dgel-based SCs by layer-by-layer (LBL)-deposited polyelectrolyte multilayers (PEM) on Dgel templates. The overall structure can be seen in Figure 8.19a. The optical structures and SEM images of Dgel and PEM-Dgel can be observed in Figure 8.19b. Dgels were fabricated by enzymatic cross-linking of X-shaped DNAs. It was particularly intriguing that the SCs functioned by utilizing physiological fluids. The obtained porous DNA template was negatively charged, allowing functionalization with polyelectrolytes. The template was coated with PEDOT:PSS and polydiallydimethyl-ammonium chloride (PDADMAC) to form electrodes. In both PBS and artificial urine, the SC device gave somewhat ideal energy storage. This research showcased that with biocompatible materials, there is no need for encapsulation, and physiological fluids can be utilized as electrolytes to power up the implanted devices.[141]

CNTs are widely used for implantable SCs.[142–144] CNTs possess a range of advantageous properties, including high electrical conductivity, high chemical stability, exceptional mechanical strength, and intrinsic energy storage capability.[145] In one example, Qian and coworkers[146] developed multi-strand SCs coupled with an energy harvester based on CNT fibers. Firstly, the anode of the SCs was constructed by coating CNT fibers with glucose oxidase (GOx) and tetrathiafulvalene (TTF), which are the glucose oxidase enzyme and electron transfer mediator, respectively. A CNT fiber coated with PANI was twisted with the previous one to form the energy harvester anode. Lastly, CNT fibers were coated with platinum on carbon (Pt/C) and re-twisted with CNT/PANI fibers to form the cathode. Twisting this cathode and anode formed the energy-harvesting, self-rechargeable SCs. The schematic structure, optical photograph, and simple working mechanism are provided in Figure 8.20a. The GOx/TTF/CNT fiber was electrochemically characterized in a glucose environment from 0–20 mM by linear sweep voltammetry (LSV). It was observed that the power density increases from 0 to 8 mM glucose concentrations, reaching up to 99.1 μW cm^{-2}. CNT/PANI fibers were also separately characterized and found to have an areal capacitance of 123 mF cm^{-2}. The overall energy module was then used as an electrical stimulation device. The fiber-based SC was connected to both ends of a rat sciatic nerve. This is schematically shown in Figure 8.20b.

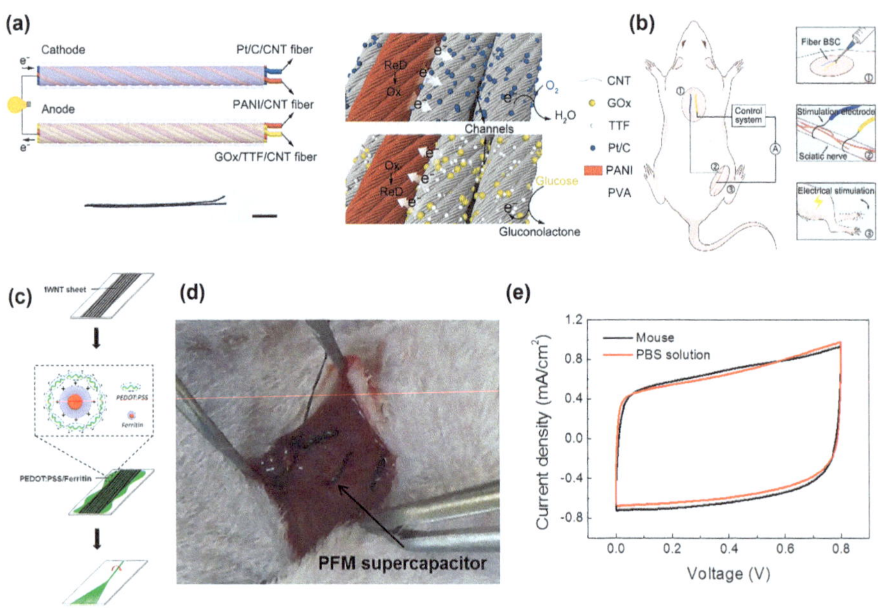

Figure 8.20 Fiber/yarn-based biocompatible implantable SCs. (a) Schematic representation of a self-charging SC and its working mechanism. (b) Schematic representation of the fibers attached to a sciatic nerve to act as a muscle stimulator. Adapted from ref. 146 with permission from John Wiley & Sons, Copyright 2023 Wiley-VCH GmbH. (c) Schematic representation of the preparation of PEDOT:PSS/Ferritin/MWCNT fibers. (d) Optical photograph of the SC fibers used to suture abdominal muscles. (e) CV comparison of the standard electrochemical characterization with PBS and *in vivo* performance. Adapted from ref. 147 with permission from Elsevier, Copyright 2018.

The fiber-based SC could deliver a discharge current of 150 μA, which was able to move the leg of the rat.[146]

In another study, Sim and coworkers[147] again utilized fibers based on multiwalled CNT (MWCNT), PEDOT:PSS, and ferritin. Ferritin was added to the system to improve the physicochemical properties and biocompatibility. MWCNT sheets were prepared by spinning on glass. Later, they were coated with PEDOT:PSS and ferritin solution. After the PEDOT:PSS/ferritin solution was fully adsorbed by the MWCNT sheets, it was twisted to obtain a fiber. A schematic representation of the preparation steps is given in Figure 8.20c. Electrochemical characterization was done using PBS as the electrolyte. Ferritin addition increased the areal capacitance from 4.3 $mF\,cm^{-2}$ to 32.9 $mF\,cm^{-2}$. An *in vivo* test was carried out with the fibers by implanting them into a mouse, specifically stitching the abdominal muscle (Figure 8.20d). The wound that was sutured with the SC fibers was observed to be healed without any infection in 4 weeks and covered with vascularized adipose tissue. Moreover, implantable SC fibers showed the same performance in mice when compared to the electrochemical characterization with a PBS solution.[147]

Besides CNTs, rGOs can also be utilized for implantable and bioresorbable energy storage devices. Lv and coworkers[148] developed a conductive aminated-rGO and methacrylic-anhydride-co-modified sericin (SrMA/A-rGO) by sequentially cross-linking polyethylene glycol succinimide carbonate (4PEGSC) and polyethylene glycol acrylate (PEGDA). Sericin is a natural protein from silkworm silk, which is highly biocompatible and can be bioresorbed without any adverse effects. Biocompatibility of the SrMA/A-rGO hydrogel electrodes was tested in human umbilical vein endothelial cells (HUVEC), rat fibroblasts (L929), and rat cardiomyoblasts (H9C2). After 24 hours of incubation, cell viability was higher than 90% for all, suggesting good cytocompatibility. The electrochemical tests with a PVA-based gel electrolyte resulted in an areal capacitance of 103.5 $\mu F\,cm^{-2}$ with an areal energy density of 0.014 $\mu W\,h\,cm^{-2}$. The bioresorbable SCs were implanted into the subcutaneous region of rats and observed for 35 days. The biodegradation of SrMA/A-rGO was much better than for sericin/A-rGO, without showing any organ staining or inflammatory response.[148]

The biodegradation of metals and metal oxides has also been studied for implantable SCs. There are already a few studies that explore the degradation of metals in terms of transiency.[149,150] Inspired by these, Sheng and coworkers[151] developed fully biodegradable and bioresorbable SCs based on amorphous molybdenum oxide (MoO_x) grown on Mo foil. Mo foil was electrochemically oxidized with 1 M NaCl as the electrolyte with CV ran at 100 $mV\,s^{-1}$. For the electrolyte, sodium alginate (Alg-Na) was used. The electrolyte was coated onto the electrodes and dried. The SCs were then assembled symmetrically and encapsulated with polylactic acid (PLA) or PVA to control the bioresorption rate. The electrochemical behavior of the electrodes was characterized in 1 M NaCl electrolyte, achieving the highest areal capacitance of 227.4 $mF\,cm^{-2}$. On the other hand, the fabricated symmetric implantable SCs gave a maximum areal capacitance of 112.5 mF cm^{-2}. A systematic degradation study was conducted in a typical PBS solution at 37 °C, the average body temperature of the human body. A complete dissolution was seen after 3 months. Moreover, assembled SCs showed no cytotoxicity in human endometrial adenocarcinoma (HEC-1-A). The cells were encapsulated with PVA and starch paper. At elevated temperatures, a complete dissolution in PBS occurred in 40 days. The biodegradability of SCs was carried out *in vivo* by implanting them subcutaneously in the dorsal side of the rats. In this case, the dissolution was completed in 6 months. Serially connected SCs were implanted into the rats and were shown to power an LED, proving that the bioresorbable SCs can power up implantable electronics and degrade over time.[151]

Shao and coworkers[152] also worked with MoO_3 due to its well-known biodegradability, yet formed a heterostructure with MoS_2 to improve the capacitive performance. Anodization was employed to obtain MoO_3 flakes on the surface of Mo foil, followed by a hydrothermal reaction for sulfidation with thiourea. An optical photograph of the electrodes in different chemical states are provided in Figure 8.21a. With three electrode measurements utilizing 1 M Na_2SO_4, the best capacitive behavior with enhanced storage was obtained with the MoO_3/MoS_2 composite structure (Figure 8.21b). The researchers constructed an asymmetric Zn-ion SC by utilizing the composite electrodes as the cathode, Zn as the anode, and gelatin/$ZnSO_4$ as a biodegradable gel electrolyte. They obtained a maximum areal capacitance of 181.86 mF cm^{-2} and an energy density of 30.56 μW h cm^{-2}. The schematic structure of the Zn-ion SC and related CV curves is provided in Figure 8.21c. The prepared SCs encapsulated with a wax layer were dissolved in 0.1 M PBS at 80 °C to prove the biodegradation. In 60 days, the SCs were almost fully dissolved, proving their reliability as a power source for implantable bioelectronics.[152]

It is also possible to use traditional pseudocapacitive materials for implantable SCs. RuO_2 is a well-known pseudocapacitive material with very high energy densities, although its high cost limits its applications.[42,153] In one study, Makino and coworkers[154] explored the performance of poorly crystalline RuO_2 and highly crystalline RuO_2 in near-neutral pH and benign electrolytes. The researchers tried various electrolytes, including lithium sulfate (Li_2SO_4), acetic acid (AcOH), lithium acetate (AcOLi), and H_2SO_4, and explored the performances in PBS and fetal bovine serum. The redox activity was similar to that in the AcOH-AcOLi system, and gravimetric capacitances of 837 F g^{-1} and 772 F g^{-1} were achieved in PBS and fetal bovine serum, respectively. Compared to the gravimetric capacitance of 1038 F g^{-1} in the AcOH-AcOLi electrolyte at pH 5.4, the capacitance loss was not drastic in physiological fluids. These findings show that RuO_2 electrodes can be powered easily by physiological fluids and offer an alternative solution for developing implantable SCs.[154] In another study led by Li and coworkers,[155] researchers utilized self-assembled zinc oxide (ZnO) layers on top of PLA nanopillar arrays and a PVA/PBS hydrogel layer to obtain bioresorbable

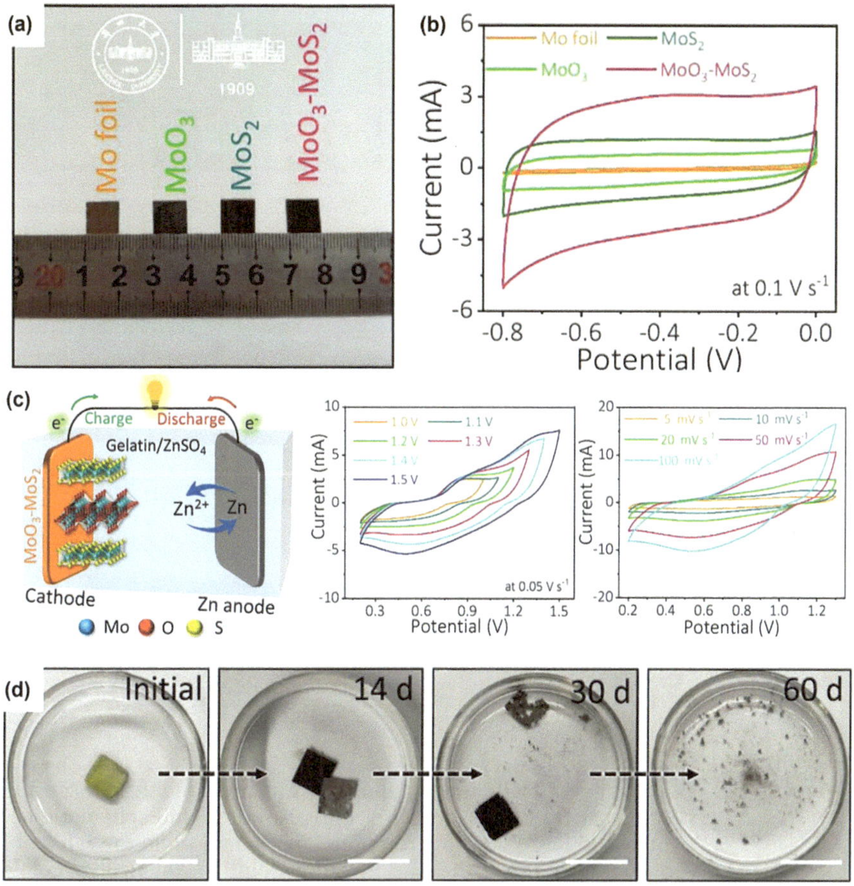

Figure 8.21 Implantable SCs with MoO_3/MoS_2 composite electrodes. (a) Photograph and (b) CV measurements of the electrodes. (c) Schematic representation of the assembled Zn-ion supercapacitor with MoO_3/MoS_2 composite electrode and relevant CV curves. (d) Dissolution of MoO_3/MoS_2-based Zn-ion supercapacitors in PBS. Adapted from ref. 152 with permission from John Wiley & Sons, Copyright 2022 Wiley-VCH GmbH.

capacitors. Although not labeled as an SC, this is a nice example of using bioresorbable metals and metal oxides for implantable energy sources. Their structure can also work in biofluids and can be fully degraded and absorbed by the body.[155]

MXenes are also employed for implantable SCs. Liu and coworkers[156] fabricated an all-hydrogel device consisting of PANI@rGO/MXene as electrodes and polyacrylamide-alginate (PAM-Alg) as the electrolyte. The overall schematic of the work is given in Figure 8.22a. The PANI solution in NMP was prepared and mixed with a GO/MXene solution with the addition of sodium ascorbate. After mixing at 95 °C for 2 hours, PANI@rGO/MXene

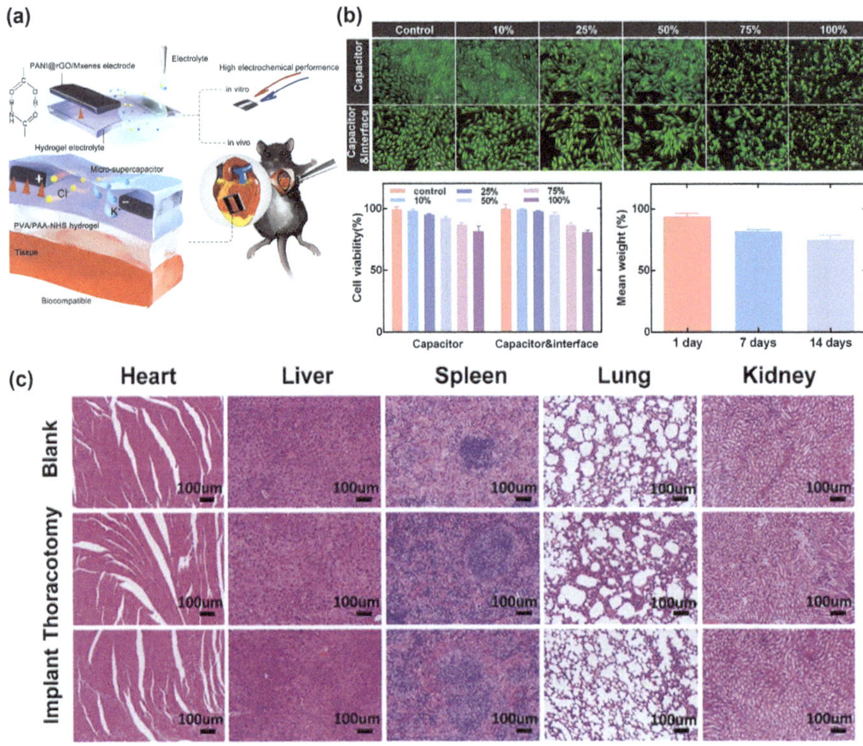

Figure 8.22 All-hydrogel-based biodegradable and implantable SCs utilizing PANI@rGO/MXene hydrogel electrodes. (a) Overall schematic of the work. (b) Fluorescent microscopic images and plots for cell viability and degradation of the SCs in saline solution. (c) H&E histological staining images for the control group's heart, liver, spleen, lung, and kidney, SC without the adhesive (thoracotomy), and fully implanted structure. Adapted from ref. 156 with permission from John Wiley & Sons, Copyright 2021 Wiley-VCH GmbH.

hydrogel electrodes were obtained. On the other hand, hydrogel electrolyte was prepared *via* mixing alginate and acrylamide solutions with the additions of ammonium persulfate (APS—as a photo-initiator), and *N*,*N*-methylenebisacrylamide (MBAA—as a cross-linker). After obtaining the PAM-Alg pregel, calcium sulfate slurry ($CaSO_4{\cdot}2H_2O$) was added to the pregel, and cured with UV light. The PAM-Alg gel was directly cured over two PANI@rGO/MXene electrodes to obtain the SCs. A maximum gravimetric capacitance of 45.62 $F\,g^{-1}$ was obtained, with an energy density of 333 $\mu W\,h\,cm^{-2}$ and 4.68 $W\,h\,kg^{-1}$. Although the all-hydrogel implantable SCs are not a new concept, the researchers added an extra wet-adhesive hydrogel glue layer for better tissue interface and biocompatibility. They fabricated a PVA/poly(acrylic acid) (PAA) hydrogel layer, which was then modified with an *N*-hydroxysuccinimide (NHS) ester bond. The PAA-NHS ester group formed within the hydrogel acts like a biomedical glue and is reported to form bonds with amine groups on most tissues. After forming the complete structure seen in Figure 8.22a, the researchers thoroughly delved into the biocompatibility and cytotoxicity tests. Rat cardiomyocytes (H9c2(2-1)) were chosen as the testing cell type because the fabricated SCs were planned to be used in heart-related applications. Fluorescent microscopic images and related cell viability graphs are provided in Figure 8.22b. The culture media was conditioned with varying extracts ranging from 1% to 100%. From the graphs obtained from the imaging, a cell viability of >90% was achieved up to 50% conditioning, showing excellent biocompatibility. Moreover, a degradation of 20% of the mean weight was observed in saline solutions just after 14 days. For the *in vivo* compatibility tests, the rats were divided into three groups: one as a control, one with the device lacking the PVA/PAA adhesive, and one with the complete structure. Hematoxylin and eosin (H&E) histological staining was used to observe the possible spread of inflammation. The images for different organs such as the heart, liver, spleen, lung, and kidney are provided in Figure 8.22c. No difference was seen between the control and other samples after 14 days, indicating no adverse effect through the implantation of the implantable SC and its degradation in the body.[156]

MSCs also offer another energy source solution for implantable energy sources. As in the ingestible Zn-based devices,[128] Zn electrode-based MSCs are also promising for implantable devices. Tian and coworkers[157] developed a three-dimensional Zn@polypyrrole (PPy) by screen-printing Zn as a slurry and then the electrodeposition of PPy. An electrolyte of NaCl/agarose was prepared as a biocompatible electrolyte. Schematics of the fabrication procedure are provided in Figure 8.23a. The fabricated MSC had an energy density of 0.39 $\mu W\,h\,cm^{-2}$. The fabricated SCs were encapsulated in silk and implanted into a rat. Related optical images are provided in

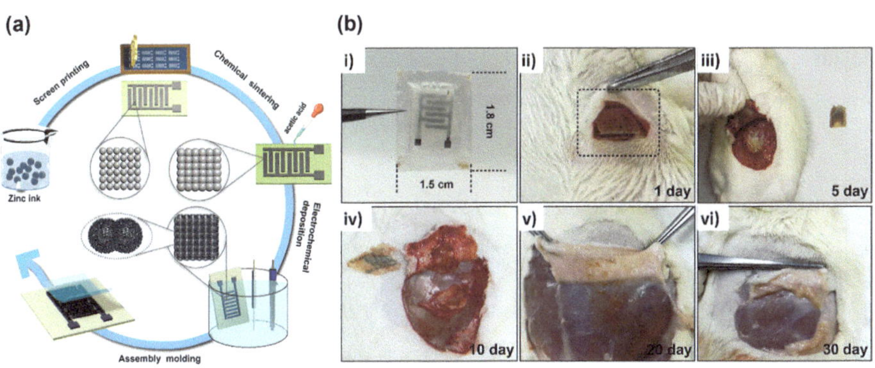

Figure 8.23 Zn/PPy-based MSCs for implantable devices. (a) Schematic representation of the fabrication of implantable Zn/PPy-based MSCs. (b) Optical images of (i) silk-encapsulated MSCs and (ii–vi) *in vivo* degradation of the MSCs. Adapted from ref. 157 with permission from American Chemical Society, Copyright 2021.

Figure 8.23b(i–vi). The MSCs were seen to degrade *in vivo* without causing any adverse effects on the rat.[157]

Despite all the advancements, the energy density of SCs still falls below that of batteries, limiting their usage in implantable devices. To take advantage of their high power density and long cycle lives, researchers have come up with a different solution rather than trying to increase the energy density: wireless charging. By incorporating a simple transmitting coil, typically made of a biocompatible metal for implantable applications, SCs can be wirelessly charged to either recharge the device or activate it on demand.[158] Sheng and coworkers[159] developed such an integrated device, using a bioresorbable Mg transmitting coil, a rectifying module, and bioresorbable Zn-ion SCs. The conceptual image and digital design of the device are given in Figure 8.24a. For the Zn-ion SCs, a 2D MoS_2/rGO composite was employed as a cathode. 2D MoS_2 was fabricated *via* a hydrothermal reaction on Mo foil with GO addition. As an electrolyte, the Alg–Zn electrolyte was prepared by immersing the Alg into 1 M $ZnCl_2$ for 24 hours. Using Zn foil as the anode, the Zn-ion SCs were fabricated. Then, Zn-ion SCs were connected with Mg foils in serpentine shapes and connected using a biodegradable tungsten (W) paste. With the connection to the rectifier and Mg coil, the implantable and rechargeable energy supply was ready. The final state of the system can be seen in Figure 8.24b. With the Alg–Zn electrolyte, the dendrite formation was prevented, prolonging the cycle-life of Zn-ion SCs. An areal capacitance of 93.5 mF cm^{-2} with an output voltage of 1.3 V was obtained. *In vitro* and *in vivo* tests showed that the prepared Zn-ion SCs were biocompatible and degraded after 2 months when implanted into rats. The researchers also developed a drug-release system that can be activated *via* charging SCs wirelessly, which controls the release of the drugs depending on the charge level. The researchers implanted the integrated drug system in rats, and then induced fever by yeast. With the wireless charging Zn-ion SCs,

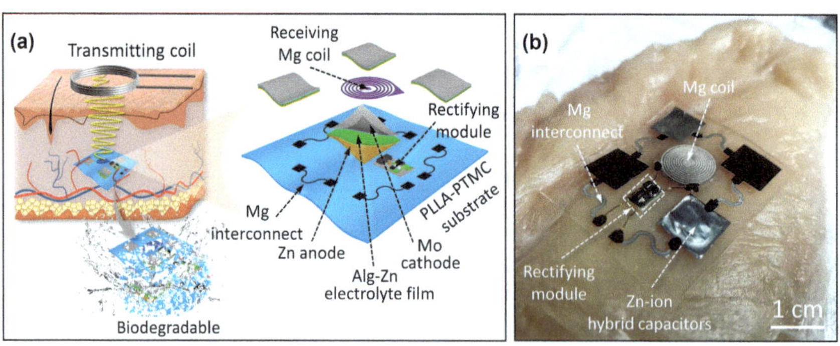

Figure 8.24 Wirelessly chargeable Zn-ion SCs for implantable devices. (a) Schematic design and (b) photograph of the integrated energy supply system consisting of bioresorbable Zn-ion SCs. Adapted from ref. 159 with permission from American Association for the Advancement of Science, Copyright 2023.

the drug was successfully released from the system of the rats, as the rectal temperature dropped.[159] This research is a great example of the usage of implantable SCs for wireless charging and expands the vision of future research on implantable and bioresorbable SCs.

In summary, implantable SCs offer significant promise for advancing implantable bioelectronics, which could pave the way for implantable and decentralized healthcare monitoring. While traditional batteries are still widely used in such devices, they face challenges related to lifespan, safety, and compatibility with biological tissues. Though still in its early stages, research into implantable supercapacitors highlights their ability to offer fast charge–discharge cycles, higher stability, and biocompatibility/bioresorbability, which could address the limitations of current power sources. As research progresses, implantable SCs could substantially advance the functionality and longevity of implantable bioelectronic devices.

8.4 Conclusions and Future Outlook

In conclusion, wearable, implantable, and ingestible SCs for healthcare applications represent a niche yet emerging field. Each application—whether wearable, implantable, or ingestible—demands unique considerations regarding biocompatibility, flexibility, energy density, and operational safety within physiological environments. Representative works that are discussed in this chapter with detailed metrics and important aspects are summarized in Table 8.3. The realization of these devices as viable components in medical technologies will allow a more decentralized health system and continuous monitoring of the patients, and lessen the risk of complications by reducing the number of required surgical procedures. The biggest challenge for SCs is their lower energy densities compared to batteries, limiting their extensive use. On the other hand, developments in self-powering SCs or wirelessly chargeable units seem to eliminate the need for excessive energy density.

8.4.1 Key Challenges

Key challenges for the future of SCs to be used in bioelectronics are as follows:

- Energy density: SCs still lack sufficient energy density to fully power most implantable and ingestible devices (Table 8.1). Ongoing research on SCs—not limited to bioelectronics—continues to focus on increasing energy density to expand their potential applications in industry. Nevertheless, SCs can still be used as transmission units in such devices to reduce overall energy consumption by providing quick bursts of power for signalling.
- Lack of flat voltage plateau: SCs, in general, lack a flat discharge voltage plateau, making them incompatible for powering long-term devices such as pacemakers (Figure 8.5). Most bioelectronic devices require a

Table 8.3 Summary of wearable, ingestible/edible, and implantable SCs.

Category	Configuration	Capacitance	Energy density	Retention	Aspect	Ref.
Wearable	Planar W, Fe, Mo@ NaCl/agarose electrolyte	1.6 mF cm^{-2}@ 0.15 mA cm^{-2}	0.14 μW h cm^{-2}	—	Bioresorbable	93
	Symmetric PPy/PEDOT:PSS on stainless-steel yarns@artificial sweat	343.1 mF cm^{-2}@ 0.5 mA cm^{-2}	30.5 μW h cm^{-2}	68% after 162 cycles	Sweat-activated	100
	Symmetric PANI/CNT@ 1 M H_2SO_4	217.7 mF cm^{-2}@ 1 mA cm^{-2}	~1.8 μW h cm^{-2}	—	Biocompatible	101
	Fabric-based CF/$NiCoO_2$/ N–C//CF/N–C@PVA/KOH electrolyte	98.5 mF cm^{-2}@ 2 mA cm^{-2}	—	95% after 8000 cycles	Integrated self-powered glucose sensor	102
	Hydrogel PAM/PrGO/ CNON@PAM-PrGO electrolyte containing KOH	252.63 F g^{-1}	126.64 W h kg^{-1}	91.3% after 1000 cycles	Self-adhesive and stretchable	104
	Hydrogel symmetric PVA/ PANI@PVA/ H_2SO_4	295 F g^{-1}@ 0.1 A g^{-1}	40.975 W h kg^{-1}	90% after 3500 cycles	Can be used as strain sensor	105
	Symmetric MSCs based on CNF/PEDOT:PSS@EMIM: ES/HEC electrolyte	7 F cm^{-3}@ 0.0125 mA cm^{-2}	0.26 mW h cm^{-3}	90% after 10 000 cycles	Skin-compatible, ultrathin	109
	Symmetric MXene/TiS_2 MSCs@PVA/ H_2SO_4	89.1 F g^{-1}@ 0.8 A g^{-1}	31.6 W h kg^{-1}	79.8% after 10 000 cycles	Can be used as strain sensor	110
Ingestible/ edible	Symmetric Au/AC@roasted seaweed/Gatorade/MSG electrolyte	129 F g^{-1}@ 1 A g^{-1}	~10 W h kg^{-1}	96.9% after 1000 cycles	Edible and antibacterial	120
	Symmetric Au/ACs@dough electrolyte with NaCl & K_2CO_3	48 F g^{-1}@ 0.1 A g^{-1}	—	—	Edible and resistant to drying	121

	Symmetric HEC/AC/CB@HEC/gelatin/soy sauce electrolyte	3.75 F g^{-1}@ 30 mV s^{-1}	0.62 W h kg^{-1}	86.5% after 10 000 cycles	Edible, resistant to freezing and drying, and promotes cell growth	82
	Symmetric Au/AC@nori algae/Na_2SO_4 electrolyte	40 F g^{-1}@ 0.1 mA g^{-1}	3.36 mW h g^{-1}	—	Electrodes are also used for energy harvesting	126
	Symmetric Au/AC MSC on gelatin@agar/MSG electrolyte	78.2 mF cm^{-2}@ 20 mV s^{-1}	10.86 µW h cm^{-2}	67.65% after 4000 cycles	Highly flexible	127
	Zinc foil//Au/AC MSC@gelatin/$ZnSO_4$ electrolyte	605 mF cm^{-2}@ 0.2 mA cm^{-2}	215.1 µW h cm^{-2}	~70% after 500 cycles	Highly flexible and can be used as a supplement for cell growth	128
Implantable	Symmetric Au/PANI/CNT/PAH/Dgel electrodes@cell culture medium	66.7 F g^{-1}@ 1 A g^{-1}	—	~92% after 1000 cycles	Biofluids can act as the electrolyte	140
	Symmetric Au/PEM/Dgel@cell culture medium	25.5 F g^{-1}@ 25 mV s^{-1}	—	6.1% after 1000 cycles	Artificial urine and PBS can also be used to operate the devices	141
	Pt/C/CNT//PANI/CNT as cathode and PANI/CNT//GOx/TTF/CNT as anode@PBS	123 mF cm^{-2}@ 1 mA cm^{-2}	—	95.8% after 10 000 cycles	Energy harvesting *via* glucose, can supply pulse discharges for nerve stimulation	146
	Symmetric MWCNT/ferritin/PEDOT:PSS@PBS	32.9 mF cm^{-2}@ 10 mV s^{-1}	0.82 µW h cm^{-2}	83.5% after 1000 cycles	Works with biofluids, can be used for suture	147
	Symmetric hydrogel Mo wire/SrMA/A-rGO@PVA/NaCl gel electrolyte	103.5 µF cm^{-2}@ 5 µA cm^{-2}	0.014 µW h cm^{-2}	88.15% after 15 000 cycles	Bioresorbable, although induces acute inflammation	148

Table 8.3 (*Continued*)

Category	Configuration	Capacitance	Energy density	Retention	Aspect	Ref.
	Symmetric Mo foil/MoO_x@Alg-Na gel electrolyte	112.5 mF cm^{-2}@ 1 mA cm^{-2}	15.64 µW h cm^{-2}	86.7% after 2000 cycles	Bioresorbable and nutritional	151
	Mo foil/MoO_3/MoS_2//Zn foil@$ZnSO_4$/gelatin gel electrolyte	181.86 mF cm^{-2}@ 0.5 mA cm^{-2}	30.56 µW h cm^{-2}	81.62% after 1500 cycles	Bioresorbable, can also be used as a Mg primary battery	152
	RuO_2 nanosheets (three-electrode measurement) @PBS	837 F g^{-1}@ 2 mV s^{-1}	—	—	Fetal bovine serum can also act as electrolyte	154
	Symmetric PLA nanopillars/Fe film/ZnO@PVA/PBS gel electrolyte	~0.35 mF cm^{-2}@ 10 mV s^{-1}	0.153 µW h cm^{-2}	70% after 3000 cycles	Long-term use *in vivo*, and work time is tunable by manipulating encapsulation	155
	Symmetric MSC hydrogel PANI/rGO/MXene@PAM-Alg/H_2SO_4 hydrogel electrolyte	45.62 F g^{-1}@ 0.1 A g^{-1}	333 µW h cm^{-2}	84% after 10 000 cycles	Highly flexible and stretchable, self-adhesive	156
	Symmetric MSC Zn/PPy@NaCl/agarose gel electrolyte	—	0.39 µW h cm^{-2}	—	Bioresorbable	157
	Mo foil/MoS_2/rGO//Zn foil@Alg–Zn electrolyte	93.5 mF cm^{-2}@ 10 mV s^{-1}	1.65 µW h cm^{-2}	94.6% after 800 cycles	Bioresorbable, can be recharged wirelessly while *in vivo*	159

constant voltage in order to operate reliably, which SCs—including pseudocapacitors—do not provide. To maintain a voltage above the required threshold, SCs still face limitations due to their low energy density.

- Self-discharge: SCs tend to lose their charge relatively quickly. The literature often focuses on suppressing self-discharge, *i.e.*, minimizing the leakage current. Nevertheless, when combined with their low energy density, self-discharge presents a significant obstacle to the use of SCs in bioelectronic devices that demand long-term functionality.
- Miniaturization: although it is possible to fabricate SCs smaller than batteries, reducing their size leads to a significant decrease in energy density. This trade-off imposes another limitation on the use of SCs in bioelectronics, where both compactness and sufficient energy supply are essential.

8.4.2 Future Perspectives

Future perspectives for the potential application of SCs in bioelectronics could be as follows:

- Enhancing energy density: this should be the primary focus to be improved in order for SCs to become reliable energy storage devices for bioelectronics. Exploring novel materials or architectures will be crucial in future years. Improving the energy densities to at least 100 $W\,h\,kg^{-1}$ should be the aim for the coming years.
- Suppressing self-discharge: reducing leakage current in SCs is a critical challenge that requires close attention. The development of novel gel or solid-state electrolytes, particularly when combined with pseudocapacitive energy storage mechanisms, holds significant promise for advancing the performance and reliability of SCs.
- Biocompatibility and bioresorption: beyond biocompatibility, the bioresorption properties of bioelectronic devices—and their integrated energy storage units—are crucial, especially for systems that require surgical removal after use. Therefore, the use of materials that are not only non-toxic but also support cellular proliferation is essential for the development of next-generation biomedical devices.
- Wirelessly rechargeable SCs: integrating wireless charging systems into bioelectronic platforms enables on-demand recharging of SCs, facilitating repeated use without the reliance on conventional batteries that often contain toxic components. This approach enhances both user safety and device sustainability.
- Coupling SCs with energy harvesters: although still a niche research area, coupling SCs with energy-harvesting technologies offers a promising route toward fully battery-free bioelectronics. The inherently low energy density and high power density of SCs make them well-suited for use with energy harvesters. Devices such as triboelectric

and piezoelectric nanogenerators can be embedded within bioelectronic systems to continuously charge SCs without the need for external power sources, enabling long-term uninterrupted operation.

- Low-power electronics: in parallel with efforts to enhance the energy density of SCs, the development of low-power bioelectronic devices is equally important. Reducing the power consumption of these devices can significantly extend their operational lifetime and make the use of SCs more feasible in practical applications. By optimizing circuit design, incorporating energy-efficient components, and minimizing standby power, future bioelectronics can better align with the energy delivery capabilities of SCs—facilitating their adoption in wearable, implantable, and ingestible healthcare systems.

References

1. H. I. Becker, Low voltage electrolytic capacitor, *US Pat.*, US2800616A, 1957.
2. B. K. Kim, S. Sy, A. Yu and J. Zhang, Electrochemical Supercapacitors for Energy Storage and Conversion, in *Handbook of Clean Energy Systems*, Chichester, UK, John Wiley & Sons, Ltd, 2015, p. 1–25.
3. F. Beguin and E. Frackowiak, *Supercapacitors*, ed. F. Béguin and E. Frąckowiak, Wiley, 2013, p. 568.
4. R. Kötz and M. Carlen, Principles and applications of electrochemical capacitors, *Electrochim. Acta*, 2000, **45**(15–16), 2483–2498.
5. B. E. Conway, *Electrochemical Supercapacitors*, Boston, MA, Springer US, 1999, p. 736.
6. P. Simon and Y. Gogotsi, Materials for electrochemical capacitors, *Nat. Mater.*, 2008, **7**(11), 845–854.
7. Maxwell Technologies, 2023. Available from: https://maxwell.com/.
8. Energy Storage Application Brief, 2023. Available from: maxwell.com.
9. Cap-XX. 2023. Available from: https://www.cap-xx.com/.
10. CAPMOD Data Sheet, 2022. Available from: https://www.cap-xx.com/.
11. SuperBattery-Sustainable and Safe, 2022. Available from: www.skeletontech.com.
12. Capacitech Energy, 2023. Available from: https://www.capacitechenergy.com/.
13. SCP Series PrizmaCap™ 2.5V Series, 2023. Available from: www.kyocera-avx.com.
14. Y. J. Gao, W. Song, J. L. Liu and S. Bashir, *Advances in sustainable energy: Policy, materials and devices*, Springer International Publishing, 2021, pp. 1–862.
15. C. J. Bettinger, Materials Advances for Next-Generation Ingestible Electronic Medical Devices, *Trends Biotechnol.*, 2015, **33**(10), 575–585.
16. A. S. Sharova, F. Melloni, G. Lanzani, C. J. Bettinger and M. Caironi, Edible Electronics: The Vision and the Challenge, *Adv. Mater. Technol.*, 2021, **6**, 2000757.

17. P. Simon, Y. Gogotsi and B. Dunn, Where Do Batteries End and Supercapacitors Begin?, *Science*, 2014, **343**(March), 1210–1211.
18. P. Simon and Y. Gogotsi, Perspectives for electrochemical capacitors and related devices, *Nat. Mater.*, 2020, **19**(11), 1151–1163.
19. M. B. Durukan, *Multifunctional and Physically Transient Supercapacitors, Triboelectric Nanogenerators and Capacitive Sensors*, Middle East Technical University, Ankara, 2023.
20. L. L. Zhang and X. S. Zhao, Carbon-based materials as supercapacitor electrodes, *Chem. Soc. Rev.*, 2009, **38**(9), 2520.
21. E. Frackowiak and F. Beguin, Carbon materials for the electrochemical storage of energy in capacitors, *Carbon*, 2001, **39**, 937–950.
22. H. Helmholtz, Ueber einige Gesetze der Vertheilung elektrischer Ströme in körperlichen Leitern mit Anwendung auf die thierisch-elektrischen Versuche, *Ann. Phys.*, 1853, **165**(6), 211–233.
23. M. Gouy, Sur la constitution de la charge électrique à la surface d'un électrolyte, *J. Phys. Theor. Appl.*, 1910, **9**(1), 457–468.
24. D. L. Chapman, A contribution to the theory of electrocapillarity, *Philos. Mag. Series 6*, 1913, **25**(148), 475–481.
25. O. Stern, Zur Theorie der Elektrolytischen Doppelschicht, *Z. Elektrochem.*, 1924, **30**, 508–516.
26. A. G. Pandolfo and A. F. Hollenkamp, Carbon properties and their role in supercapacitors, **157***J. Power Sources*, 2006, 11–27.
27. Y. Zhu, S. Murali, M. D. Stoller, K. J. Ganesh, W. Cai and P. J. Ferreira, *et al.*, Carbon-based supercapacitors produced by activation of graphene, *Science*, 2011, **332**(June), 1537–1542.
28. C. Liu, Z. Yu, D. Neff, A. Zhamu and B. Z. Jang, Graphene-based supercapacitor with an ultrahigh energy density, *Nano Lett.*, 2010, **10**(12), 4863–4868.
29. M. Kim, Y. Kim, K. M. Lee, S. Y. Jeong, E. Lee and S. H. Baeck, *et al.*, Electrochemical improvement due to alignment of carbon nanofibers fabricated by electrospinning as an electrode for supercapacitor, *Carbon*, 2016, **99**, 607–618.
30. C. Kim and K. S. Yang, Electrochemical properties of carbon nanofiber web as an electrode for supercapacitor prepared by electrospinning, *Appl. Phys. Lett.*, 2003, **83**(6), 1216–1218.
31. V. T. Le, H. Kim, A. Ghosh, J. Kim, J. Chang and Q. A. Vu, *et al.*, Coaxial fiber supercapacitor using all-carbon material electrodes, *ACS Nano*, 2013, 7(7), 5940–5947.
32. R. R. Salunkhe, J. Tang, Y. Kamachi, T. Nakato, J. H. Kim and Y. Yamauchi, Asymmetric Supercapacitors Using 3D Nanoporous Carbon and Cobalt Oxide Electrodes Synthesized from a Single Metal-Organic Framework, *ACS Nano*, 2015, **9**(6), 6288–6296.
33. E. Frackowiak, K. Metenier, V. Bertagna and F. Beguin, Supercapacitor electrodes from multiwalled carbon nanotubes, *Appl. Phys. Lett.*, 2000, **77**(15), 2421.

34. Y. Gao, Y. S. Zhou, M. Qian, X. N. He, J. Redepenning and P. Goodman, *et al.*, Chemical activation of carbon nano-onions for high-rate supercapacitor electrodes, *Carbon*, 2013, **51**(1), 52–58.
35. M. Zeiger, N. Jäckel, V. N. Mochalin and V. Presser, Review: carbon onions for electrochemical energy storage, *J. Mater. Chem. A*, 2016, **4**(9), 3172–3196.
36. P. Lv, X. W. Tan, K. H. Yu, R. L. Zheng, J. J. Zheng and W. Wei, Superelastic graphene/carbon nanotube aerogel: A novel thermal interface material with highly thermal transport properties, *Carbon*, 2016, **99**, 222–228.
37. J. Li, X. Wang, J. Li, Q. Huang, X. Wang and S. Gamboa, *et al.*, Studies on preparation and performances of carbon aerogel electrodes for the application of supercapacitor, *J. Power Sources*, 2006, **158**, 784.
38. B. Yuan, C. Xu, D. Deng, Y. Xing, L. Liu and H. Pang, *et al.*, Graphene oxide/nickel oxide modified glassy carbon electrode for supercapacitor and nonenzymatic glucose sensor, *Electrochim. Acta*, 2013, **88**, 708–712.
39. K. Ge, H. Shao, P. L. Taberna and P. Simon, Understanding Ion Charging Dynamics in Nanoporous Carbons for Electrochemical Double Layer Capacitor Applications, *ACS Energy Lett.*, 2023, **8**(6), 2738–2745.
40. B. E. Conway and E. Gileadi, Kinetic theory of pseudo-capacitance and electrode reactions at appreciable surface coverage, *Trans. Faraday Soc.*, 1962, **58**, 2493–2509.
41. T. Brousse, D. Belanger and J. W. Long, To Be or Not To Be Pseudocapacitive?, *J. Electrochem. Soc.*, 2015, **162**(5), A5185–A5189.
42. V. Augustyn, P. Simon and B. Dunn, Pseudocapacitive oxide materials for high-rate electrochemical energy storage, *Energy Environ. Sci.*, 2014, **7**, 1597–1614.
43. B. O. Park, C. D. Lokhande, H. S. Park, K. D. Jung and O. S. Joo, Performance of supercapacitor with electrodeposited ruthenium oxide film electrodes—effect of film thickness, *J. Power Sources*, 2004, **134**(1), 148–152.
44. D. Xuan, W. Chengyang, C. Mingming, J. Yang and W. Jin, Electrochemical performances of nanoparticle Fe_3O_4/activated carbon supercapacitor using KOH electrolyte solution, *J. Phys. Chem. C*, 2009, **113**(6), 2643–2646.
45. R. Kumar, R. K. Singh, A. R. Vaz and S. A. Moshkalev, Microwave-assisted synthesis and deposition of a thin ZnO layer on microwave-exfoliated graphene: optical and electrochemical evaluations, *RSC Adv.*, 2015, **5**(83), 67988–67995.
46. W. Wei, X. Cui, W. Chen and D. G. Ivey, Manganese oxide-based materials as electrochemical supercapacitor electrodes, *Chem. Soc. Rev.*, 2011, **40**(3), 1697–1721.
47. L. Yuan, X. H. Lu, X. Xiao, T. Zhai, J. Dai and F. Zhang, *et al.*, Flexible Solid-State Supercapacitors Based on Carbon Nanoparticles/MnO_2 Nanorods Hybrid Structure, *ACS Nano*, 2011, **6**(1), 656–661.

48. T. C. Liu, W. G. Pell and B. E. Conway, Stages in the development of thick cobalt oxide films exhibiting reversible redox behavior and pseudocapacitance, *Electrochim. Acta*, 1999, **44**(17), 2829–2842.
49. R. Kumar, R. K. Singh, P. K. Dubey, D. P. Singh and R. M. Yadav, Self-assembled hierarchical formation of conjugated 3D cobalt oxide nanobead-CNT-graphene nanostructure using microwaves for high-performance supercapacitor Electrode, *ACS Appl. Mater. Interfaces*, 2015, **7**(27), 15042–15251.
50. R. Kumar, R. K. Singh, R. Savu, P. K. Dubey, P. Kumar and S. A. Moshkalev, Microwave-assisted synthesis of void-induced graphene-wrapped nickel oxide hybrids for supercapacitor applications, *RSC Adv.*, 2016, **6**(32), 26612–26620.
51. P. Kumari, K. Khawas, S. Nandy and B. K. Kuila, A supramolecular approach to Polyaniline graphene nanohybrid with three dimensional pillar structures for high performing electrochemical supercapacitor applications, *Electrochim. Acta*, 2016, **190**, 596–604.
52. H. Park, J. W. Kim, S. Y. Hong, G. Lee, D. S. Kim and J. H. Oh, *et al.*, Microporous Polypyrrole-Coated Graphene Foam for High-Performance Multifunctional Sensors and Flexible Supercapacitors, *Adv. Funct. Mater.*, 2018, **28**(33), 1–11.
53. G. A. Snook, P. Kao and A. S. Best, Conducting-polymer-based supercapacitor devices and electrodes, *J. Power Sources*, 2011, **196**(1), 1–12.
54. S. Trasatti and G. Buzzanca, Ruthenium dioxide: A new interesting electrode material. Solid state structure and electrochemical behaviour, *J. Electroanal. Chem. Interfacial Electrochem.*, 1971, **29**(2), A1–A5.
55. T. Brousse, D. Belanger and J. W. Long, To Be or Not To Be Pseudocapacitive?, *J. Electrochem. Soc.*, 2015, **162**(5), A5185–A5189.
56. M. R. Lukatskaya, B. Dunn and Y. Gogotsi, Multidimensional materials and device architectures for future hybrid energy storage, *Nat. Commun.*, 2016, 7, 12647.
57. T. Brousse, D. Bélanger and J. W. Long, To Be or Not To Be Pseudocapacitive?, *J. Electrochem. Soc.*, 2015, **162**(5), A5185–A5189.
58. S. Fleischmann, J. B. Mitchell, R. Wang, C. Zhan, D. Jiang and V. Presser, *et al.*, Pseudocapacitance: From Fundamental Understanding to High Power Energy Storage Materials, *Chem. Rev.*, 2020, **120**(14), 6738–6782.
59. D. P. Chatterjee and A. K. Nandi, A review on the recent advances in hybrid supercapacitors, *J. Mater. Chem. A*, 2021, **9**, 15880–15918.
60. S. Zhang and N. Pan, Supercapacitors performance evaluation, *Adv. Energy Mater.*, 2015, **5**, 1401401.
61. T. S. Mathis, N. Kurra, X. Wang, D. Pinto, P. Simon and Y. Gogotsi, Energy Storage Data Reporting in Perspective—Guidelines for Interpreting the Performance of Electrochemical Energy Storage Systems, *Adv. Energy Mater.*, 2019, **9**, 1902007.
62. A. Sarem-Aslani and K. Mullett, Industrial perspective on deep brain stimulation: History, current state, and future developments, *Front. Integr. Neurosci.*, 2011, **5**, 46.

63. Eterna™ SCS System, 2025. Available from: https://www.neuromodulation.abbott/.
64. Intellis™ Platform, 2025. Available from: https://www.medtronic.com/.
65. J. Parker, P. Single and J. L. Shils, Power use in neurostimulators, in *Essential Neuromodulation*, Elsevier, 2022, pp. 231–253.
66. W. Lemaire, M. Benhouria, K. Koua, W. Tong, G. Martin-Hardy, M. Stamp, *et al.* Feasibility Assessment of an Optically Powered Digital Retinal Prosthesis Architecture for Retinal Ganglion Cell Stimulation, *ArXiv*, 2020, arXiv:2010.12600, DOI: 10.48550/arXiv.2010.12600.
67. S. K. Kelly, D. B. Shire, J. Chen, P. Doyle, M. D. Gingerich and S. F. Cogan, *et al.*, Communication and control system for a 15-channel hermetic retinal prosthesis, *Biomed. Signal Process. Control*, 2011, **6**(4), 356–363.
68. F. G. Zeng, S. Rebscher, W. Harrison, X. Sun and H. Feng, Cochlear Implants: System Design, Integration, and Evaluation, *IEEE Rev. Biomed. Eng.*, 2008, **1**, 115–142.
69. H. A. Yigit, H. Ulusan, M. B. Yuksel, S. Chamanian, B. Ciftci, A. Koyuncuoglu, *et al.* A Pulse-Width Modulated Cochlear Implant Interface Electronics with 513 μW Power Consumption, in *2019 IEEE/ACM International Symposium on Low Power Electronics and Design (ISLPED)*, IEEE, 2019, p. 1–5.
70. A. Ghafari, E. Mohammadi, M. Dastjerdi, S. Honarmand, Z. A. Radmoghadam and S. Akbari, Batteries applications in the biomedical industry: A review, *Journal of Nanoscience and Technology*, 2023, **10**, 33–52.
71. Q. Zheng, Y. Wei, D. Li, Z. Zeng and Z. Li, Choosing energy sources for battery-free pacemakers, *Device*, 2025, **3**, 100746.
72. K. M. Huang, H. B. Qiu, Y. Deng, L. H. Wu and H. B. Chen, Capsule Endoscopy Technology: A New Era in Digestive Tract Examination, *J. Dig. Endosc.*, 2024, **15**(04), 243–249.
73. Q. Cao, R. Deng, Y. Pan, R. Liu, Y. Chen and G. Gong, *et al.*, Robotic wireless capsule endoscopy: recent advances and upcoming technologies, *Nat. Commun.*, 2024, **15**, 4597.
74. Dexcom G7, 2025. Available from: https://www.dexcom.com/.
75. FreeStyle Libre 3 Plus, 2025. Available from: https://www.freestyle.abbott/.
76. Galaxy Watch7, 2025. Available from: https://www.samsung.com/.
77. Apple Watch Series 10, 2025. Available from: https://www.apple.com/.
78. Garmin Venu 3, 2025. Available from: https://www.garmin.com/.
79. Z. Gao, Y. Zhou, J. Zhang, J. Foroughi, S. Peng and R. H. Baughman, *et al.*, Advanced Energy Harvesters and Energy Storage for Powering Wearable and Implantable Medical Devices, *Adv. Mater.*, 2024, **36**, 2404492.
80. Y. Jiao, E. He, T. Ye, Y. Wang, H. Yin and Y. Zhang, Implantable Batteries for Bioelectronics, *Acc. Mater. Res.*, 2025, **6**(2), 221–232.
81. C. V. V. Muralee Gopi, S. Alzahmi, V. Narayanaswamy, K. V. G. Raghavendra, B. Issa and I. M. Obaidat, A review on electrode materials

of supercapacitors used in wearable bioelectronics and implantable biomedical applications, *Mater. Horiz.*, 2025, **12**, 4092–4132.
82. M. B. Durukan, D. Keskin, Y. Tufan, O. Dincer, M. O. Cicek and B. Yildiz, *et al.*, An Edible Supercapacitor Based on Zwitterionic Soy Sauce-Based Gel Electrolyte, *Adv. Funct. Mater.*, 2024, **34**(6), 2307051.
83. Q. He, Z. C. Ling, D. H. Li, K. P. Yang, H. B. Yang and Z. K. Yan, *et al.*, Sargassum Nanocellulose-Based Fully Ingestible Supercapacitor, *Adv. Mater.*, 2025, **37**(9), 2416307.
84. E. Wang, M. Wu, L. Luo, X. Cui, L. Xu and R. Luo, *et al.*, Symbiotic biodegradable flexible supercapacitor in vivo, *Device*, 2025, 100724.
85. P. Kumar, E. Di Mauro, S. Zhang, A. Pezzella, F. Soavi and C. Santato, *et al.*, Melanin-based flexible supercapacitors, *J. Mater. Chem. C*, 2016, **4**(40), 9516–9525.
86. M. Pantrangi, E. Ashalley, M. K. Hadi, H. Xiao, Y. Zhang and W. Ahmed, *et al.*, Flexible micro-supercapacitors: Materials and architectures for smart integrated wearable and implantable devices, *Energy Storage Mater.*, 2024, **73**, 103791.
87. P. Luo, Q. Liu, R. Chen, H. Shao, Y. Ma and Y. Zhao, Recent progress in biocompatible miniature supercapacitors, *Energy Mater.*, 2025, **5**(7), 500070.
88. C. Meng, O. Z. Gall and P. P. Irazoqui, A flexible super-capacitive solid-state power supply for miniature implantable medical devices, *Biomed. Microdevices*, 2013, **15**(6), 973–983.
89. H. W. Park and K. C. Roh, Recent advances in and perspectives on pseudocapacitive materials for Supercapacitors–A review, *J. Power Sources*, 2023, **557**, 232558.
90. K. Liu, C. Yu, W. Guo, L. Ni, J. Yu and Y. Xie, *et al.*, Recent research advances of self-discharge in supercapacitors: Mechanisms and suppressing strategies, *J. Energy Chem.*, 2021, **58**, 94–109.
91. G. Liu, Z. Lv, S. Batool, M. Z. Li, P. Zhao and L. Guo, *et al.*, Biocompatible Material-Based Flexible Biosensors: From Materials Design to Wearable/Implantable Devices and Integrated Sensing Systems, *Small*, 2023, **19**, 2207879.
92. R. Ghaffari, J. A. Rogers and T. R. Ray, Recent progress, challenges, and opportunities for wearable biochemical sensors for sweat analysis, *Sens. Actuators, B*, 2021, **332**, 129447.
93. G. Lee, S. K. Kang, S. M. Won, P. Gutruf, Y. R. Jeong and J. Koo, *et al.*, Fully Biodegradable Microsupercapacitor for Power Storage in Transient Electronics, *Adv. Energy Mater.*, 2017, **7**(18), 1700157.
94. L. Yin, H. Cheng, S. Mao, R. Haasch, Y. Liu and X. Xie, *et al.*, Dissolvable metals for transient electronics, *Adv. Funct. Mater.*, 2014, **24**(5), 645–658.
95. M. Irimia-Vladu, E. D. Glowacki, N. S. Sariciftci and S. Bauer, *Green Materials for Electronics*, Wiley-VCH Verlag GmbH & Co. KGaA, Weinheim, Germany, 2017.
96. M. B. Durukan, M. O. Cicek, D. Doganay, M. C. Gorur, S. Çınar and H. E. Unalan, Multifunctional and Physically Transient Supercapacitors,

Triboelectric Nanogenerators, and Capacitive Sensors, *Adv. Funct. Mater.*, 2022, **32**(1), 1–12.
97. S. Greengard, Transient electronics take shape, *Commun. ACM*, 2018, **61**(10), 14–16.
98. K. K. Fu, Z. Wang, J. Dai, M. Carter and L. Hu, Transient Electronics: Materials and Devices, *Chem. Mater.*, 2016, **28**(11), 3527–3539.
99. B. Tawiah, R. K. Seidu, B. K. Asinyo and B. Fei, A review of fiber-based supercapacitors and sensors for energy-autonomous systems, *J. Power Sources*, 2024, **595**, 234069.
100. G. Xiao, J. Ju, M. Li, H. Wu, Y. Jian and W. Sun, *et al.*, Weavable yarn-shaped supercapacitor in sweat-activated self-charging power textile for wireless sweat biosensing, *Biosens. Bioelectron.*, 2023, **235**, 115389.
101. Z. Luo, Y. Wang, B. Kou, C. Liu, W. Zhang and L. Chen, "Sweat-chargeable" on-skin supercapacitors for practical wearable energy applications, *Energy Storage Mater.*, 2021, **38**, 9–16.
102. T. Sun, L. Shen, Y. Jiang, J. Ma, F. Lv and H. Ma, *et al.*, Wearable Textile Supercapacitors for Self-Powered Enzyme-Free Smartsensors, *ACS Appl. Mater. Interfaces*, 2020, **12**(19), 21779–21787.
103. J. A. Chiong, H. Tran, Y. Lin, Y. Zheng and Z. Bao, Integrating Emerging Polymer Chemistries for the Advancement of Recyclable, Biodegradable, and Biocompatible Electronics, *Adv. Sci.*, 2021, **8**, 2101233.
104. L. Jiang, Y. Li, F. Zou, D. Gan, M. Gao and L. Yuan, *et al.*, Highly self-adhesive, compressible, stretchable, all hydrogel-based supercapacitor for wearable/portable electronics, *Mater. Today Phys.*, 2023, **33**, 101046.
105. M. Xu, J. Zhu, J. Xie, Y. Mao and W. Hu, Dynamically Cross-Linked, Self-Healable, and Stretchable All-Hydrogel Supercapacitor with Extraordinary Energy Density and Real-Time Pressure Sensing, *Small*, 2024, **20**(10), 2305448.
106. S. J. Arbaz, B. Ramulu and J. S. Yu, Micro-Supercapacitors Based on Fungi-Derived Biocarbon Microfibers Infused with NiMoO Nanoparticles for Biomedical and E-Skin Applications, *Adv. Fiber Mater.*, 2024, **6**(4), 1008–1025.
107. M. L. Hammock, A. Chortos, B. C. K. Tee, J. B. H. Tok and Z. Bao, 25th anniversary article: The evolution of electronic skin (E-Skin): A brief history, design considerations, and recent progress, *Adv. Mater.*, 2013, **25**(42), 5997–6038.
108. J. Zhang, G. Zhang, T. Zhou and S. Sun, Recent Developments of Planar Micro-Supercapacitors: Fabrication, Properties, and Applications, *Adv. Funct. Mater.*, 2020, **30**, 1910000.
109. M. G. Say, I. Sahalianov, R. Brooke, L. Migliaccio, E. D. Głowacki and M. Berggren, *et al.*, Ultrathin Paper Microsupercapacitors for Electronic Skin Applications, *Adv. Mater. Technol.*, 2022, **7**(8), 2101420.
110. L. Wang, Y. Tang, Y. Li, C. Liu, N. Wei and W. Zeng, *et al.*, Multifunctional Integrated Interdigital Microsupercapacitors and Self-Powered Iontronic Tactile Pressure Sensor for Wearable Electronics, *ACS Appl. Mater. Interfaces*, 2022, **14**(41), 47136–47147.

111. C. Steiger, A. Abramson, P. Nadeau, A. P. Chandrakasan, R. Langer and G. Traverso, Ingestible electronics for diagnostics and therapy, *Nat. Rev. Mater.*, 2019, **4**, 83–98.
112. R. S. Mackay and B. Jacobson, Endoradiosonde, *Nature*, 1957, **179**(4572), 1239–1240.
113. J. T. Farrar, V. K. Zworykin and J. Baum, Pressure-Sensitive Telemetering Capsule for Study of Gastrointestinal Motility, *Science*, 1957, **126**(3280), 975–976.
114. P. N. Cutchis, A. F. Hogrefe and J. C. Lesho, The Ingestible Thermal Monitoring System, *Johns Hopkins APL Tecnical Dig.*, 1988, **9**(1), 16–21.
115. A. Abdigazy, M. Arfan, G. Lazzi, C. Sideris, A. Abramson and Y. Khan, End-to-end design of ingestible electronics, *Nat. Electron.*, 2024, **7**, 102–118.
116. I. K. Ilic, V. Galli, L. Lamanna, P. Cataldi, L. Pasquale and V. F. Annese, *et al.*, An Edible Rechargeable Battery, *Adv. Mater.*, 2023, **35**(20), 2211400.
117. Y. Wu, D. Ye, Y. Shan, S. He, Z. Su and J. Liang, *et al.*, Edible and Nutritive Electronics: Materials, Fabrications, Components, and Applications, *Adv. Mater. Technol.*, 2020, **5**, 2000100.
118. J. Kim, I. Jeerapan, B. Ciui, M. C. Hartel, A. Martin and J. Wang, Edible Electrochemistry: Food Materials Based Electrochemical Sensors, *Adv. Healthcare Mater.*, 2017, **6**(22), 1700770.
119. S. Y. Yang, V. Sencadas, S. S. You, N. Z. X. Jia, S. S. Srinivasan and H. W. Huang, *et al.*, Powering Implantable and Ingestible Electronics, *Adv. Funct. Mater.*, 2021, **31**(44), 2009289.
120. X. Wang, W. Xu, P. Chatterjee, C. Lv, J. Popovich and Z. Song, *et al.*, Food-Materials-Based Edible Supercapacitors, *Adv. Mater. Technol.*, 2016, **1**(3), 1600059.
121. R. Wang, M. Yao, S. Huang, J. Tian and Z. Niu, Sustainable Dough-Based Gel Electrolytes for Aqueous Energy Storage Devices, *Adv. Funct. Mater.*, 2021, **31**(14), 2009209.
122. X. Peng, H. Liu, Q. Yin, J. Wu, P. Chen and G. Zhang, *et al.*, A zwitterionic gel electrolyte for efficient solid-state supercapacitors, *Nat. Commun.*, 2016, **7**, 11782.
123. K. Fang, R. Wang, H. Zhang, L. Zhou, T. Xu and Y. Xiao, *et al.*, Mechano-Responsive, Tough, and Antibacterial Zwitterionic Hydrogels with Controllable Drug Release for Wound Healing Applications, *ACS Appl. Mater. Interfaces*, 2020, **12**(47), 52307–52318.
124. V. Galli, V. F. Annese, G. Coco, P. Cataldi, V. Scribano and I. K. Ilic, *et al.*, A Coplanar Edible Rechargeable Battery with Enhanced Capacity, *Adv. Mater. Technol.*, 2024, **9**(23), 2400715.
125. P. Cataldi, L. Lamanna, C. Bertei, F. Arena, P. Rossi and M. Liu, *et al.*, An Electrically Conductive Oleogel Paste for Edible Electronics, *Adv. Funct. Mater.*, 2022, **32**(23), 2113417.
126. L. Lamanna, G. Pace, I. K. Ilic, P. Cataldi, F. Viola and M. Friuli, *et al.*, Edible cellulose-based conductive composites for triboelectric nanogenerators and supercapacitors, *Nano Energy*, 2023, **108**, 108168.

127. C. Gao, C. Bai, J. Gao, Y. Xiao, Y. Han and A. Shaista, *et al.*, A directly swallowable and ingestible micro-supercapacitor, *J. Mater. Chem. A*, 2020, **8**(7), 4055–4061.
128. K. Chen, L. Yan, Y. Sheng, Y. Ma, L. Qu and Y. Zhao, An Edible and Nutritive Zinc-Ion Micro-supercapacitor in the Stomach with Ultrahigh Energy Density, *ACS Nano*, 2022, **16**(9), 15261–15272.
129. K. Jeffrey and V. Parsonnet, Cardiac Pacing, 1960-1985 A Quarter Century of Medical and Industrial Innovation, *Circulation*, 1998, **97**, 1978–1991.
130. L. K. Altman, A. H. W. Larsson, 86; Had First Internal Pacemaker, The New York Times, 2002.
131. M. Yu, Y. Peng, X. Wang and F. Ran, Emerging Design Strategies Toward Developing Next-Generation Implantable Batteries and Supercapacitors, *Adv. Funct. Mater.*, 2023, **33**, 2301877.
132. X. Huang, L. Wang, H. Wang, B. Zhang, X. Wang and R. Y. Z. Stening, *et al.*, Materials Strategies and Device Architectures of Emerging Power Supply Devices for Implantable Bioelectronics, *Small*, 2020, **16**, 1902827.
133. S. K. Kang, J. Koo, Y. K. Lee and J. A. Rogers, Advanced Materials and Devices for Bioresorbable Electronics, *Acc. Chem. Res.*, 2018, **51**(5), 988–998.
134. R. Singh, M. J. Bathaei, E. Istif and L. Beker, A Review of Bioresorbable Implantable Medical Devices: Materials, Fabrication, and Implementation, *Adv. Healthcare Mater.*, 2020, **9**, 2000790.
135. R. A. Huggins, Pulse Applications of Electrochemical Cells-Materials, *Lonics*, 1995, **1**, 5.
136. M. Block, D. Hammel, D. Bocker, M. Borggrefe, T. Seifert and H. H. Scheld, *et al.*, Internal Defibrillation with Smaller Capacitors: A Prospective Randomized Cross-Over Comparison of Defibrillation Efficacy Obtained with 90-/xF and 125-//F Capacitors in Humans, *J. Cardiovasc. Electrophysiol.*, 1995, **6**, 333–342.
137. R. J. Damiano, Implantable Cardioverter Defibrillators: Current Status and Future Directions, *J. Cardiovasc. Surg.*, 1992, **7**(1), 36–57.
138. C. D. Swerdlow, R. M. Kass, P. S. Chen, C. Hwang and S. Raissi, Effect of Capacitor Size and Pathway Resistance on Defibrillation Threshold for Implantable Defibrillators, *Circulation*, 1994, **90**, 1840–1846.
139. E. Alt, F. Evans, P. D. Wolf, J. Pasquantonio, L. Stotts and S. Heinz, *et al.*, Does reducing capacitance have potential for further miniaturisation of implantable defibrillators?, *Heart*, 1997, **77**, 234–237.
140. J. Hur, K. Im, S. W. Kim, U. J. Kim, J. Lee and S. Hwang, *et al.*, DNA hydrogel templated carbon nanotube and polyaniline assembly and its applications for electrochemical energy storage devices, *J. Mater. Chem. A*, 2013, **1**(46), 14460–14466.
141. J. Hur, K. Im, S. Hwang, B. Choi, S. Kim and S. Hwang, *et al.*, DNA hydrogel-based supercapacitors operating in physiological fluids, *Sci. Rep.*, 2013, **3**, 1282.
142. M. Xu, Y. Liu, K. Yang, S. Li, M. Wang and J. Wang, *et al.*, Minimally invasive power sources for implantable electronics, *Exploration*, 2024, **4**(1), 20220106.

143. Y. Liu, Q. Yu, L. Yang and Y. Cui, Materials and Biomedical Applications of Implantable Electronic Devices, *Adv. Mater. Technol.*, 2023, **8**, 2200853.
144. S. He, Y. Hu, J. Wan, Q. Gao, Y. Wang and S. Xie, *et al.*, Biocompatible carbon nanotube fibers for implantable supercapacitors, *Carbon*, 2017, **122**, 162–167.
145. J. S. Chae, H. Lee, S. H. Kim, N. R. Chodankar, S. M. Kang and S. Lee, *et al.*, A durable high-energy implantable energy storage system with binder-free electrodes useable in body fluids, *J. Mater. Chem. A*, 2022, **10**(9), 4611–4620.
146. Z. Qian, Y. Yang, L. Wang, J. Wang, Y. Guo and Z. Liu, *et al.*, An Implantable Fiber Biosupercapacitor with High Power Density by Multi-Strand Twisting Functionalized Fibers, *Angew. Chem., Int. Ed.*, 2023, **62**(28), e202303268.
147. H. J. Sim, C. Choi, D. Y. Lee, H. Kim, J. H. Yun and J. M. Kim, *et al.*, Biomolecule based fiber supercapacitor for implantable device, *Nano Energy*, 2018, **47**, 385–392.
148. Q. Lv, X. Li, X. Tian, D. A. Fu, H. Liu and J. Liu, *et al.*, A Degradable and Biocompatible Supercapacitor Implant Based on Functional Sericin Hydrogel Electrode, *Adv. Energy Mater.*, 2023, **13**(16), 2203814.
149. J. S. Shim, J. A. Rogers and S. K. Kang, Physically transient electronic materials and devices, *Mater. Sci. Eng., R*, 2021, **145**, 100624.
150. G. Lee, S. K. Kang, S. M. Won, P. Gutruf, Y. R. Jeong and J. Koo, *et al.*, Fully Biodegradable Microsupercapacitor for Power Storage in Transient Electronics, *Adv. Energy Mater.*, 2017, **7**(18), 1700157.
151. H. Sheng, J. Zhou, B. Li, Y. He, X. Zhang and J. Liang, *et al.*, A thin, deformable, high-performance supercapacitor implant that can be biodegraded and bioabsorbed within an animal body, *Sci. Adv.*, 2021, **7**, eabe3097.
152. M. Shao, H. Sheng, L. Lin, H. Ma, Q. Wang and J. Yuan, *et al.*, High-Performance Biodegradable Energy Storage Devices Enabled by Heterostructured MoO3–MoS2 Composites, *Small*, 2023, **19**(10), 2205529.
153. B. E. Conway, *Electrochemical Supercapacitors*, Springer US, Boston, MA, 1999.
154. S. Makino, T. Ban and W. Sugimoto, Towards Implantable Bio-Supercapacitors: Pseudocapacitance of Ruthenium Oxide Nanoparticles and Nanosheets in Acids, Buffered Solutions, and Bioelectrolytes, *J. Electrochem. Soc.*, 2015, **162**(5), A5001–A5006.
155. H. Li, C. Zhao, X. Wang, J. Meng, Y. Zou and S. Noreen, *et al.*, Fully Bioabsorbable Capacitor as an Energy Storage Unit for Implantable Medical Electronics, *Adv. Sci.*, 2019, **6**(6), 1801625.
156. Y. Liu, H. Zhou, W. Zhou, S. Meng, C. Qi and Z. Liu, *et al.*, Biocompatible, High-Performance, Wet-Adhesive, Stretchable All-Hydrogel Supercapacitor Implant Based on PANI@rGO/Mxenes Electrode and Hydrogel Electrolyte, *Adv. Energy Mater.*, 2021, **11**(30), 2101329.

157. W. Tian, Y. Li, J. Zhou, T. Wang, R. Zhang and J. Cao, *et al.*, Implantable and Biodegradable Micro-Supercapacitor Based on a Superassembled Three-Dimensional Network Zn@PPy Hybrid Electrode, *ACS Appl. Mater. Interfaces*, 2021, **13**(7), 8285–8293.
158. X. Huang, Y. Liu, G. W. Kong, J. H. Seo, Y. Ma and K. I. Jang, *et al.*, Epidermal radio frequency electronics for wireless power transfer, *Microsyst. Nanoeng.*, 2016, **2**, 16052.
159. H. Sheng, L. Jiang, Q. Wang, Z. Zhang, Y. Lv and H. Ma, *et al.*, A soft implantable energy supply system that integrates wireless charging and biodegradable Zn-ion hybrid supercapacitors, *Sci. Adv.*, 2023, **9**(46), eadh8083.

CHAPTER 9

Electromagnetic Waves and Biological Tissue Interactions: Risks and Opportunities in Wireless Power Transfer

H. DINIS*[a,b] AND P. M. MENDES[a,b]

[a] Center for MicroElectromechanical Systems (CMEMS-Uminho), University of Minho, 4800-058 Guimarães, Portugal; [b] LABBELS–Associate Laboratory, 4710-057 Braga, Portugal
*Email: hdinis@dei.uminho.pt

9.1 Introduction

Electromagnetic (EM) fields in the GHz range produce low-energy photons and, therefore, under ordinary conditions, their energy is too low to break chemical bonds, excite neutrons, or ionize biological molecules. Consequently, they are often referred to as nonionizing radiation.[1]

Although living organisms have thrived in natural electromagnetic environments, they are nowadays subjected to a plethora of human-made EM fields, which are used for communication, industry, and medical applications, among others. Besides their primary goals, these fields can also produce secondary effects that may influence the vital activities of an organism. These effects may or may not manifest themselves immediately after exposure and depend on many physical and biological factors.[1]

The interaction of EM fields with biological tissues has long been a subject of scientific study, ever since their use in therapeutic applications. Since

Detection Science Series No. 33
Energy Materials for Ingestible and Implantable Healthcare Devices
Edited by Santanu Patra and Anja Boisen

Published by the Royal Society of Chemistry, www.rsc.org

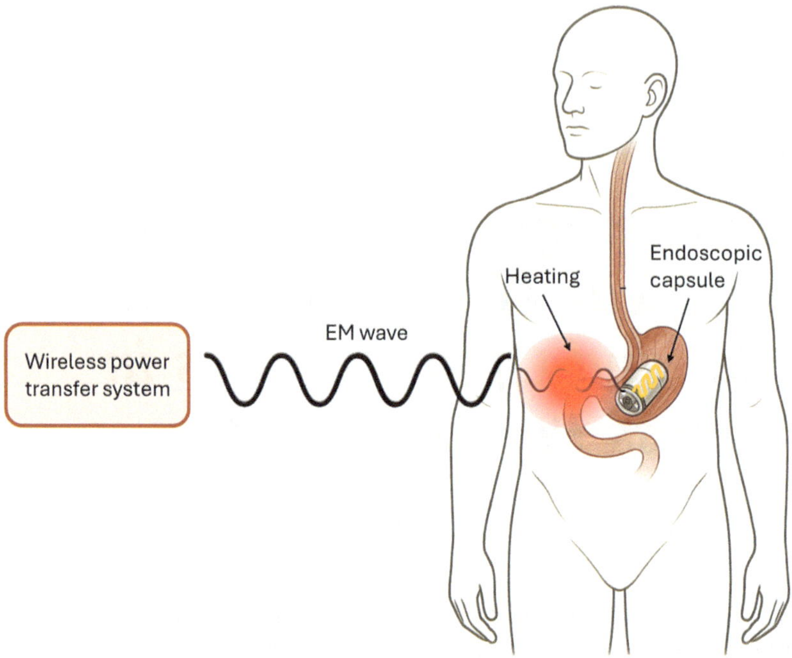

Figure 9.1 Illustration of the interaction between an EM wave used for wireless power transfer and the human body. It represents the attenuation of the EM wave as it propagates in the biological tissue to reach an endoscopic capsule, and the wave's lost energy being converted to heat.

then, the knowledge on these interactions has increased massively, but this field is still far from being fully understood. The widespread use of EM sources in telecommunications (Wi-Fi, Bluetooth, 5G, WPT technologies, *etc.*) has increased attention to the lack of understanding of the mechanism of interaction between EM waves and biological tissues.[1]

With the increased miniaturization of implantable and ingestible devices, which grants them new functionalities and applications, comes the trade-off of having smaller batteries. As such, wireless power transfer (WPT) becomes an interesting technology, as it can be used to meet the energy requirements of these devices and, potentially, allow for further miniaturization.[2] Since WPT often relies on EM waves, understanding the interaction of these waves with biological tissue, with one type of interaction being exemplified in Figure 9.1, becomes a necessity both for safety and optimization purposes.

9.1.1 Maxwell's Equations

Maxwell's work was based on the vast knowledge developed by Gauss, Ampere, Faraday, and others. Maxwell's equations describe the characteristics and behaviour of EM fields at the macroscopic levels. They provided a unification of the state of EM science at that time and hypothesized the existence of the electrical displacement current, which led Hertz to discover EM propagation.[3]

EM waves propagate in biological tissues, and this interaction can be described by this set of four equations. Maxwell's equations, in their differential form, are presented as:[3]

$$\nabla \times H = J + \frac{\partial D}{\partial t} \tag{9.1}$$

$$\nabla \times E = -\frac{\partial B}{\partial t} \tag{9.2}$$

$$\nabla \cdot B = 0 \tag{9.3}$$

$$\nabla \cdot D = \rho \tag{9.4}$$

The quantities in these equations are defined as follows:

- H is the magnetic field, in Ampere per meter ($\mathrm{A\,m^{-1}}$);
- J is the electric current density, in Ampere per meter squared ($\mathrm{A\,m^{-2}}$);
- D is the electric flux density, in Coulomb per meter squared ($\mathrm{C\,m^{-2}}$);
- E is the electric field, in Volt per meter ($\mathrm{V\,m^{-1}}$);
- B is the magnetic flux density, in Weber per meter squared ($\mathrm{Wb\,m^{-2}}$); and
- ρ the electric charge density, in Coulomb per meter cubed ($\mathrm{C\,m^{-3}}$).

Eqn (9.1) describes that a circulating magnetic field is produced by an electric field that changes with time or by an electric current. Eqn (9.2) states that variations of a magnetic field over time generate a circulating electric field. Eqn (9.3) illustrates that there is no divergence of the magnetic field, as no magnetic monopoles (therefore sources or sinks) exist. Eqn (9.4) defines that the divergence of the electric field is caused by the presence of charges (which act as sources or sinks).[4]

Maxwell's constitutive equations are defined as:[3]

$$B = \mu H \tag{9.5}$$

$$D = \varepsilon E \tag{9.6}$$

$$J = \sigma E \tag{9.7}$$

where μ is the permeability in Henry per meter ($\mathrm{H\,m^{-1}}$), ε the permittivity in Farad per meter ($\mathrm{F\,m^{-1}}$), and σ the conductivity ($\Omega^{-1}\,\mathrm{m^{-1}}$) of the propagation medium.

The conductivity of free space (vacuum) is 0 $\Omega^{-1}\,\mathrm{m^{-1}}$, and its permittivity and permeability are, respectively, given by:[1]

$$\varepsilon_0 = 8.854 \times 10^{-12} \tag{9.8}$$

$$\mu_0 = 4\pi \times 10^{7} \tag{9.9}$$

The relative permittivity and permeability of free space are both 1. Biological entities such as cells, tissues, or organs have a relative permeability of 1, except for those containing ferromagnetic particles, such as red blood cells and the liver. However, the relative permittivity varies significantly between tissue types and with the temperature of the tissue and the frequency of the EM wave.[1]

In a lossy material, such as biological tissue, the wave is attenuated as it propagates through the tissue. The magnitude of the propagating wave as a function of distance and time can be described as:[5]

$$E(z,t) = E_0 e^{-\alpha z} \sin(\omega t - \beta z - \phi) \tag{9.10}$$

The attenuation constant α (Np m^{-1}) shows how quickly the wave is attenuated as it propagates through the medium. The electric field inside the medium can be expressed as a function of the propagation distance by $E_0 e^{-\alpha z}$, where E_0 is the magnitude of the incident field at the body surface, and z is the propagation distance (see Figure 9.2).

The attenuation constant α is dependent on the field's frequency and the material's permittivity and conductivity, and can be calculated as:[6]

$$\alpha = \omega \left\{ \frac{\mu \varepsilon'}{2} \left[\sqrt{1 + \left(\frac{\varepsilon''}{\varepsilon'} \right)^2} - 1 \right] \right\}^{\frac{1}{2}} \tag{9.11}$$

and the phase constant β is given by:

$$\beta = \omega \left\{ \frac{\mu \varepsilon'}{2} \left[\sqrt{1 + \left(\frac{\varepsilon''}{\varepsilon'} \right)^2} + 1 \right] \right\}^{\frac{1}{2}} \tag{9.12}$$

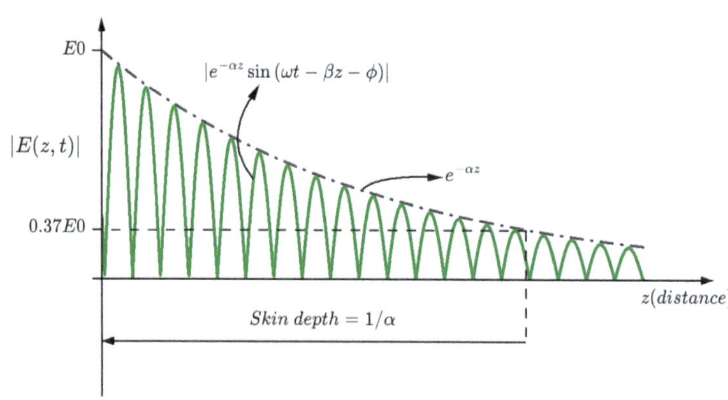

Figure 9.2 Magnitude of the wave $E(z, t)$ travelling through a medium. The peak magnitude of this wave in a lossy material decreases exponentially as a function of distance travelled z. Its envelope is given by $e^{-\alpha z}$.

From the attenuation constant, α, it is possible to determine the penetration depth, or skin depth, δ, which is the depth of tissue by which the electric field amplitude is reduced by e^{-1} (63%):[6]

$$\delta = \frac{1}{\alpha} \tag{9.13}$$

For the propagation of a wave, it is also possible to define a phase velocity, which is the velocity at which a fixed phase point on the wave travels in a medium with relative permittivity ε_r and relative permeability μ_r, as:

$$v_p = \frac{1}{\sqrt{\mu\varepsilon}} = \frac{1}{\sqrt{\mu_r\mu_0\varepsilon_r\varepsilon_0}} = \frac{1}{\sqrt{\mu_r\varepsilon_r}\sqrt{\mu_0\varepsilon_0}} = \frac{c}{\sqrt{\mu_r\varepsilon_r}} \tag{9.14}$$

where $c = 2.998 \times 10^8$ m s^{-1} is the speed of light.

Finally, the wavelength of the wave is given by:

$$\lambda = \frac{c}{f\sqrt{\mu_r\varepsilon_r}} \tag{9.15}$$

Nevertheless, given the complexity and nonhomogeneous nature of biological tissues, it is not trivial to entirely characterize the propagation of EM waves in this medium. Fortunately, modern computation techniques allow for very accurate approximations to be made with complex simulation methods.

9.2 Dielectric Properties of Biological Tissues

The permeability of biological media is, for all practical purposes, equal to that of free space, as previously mentioned. Therefore, the interaction of microwaves with these media may be considered solely as the influence of electric fields. In that scenario, an energy exchange occurs between the microwave and the medium's free charges and structures with asymmetrical charge distributions (dipoles). The energy exchange can also induce dipoles.

The relative permittivity ε_r of a medium is a complex quantity:[7]

$$\varepsilon_r = \frac{\varepsilon}{\varepsilon_0} = \varepsilon_r' - j\varepsilon_r'' \tag{9.16}$$

As an electric field oscillates in the tissue, the dipoles rotate to keep aligned with the field. The rotation of the dipoles is hindered by the frictional resistance of the viscous biological medium, and it relates to the imaginary part of the permittivity, ε''_r. The larger the imaginary part, the higher the losses in the medium, since more energy is being expended by the dipole through motions, resulting in a temperature increase of the medium and greater attenuation of the electric field.[7]

The conductivity σ of the medium can be related to ε_r'':[7]

$$\varepsilon_r'' = \frac{\sigma}{\omega\varepsilon_0} \tag{9.17}$$

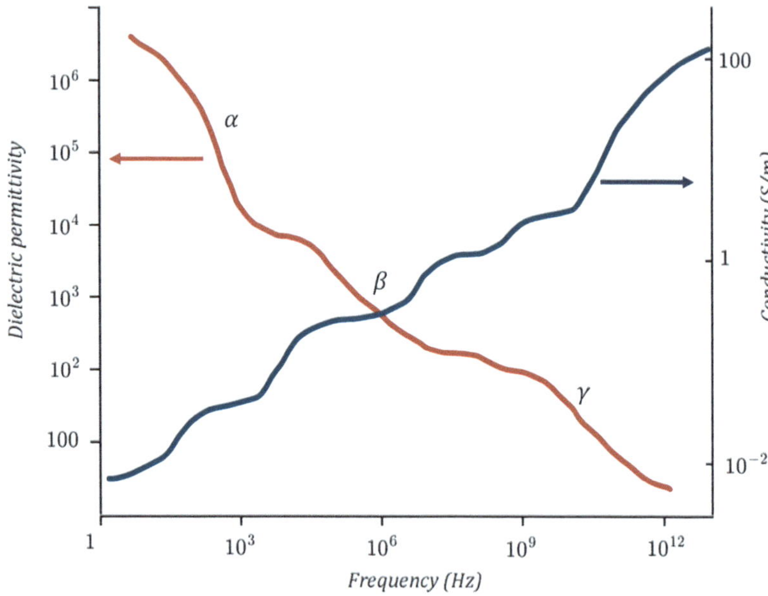

Figure 9.3 Dielectric constant and conductivity over a wide frequency range of a tissue with muscle-like properties, where different dispersion regions can be identified.

and the loss tangent is given by:[7]

$$\tan \delta = \frac{\varepsilon_r''}{\varepsilon_r'} = \frac{\sigma}{\omega \varepsilon_0 \varepsilon_r'} \tag{9.18}$$

Databases of tissue electrical properties, such as IT'IS Foundation,[8] usually provide the loss tangent ($\tan \delta$) and the real part of the relative permittivity ε_r'.

A biological tissue's permittivity and conductivity vary with frequency, and this variation is exemplified with a muscle-like tissue's properties (see Figure 9.3).

At low frequencies, biological tissues exhibit very high dielectric constants due to the cell's membrane capacitance. An applied electric field at these frequencies causes charges to accumulate at the boundaries of the membranes. As frequency increases, the membrane becomes progressively more short-circuited, since it acts as a capacitor, allowing for more current conduction. As such, the conductivity increases with frequency. For even higher frequencies, water molecules' rotational and vibrational properties become the most significant factor and viscous loss occurs, thus leading to higher conductivities.[1] It is, therefore, possible to distinguish three dispersion regions in the graphic:

- Alpha dispersion – variability of frequency of the cell membrane capacitance; very high dielectric constant and low conductivity;

- Beta dispersion – relaxation due to the charging of cell membranes; and
- Gamma dispersion – dielectric relaxation of free water; high conductivity and lower dielectric constant.[9]

9.3 Radiofrequency Field Coupling to Biological Systems

The interaction of time-varying EM fields with biological systems depends on the geometry and dielectric properties of the tissue, but it is also influenced by the EM wave source's configuration and the field's frequency. These factors dictate the penetration and deposition of power, the induction of electric and magnetic fields, and the absorption of energy from the EM wave in biological tissue. The location of the radiation source relative to the biological medium is also important, as it dictates what kind of interaction the radiated waves have with the tissue.[1]

9.3.1 Coupling in the Near Field

The use of RF radiation for wireless communication leads to frequent exposure of human bodies to RF waves in the near-field zone, especially with smartphones, as the reactive near-field distance is approximately 5 cm for a dipole antenna at 2400 MHz.[1] The fields in the near-field of an antenna are quasistatic,[10] therefore they must be analysed as such.

A sphere of radius r, permittivity ε, and conductivity σ is in the near field of a source that generates an external applied uniform electric field E polarized in the x direction. As such, a uniform induced electric field E_E, also polarized in the x direction, appears inside the sphere due to the incident electric field E_1:[1]

$$E_E = \left(\frac{\omega \varepsilon_0}{\sigma}\right) E_1 \tag{9.19}$$

The interaction of time-varying electric fields with biological tissues results in the flow of current in the tissue, the generation of electric dipoles, and the reorientation of the dipoles already present in the tissue.[11] As seen in the previous equation, the induced electric field inside the sphere is independent of its size r. On the other hand, the amplitude of the induced electric field relates to the amplitude of the incident field by a factor that depends on the conductivity of the tissue, σ.

The magnetic permeability of biological tissues is generally 1, as previously mentioned, and the magnetic field inside the body is equal to the one outside. As such, an incident magnetic field H_1 generates an electric field E_H inside the body whose magnitude is given by:[1]

$$E_H = \left(\frac{\omega r \mu}{2}\right) H_1 \tag{9.20}$$

The incident magnetic field induces an electric field that varies with the distance from the centre of the body, r, and results in circulating electric currents.[11]

When both quasistatic electric and magnetic fields are present, the ratio between the electrically and magnetically induced electric fields is a function of body size r and the incident field ratio (E_1/H_1):[1]

$$\frac{E_E}{E_H} = \left(\frac{2\varepsilon_0}{\sigma\mu_0 r}\right)\cdot\left(\frac{E_1}{H_1}\right) \tag{9.21}$$

9.3.2 Coupling in the Far Field

In the far field, the propagating radiated RF wave has a planar wavefront and the electric and magnetic fields are related through a factor η, the intrinsic impedance of the medium.[1]

At boundaries between different propagation mediums, an RF wave is either reflected or transmitted. For a plane wave that is travelling in medium A with intrinsic impedance η_A that impinges normally on medium B with intrinsic impedance η_B, the reflection and transmission coefficients, R and T, are given by:[1]

$$R = \frac{\eta_B - \eta_A}{\eta_B + \eta_A} \tag{9.22}$$

$$T = \frac{2\eta_B}{\eta_B + \eta_A} \tag{9.23}$$

$$T = 1 + R \tag{9.24}$$

In terms of power, the fraction of the incident wave's power that is reflected is R^2, and the fraction of transmitted power is:[1]

$$T^2 = \left(1 - R^2\right) \tag{9.25}$$

For media with similar intrinsic impedances, most of the power is transmitted, while for media with very different intrinsic impedances, most of the power is reflected at the interface.

As the RF wave propagates through a tissue, part of its energy is lost and deposited in the tissue, resulting in a power density reduction as the field propagates. This reduction is quantified by the penetration depth which, in terms of power, is the depth at which the power has been reduced by a factor of e^{-2}. Penetration depth is dependent on the permittivity of the medium and the frequency of the RF wave, as seen in eqn (9.11) and (9.13).

As previously mentioned, the geometry of the biological tissue influences the way EM waves propagate in the tissue. Suppose the largest dimension of the medium is similar to the wavelength of the incident RF field. In that case, energy deposition and distribution will be influenced by the surface

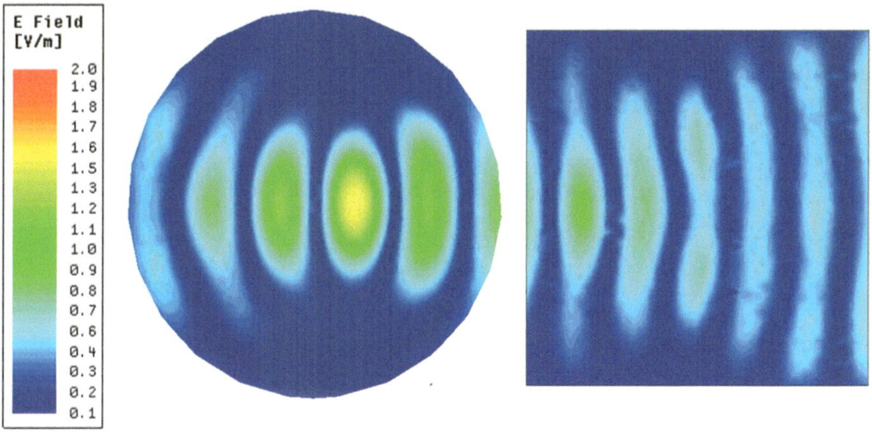

Figure 9.4 Cross-section of the electric field component of the EM wave inside a sphere and a cube with permittivity of 9. Both bodies are radiated by a 2 GHz EM wave with a planar wavefront. It is possible to distinguish a hotspot that appears in the middle of the spherical volume due to its shape, which does not happen in the object with a planar surface, where the wave propagates in a planar manner.

curvature of the body.[1] An example of this is provided in Figure 9.4, where a planar wavefront EM wave impinges on a sphere and a cube with permittivity 9. As can be seen, the electric field distribution inside the sphere generates hotspots inside the homogeneous tissue which do not occur with planar surfaces.

9.4 Interference Between Co-polar Coherent Fields

Assuming two sources in free-space at distances r_1 and r_2 from a point P, the electric field components of the two waves in P after travelling for r_1 and r_2 are:[12]

$$E_1 = E_{01} \sin\left(k_w r_1 - \omega t\right) \tag{9.26}$$

$$E_2 = E_{02} \sin\left(k_w r_2 - \omega t\right) \tag{9.27}$$

where $k_w = \dfrac{2\pi}{\lambda}$.

The amplitude E_{0r} of the electric field component of the resultant wave in P is:[12]

$$E_{0r} = \sqrt{E_{01}{}^2 + E_{02}{}^2 + 2E_{01}E_{02}\cos\varphi} \tag{9.28}$$

$$\varphi = \frac{2\pi}{\lambda}\left(r_1 - r_2\right) \tag{9.29}$$

At points P where $\varphi = 2\pi n$ $(n = 0, 1, 2, 3, \ldots)$:[12]

$$E_{0r} = \left|E_{01} + E_{02}\right| \tag{9.30}$$

When N polarized coherent electromagnetic sources with the same frequency and amplitude are considered, and knowing that the electric field amplitude of the waves emitted by these sources are $E_{01}, E_{02}, \ldots E_{0N}$ at the location of constructive interference, the resulting electric field becomes:[12]

$$E_{0r} = E_{01} + E_{02} + E_{03} + \cdots + E_{0N} \tag{9.31}$$

Consequently, for N identical sources:[12]

$$E_{0r} = NE_{0s} \tag{9.32}$$

Applying the same reasoning for magnetic field components, we arrive at:[12]

$$H_{0r} = NH_{0s} \tag{9.33}$$

The average power of the EM wave from each of the identical sources at the point of constructive interference is calculated as:

$$S_{0s} = \frac{1}{2}\left(E_{0s} \times H_{0s}^{\star}\right) \tag{9.34}$$

and the average power of the resulting wave is:

$$S_{0r} = \frac{1}{2}\left(E_{0r} \times H_{0r}^{\star}\right) = \frac{1}{2}\left(NE_{0s} \times NH_{0s}^{\star}\right) = \frac{N^2}{2}\left(E_{0s} \times H_{0s}^{\star}\right) = N^2 S_{0s} \tag{9.35}$$

In conclusion, when N identical sources are considered, the average power at the location of constructive interference between these source's EM waves increases by the Nth power.

9.5 Biological Effects of Electromagnetic Exposure

Before presenting this section's bibliographical review, it is important to distinguish between the biological and health effects of EM exposure. A biological effect occurs when the external EM field causes a physiological change in a living system. Such a change may or may not lead to a health effect. This happens when physiological change is outside the normal range that the human body can compensate for and correct. Health effects are, usually, the result of accumulated biological effects over time.[6]

The interaction of EM waves with biological systems can be studied at two levels: macroscopically (objects and whole body) and microscopically (cells, membranes and molecules). Nevertheless, interaction at the two levels cannot be considered independently. The effects of the interaction of EM energy with biological tissues is a result of three phenomena:[6]

- The penetration of EM waves into the biological medium and its propagation inside it;
- The interaction of the waves with the tissues (primary interaction); and
- The secondary effects due to the primary interaction.

9.5.1 Interactions at the Microscopic Level

Examples of the interaction of EM waves with biological tissue at the molecular and cellular levels will be described here, along with the effect this interaction bears in tissues and living organisms.

Critical biomolecules for biological function are either electrically charged or polarized.[12] When exposed to an external polarized EM field, a coherent oscillation will be induced on each of these molecules. This oscillation is most noticeable on free ions, resulting in irregular gating of electrosensitive ion channels on the cell membranes.[12] Panagopoulos *et al.*[12] theorized that the oscillation of free ions located in the vicinity of ion channels could exert enough force on said channels to disrupt normal functioning.

In the case of extremely low-frequency waves, such as power line signals (50 Hz), a field with 5 $mV m^{-1}$ can potentially disrupt normal cell function. For high-frequency signals, extremely high field intensities would be required to produce the same effect (for a 1 GHz signal, E_0 would have to be 100.000 $V m^{-1}$).[12] Nevertheless, when high-frequency fields are pulsed in telecommunication applications, the E_0 may also be very low. For example, for a pulse frequency of 217 Hz, used in GSM communication, E_0 is around 20 $mV m^{-1}$.[12] The result of the irregular gating of the ion channels is the disruption of the cell's electrochemical balance, which can potentially lead to a plethora of secondary effects.[12]

The collision rates of molecules inside the biological medium can also be affected by a polarized electric field. Considering a chemical reaction that occurs, for example, in the cell membrane between a protein and an ion, the rate at which this reaction occurs is proportional to the flow of ions that collide with the protein and carry enough energy to start the reaction. By introducing an external influence on the flow of the ions, the reaction rate is tampered with, which influences the normal behaviour of the cell.[13] Additionally, the collision angle between two molecules often dictates whether a reaction does or does not happen. In the case of proteins, this angle can be minimal. By orienting all molecules according to an external polarized field, this mechanism is influenced.[13]

9.5.2 Interactions at the Macroscopic Level

An example of a macroscopic effect of RF exposure is the microwave-induced auditory effect, where humans perceive pulse-modulated fields with frequencies between 200 MHz and 6.5 GHz. The pulsed signals have been described as clicks, pops, or buzzes, depending on the modulation characteristics.[11] In ref. 14, microwave heating is attributed to the rapid expansion of water present in the tissue, and consequent acoustic wave generation, due to the local heating of the tissue; even though the amount of heating is very little (in the order of microdegrees), the rate of heating is high (in the order of 1 $^{\circ}C s^{-1}$). Behaviour changes have also been extensively reported, but these were attributed to thermal effects.[15]

9.5.3 Uncertainty Regarding the Biological Effects of EM Waves

After a literature scan looking for biological and health effects of EM exposure, it would be expected that a conclusion could be made regarding the topic. However, such is not the case here. Different studies have produced contrasting results. While some authors argue that EM exposure is responsible for several ailments and biological alterations, other authors claim that no convincing and accurate data exist to lead to such conclusions and that the observed effects are mainly due to thermal mechanisms.

In an extensive literature review, Sage[16] summarized a series of scientific articles in which several biological effects occur under continuous and pulsed RF radiation, such as damage to DNA, chromosome aberrations, the expression of specific genes, induced responses in nervous tissue cells, reduced immune function, changes in the integrity of the blood–brain barrier, reduced fertility, higher cancer incidence, learning deficit, and sleep disruption.

Others question such results, especially when considering the low intensity of the fields. Chou[17] collected a series of papers where reported non-thermal effects were related to a temperature increase due to artefacts in the experiments. For example, behaviour studies in rats concluded that RF exposure at a low level induced behaviour changes due to non-thermal effects. It was later shown that the local heating in the rats' legs was very high when they were drinking from a metallic nozzle, thus provoking the behaviour changes. Adair *et al.*[15] state that tissue heating is an important effect of RF exposure of biological organisms and that it has been unequivocally demonstrated. They then also agree with the previous author, in the regards that most papers claiming non-thermal effects are reporting on artefacts and not accurate data, further adding that, at the time of writing of the paper, controversial claims of non-thermal effects of modulated RF signals are not substantiated experimentally nor have they been independently replicated.

The best example of the duality of available results was found in Kato's book,[18] where the author reviewed the literature to find that, for example: carcinogenesis does not seem to be affected by mobile phone signals; no concrete evidence shows that microwave exposure changes the microscopic anatomy of regions of the brain; contradicting results have been obtained regarding non-thermal effects on the blood–brain barrier, even though studies reporting an effect were later reproduced and did not observe that same effect; neurotransmitter function in the CNS is only altered under high exposure levels, *i.e.* due to the thermal impact; and the effects on nerves are caused primarily by heating, similarly to hormones.

The International Committee on Electromagnetic Safety (ICES) Technical Committee TC95, which was responsible for the IEEE C95.1 EM exposure guidelines,[19] reports that they have considered studies that link EM waves with adverse health effects when developing the safety guidelines. Some of their conclusions can be highlighted:

- The World Health Organization (WHO) analysed the 2002 International Agency for Research on Cancer report that linked exposure to ELF magnetic fields to childhood leukaemia, and concluded that the evidence is not strong enough to be considered causal;
- At higher frequencies, brain cancer has been associated with the use of mobile phones, but the epidemiologic evidence did not provide sufficient basis for concluding that brain cancer in adults is positively associated with mobile phone use and, consequently, RF exposure;
- The WHO indicated that no adverse health effects have been linked to mobile phone use;
- The International Agency for Research on Cancer (IARC) classified RF EM fields as possibly carcinogenic to humans, under Group 2B, which is used when causal association is considered credible, but when chance, bias, or inaccuracies cannot be ruled out with confidence; and
- The International Commission on Non-Ionizing Radiation Protection (ICNIRP) Standing Committee on Epidemiology concluded that evidence accumulates against the hypothesis that mobile phone use causes brain tumours in adults, and that no plausible biophysical mechanisms have explained the hypothesized effects at low-level exposure.

Consequently, ICES concluded that only the established mechanisms of electrostimulation for signals of frequencies between 0 Hz and 100 kHz and tissue heating between 100 kHz to 300 GHz needed to be considered to provide a valid basis for the exposure safety guidelines they developed,[19] which will be further explained in a following section of this chapter.

As a closing remark for this section, to the best of our knowledge it is impossible to present a clear and scientifically bulletproof conclusion to this topic. The mechanisms that can possibly yield non-thermal effects were presented, and examples of such effects were provided. On the other hand, claims that such effects are not proven at low exposure levels have also been reported. The only topic in which there is a consensus is the occurrence of thermal effects in biological tissue due to high-frequency exposure, and electrostimulation in low-frequency exposure, which are the parameters by which safety exposure limits and guidelines base themselves. Thus, following these safety guidelines is the best one can do when designing a system that exposes biological tissue to EM radiation.

9.6 Safety Electromagnetic Exposure Standards

Exposure standards are adopted standards of personal protection that refer to maximum levels to which whole or partial body exposure is permitted from EM field sources. Examples of such standards are given by the Institute of Electrical and Electronics Engineers (IEEE) (Standard C95.1-2019),[19] International Commission on Non-Ionizing Radiation Protection (ICNIRP),[11] and US Federal Communications Commission (FCC).[20] These standards were developed independently and constitute the foundations to most national exposure limits, except for China, Russia, and some Eastern European countries.[13]

As stated in the IEEE C95.1-2019 documentation:[19] "The purpose of this standard is to provide science-based exposure criteria to protect against established adverse health effects in humans associated with exposure to electric, magnetic, and electromagnetic fields; induced and contact currents; and contact voltages, over the frequency range of 0 Hz to 300 GHz", while ICNIRP (1998) states that it aims to "establish guidelines for limiting EMF exposure that will provide protection against known adverse health effects".[11]

The exposure standards began with a comprehensive and critical scientific review of available literature by a panel of experts who evaluated the literature, determined threshold levels, identified safety factors, and derived the exposure limits. For example, IEEE C95.1-2019 is a 310-page document with 1550 references. ICNIRP was developed with a similar extensive review.[13]

IEEE C95.1-2019 and ICNIRP cover the frequency range up to 300 GHz. Updates have been issued to both standards as time has gone by. For example, in 2010 ICNIRP applied a new set of guidelines to electric and magnetic fields from 1 Hz to 100 kHz and revised the existing guidelines up to 10 MHz.[21]

The institutions that create the guidelines keep a close watch on studies regarding the safety of EM exposure, to keep the standards up to date. For example, the ICNIRP reviewed two major studies regarding the carcinogenicity of mobile phones and base stations in rats.[22] Both studies concluded that male rats had increased cancer rates under exposure, and one of them also concluded that the effect was not registered in female rats or mice. In both cases, the ICNIRP found significant weaknesses such as "lack of blinding, difficulty interpreting statistical data and a failure to account for chance", thus ruling out the results as meaningful for concluding on RF fields and carcinogenesis.

There have also been efforts towards the creation of reports, guidelines, recommendations, and standards specifically about the assessment of human exposure to wireless power transfer, of which examples are ITU-R Recommendation SM.2151,[23] that delves into WPT *via* RF waves to portable devices and sensor networks, ITU-R Report ITU-R-REP-SM-2505,[24] about human hazards of RF WPT, and IEEE Standard for Assessment of Human Exposure to Electromagnetic Fields from Radiative Wireless Power Transfer Systems.[25]

To understand safety exposure guidelines, it is essential to distinguish between exposure (fields measured outside the body) and dose (induced fields within the body) when assessing exposure to RF waves.

Exposure measurements are:[13]

- External field strength – electric field ($V\,m^{-1}$), magnetic field ($A\,m^{-1}$), or magnetic flux density B (T). External field strength is usually more useful in the lower end of the frequency range, where exposures occur in the near field of the sources and the electric and magnetic fields are essentially independent; and

- Incident power density, S ($\mathrm{W\,m^{-2}}$) – useful in the far field of radiators, where exposure consists of EM fields propagating through space.

Dose measurements are:[13]

- Current density within the body J_i ($\mathrm{A\,m^{-2}}$) – mainly useful at low frequencies. It is related to the *in situ* internal electric field strength E_i ($\mathrm{V\,m^{-1}}$) by Ohm's law (eqn (9.7));
- Contact current J (A) – current passed into the body through contact with an external charged objected; and
- Specific absorption rate (SAR) ($\mathrm{W\,kg^{-1}}$) – SAR is a measure of the energy deposition rate in the body. Given its importance, it will be detailed in a separate section, which follows.

9.7 Specific Absorption Rate

SAR is a measurement of how much energy from the RF wave is absorbed by the tissue, therefore it is a great way to assess thermal effects. SAR cannot be measured directly, but it is related to the RMS of the local electric field strength E:[13]

$$\mathrm{SAR} = \frac{\sigma E^2}{\rho} \tag{9.36}$$

where σ ($\mathrm{S\,m^{-1}}$) is the conductivity and ρ is the mass density of the tissue.

SAR can be defined at single points in space, and it is averaged over a defined mass (for example, a cube of 1 gram of tissue), or over the whole body. Whole-body SAR is significant at estimating elevations of the core body temperature. The whole-body SAR of an organism will be highest at a particular resonant frequency range, which is a function of the body size and its orientation relative to the incident fields. For an average human, peak whole-body SAR occurs at 60–80 MHz. Local (or spatial) peak SAR is used when absorption takes place in a confined body region (*e.g.*, when estimating head exposure to a mobile phone's radiation).[6]

The biological effects of RF exposure are typically correlated to induced increases in tissue temperature, and SAR allows evaluating that parameter. Considering that SAR is a measurement of the power absorbed by a given mass of tissue:[26]

$$\mathrm{SAR} = \frac{\mathrm{d}P}{\mathrm{d}m} \tag{9.37}$$

and, considering and homogeneous medium with specific heat c ($\mathrm{J\,kg^{-1}\,K^{-1}}$), a form of the calorimetry law comes as:[26]

$$\frac{\mathrm{d}Q}{\mathrm{d}t} = mc\frac{\mathrm{d}T}{\mathrm{d}t} \tag{9.38}$$

The term dQ/dt is the wave power transformed into an incremental amount of heat dQ within the tissue of mass m that produces a temperature increase dT during time dt. Combining the two previous equations:[26]

$$\mathrm{SAR} = c\frac{\mathrm{d}T}{\mathrm{d}t} \tag{9.39}$$

However, the use of SAR for the evaluation of the effects of human exposure to EM fields is not free of criticism, and some authors point to issues and considerations they believe should be addressed that are not covered by this parameter. For example, Panagopoulos *et al.*[26] raise some concerns about the sole use of SAR as a criterion to decide whether or not an RF signal will cause harm, as the authors base themselves on the studies indicating that the interaction of specific biomolecules and EM waves is what dictates the biological effect of the exposure. Nevertheless, this discussion dwells on the subject previously discussed about the uncertainty in this matter. In their excellent bibliographical reviews, independent organizations (IEEE and ICNIRP) came up with very similar exposure safety guidelines, and until solid evidence is provided against them, such guidelines should be followed.

9.8 Guidelines

The exposure limits are defined differently according to the frequency of the applied EM field. In the most recent IEEE standard on safety levels for human exposure, at the time of writing of this chapter, IEEE C95.1-2019:[19]

- 0 Hz to 5 MHz: electrostimulation is the primary concern, therefore the maximum *in situ* electric field or magnetic field are the controlling criteria, which vary with body region.
- 5 MHz to 300 GHz: only thermal effects apply, and SAR and power density incident on the body are the controlling parameters; and
- 100 kHz to 5 MHz: both electrostimulation and thermal effects can occur, therefore protection against both is provided by two separate sets of limits.

The tables that follow were adapted from IEEE C.95.1-2019,[19] and they provide the dosimetric reference limits (DRLs) (see Table 9.1) and exposure reference levels (ERLs) (see Tables 9.2–9.5) for whole-body and local exposures in restricted and unrestricted environments for the specified frequency ranges.

Whole-body Exposure ERLs

It is important to note that for frequencies between 100 kHz and 6 GHz, compliance with the limits defined in Tables 9.2 and 9.3 guarantees compliance with the DRL limits defined in Table 9.1. However, failing to meet the requirements set in Tables 9.2 and 9.3 does not imply a failure to comply with the DRLs of Table 9.1, for which further assessment is required.[19] Additionally, as long as the DRLs are not exceed, the ERLs can be exceeded.

Table 9.1 Dosimetric reference limits (100 kHz to 6 GHz).[19]

Conditions	Unrestricted environment SAR ($W\,kg^{-1}$)[a]	Restricted environment SAR ($W\,kg^{-1}$)[a]
Whole-body exposure	0.08	0.4
Local exposure (head and torso)[b]	2	10
Local exposure (limbs and pinnae)[b]	4	20

[a] SAR is averaged over 30 min for whole-body exposure and 6 min for local exposure.
[b] Averaged over a 10 g mass of tissue in the shape of a cube.

Table 9.2 Whole-body exposure ERLs (100 kHz to 300 GHz) in unrestricted environments.[19a]

Frequency range (MHz)	Electric field strength [b,c] ($V\,m^{-1}$)	Magnetic field strength [b,c] ($A\,m^{-1}$)	Power density $S^{b,c}$ ($W\,m^{-2}$)	
			S_E	S_H
0.1–1.34	614	$16.3/f_M$	1000	$100\,000/f_M^2$
1.34–30	$823.8/f_M$	$16.3/f_M$	$1800/f_M^2$	$100\,000/f_M^2$
30–100	27.5	$158.3/f_M^{1.668}$	2	$9\,400\,000/f_M^{3.336}$
100–400	27.5	0.0729	2	
400–2000	—	—	$f_M/200$	
2000–300 000	—	—	10	

[a] Note – S_E and S_H are plane-wave-equivalent power density values based on electric or magnetic field strength, respectively.
[b] Averaging time 30 min.
[c] f_M is the frequency in MHz.

Table 9.3 Whole-body exposure ERLs (100 kHz to 300 GHz) in restricted environments.[19a]

Frequency range (MHz)	Electric field strength [b,c] ($V\,m^{-1}$)	Magnetic field strength [b,c] ($A\,m^{-1}$)	Power density $S^{b,c}$ ($W\,m^{-2}$)	
			S_E	S_H
0.1–1.34	1842	$16.3/f_M$	1000	$100\,000/f_M^2$
1.34–30	$1842/f_M$	$16.3/f_M$	$1800/f_M^2$	$100\,000/f_M^2$
30–100	61.4	$16.3/f_M$	2	$100\,000/f_M^2$
100–400	61.4	0.163	10	
400–2000	—	—	$f_M/40$	
2000–300 000	—	—	50	

[a] Note – S_E and S_H are plane-wave-equivalent power density values based on electric or magnetic field strength, respectively.
[b] Averaging time 30 min.
[c] f_M is the frequency in MHz.

Therefore, it suffices to demonstrate compliance with either the DRLs of Table 9.1 or the ERLs of Table 9.2 or 9.3.[19]

It is noteworthy that the limits above 6 GHz are given in terms of field strength and power density, as SAR does not apply.[19]

9.9 Additional Considerations

In scenarios where medical implants are present in the body, extra caution is advised, as extremely high local fields can occur, which may not be detectable in a 10-g averaged SAR.[19] ICNIRP even suggest that, in such cases, the time average of the SAR should be under 1 minute, and with a smaller averaging mass to try to detect the localized exposure.[27]

When the exposure happens in the near field of the EM radiation source, SAR is also well correlated with temperature increase, as stated in ref 13.

In the presence of different fields at several frequencies for which there are different values of the ERL, the percentage of the ERL incurred within each frequency interval must be determined, and the sum of all percentages must not exceed 100%.

9.9.1 Exposure to Pulsed RF Waves

Local exposures to pulsed RF waves of frequencies from 100 kHz to 300 GHz have a specific set of limits in both restricted and unrestricted environments. Peak power density limits are provided as follows to prevent high local exposure.[19]

The total incident energy density of a pulse train in any period of 100 ms is:[19]

$$\sum_{i=1}^{n} \left(S_{i\ \mathrm{peak}} \times \tau_i\right) \tag{9.40}$$

where:

- n is the number of pulses in the 100 ms period;
- $S_{i\,\mathrm{peak}}$ is the average (temporal) power density, averaged over τ_i, of the ith pulse ($\mathrm{W\,m^{-2}}$); and
- τ_i is the width (s) of the ith pulse (or a partial pulse width if any 100 ms period captures only a portion of a pulse).

Compliance is established with:[19]

$$\sum_{i=1}^{n} \left(S_{i\ \mathrm{peak}} \times \tau_i\right) \leq \frac{\mathrm{ERL_{local}} \times t_{\mathrm{avg}}}{5} \tag{9.41}$$

where:

- t_{avg} is 6 min; and
- $\mathrm{ERL_{local}}$ is the ERL for a local exposure at a given frequency (see Tables 9.4 and 9.5). The summation on the left side of eqn (9.41) applies to any 100 ms of exposure.

Table 9.4 Local exposure ERLs (100 kHz to 6 GHz) in unrestricted environments.[19a,b]

Frequency range (MHz)	Electric field strength [c,d,e] (V m^{-1})	Magnetic field strength [c,d,e] (A m^{-1})	Power density S [c,d,e] (W m^{-2})	
			S_E	S_H
0.1–1.34	1373	$36.4/f_M$	5000	$500\,000/f_M^2$
1.34–30	$1842/f_M$	$36.4/f_M$	$9000/f_M^2$	$500\,000/f_M^2$
30–100	61.4	$353/f_M^{1.668}$	10	$47\,000\,000/f_M^{3.336}$
100–400	$21.2/f_M^{0.232}$	$0.056/f_M^{0.232}$	$1.19/f_M^{0.463}$	
400–2000	—	—	$1.19/f_M^{0.463}$	
2000–6000	—	—	40	

[a] Note 1 – Below 6 GHz, portable devices are typically tested for DRL compliance (SAR).
[b] Note 2 – S_E and S_H are plane-wave-equivalent power density values based on electric or magnetic field strength, respectively.
[c] Determined in air at the location of the body surface.
[d] Spatial and temporal peaks averaged over 6 min.
[e] f_M is the frequency in MHz.

Table 9.5 Local exposure ERLs (100 kHz to 6 GHz) in restricted environments.[19a,b]

Frequency range (MHz)	Electric field strength [c,d,e] (V m^{-1})	Magnetic field strength [c,d,e] (A m^{-1})	Power density S [c,d,e] (W m^{-2})	
			S_E	S_H
0.1–1	4119	$36.4/f_M$	45 000	$500\,000/f_M^2$
1–30	$4119/f_M$	$36.4/f_M$	$9000/f_M^2$	$500\,000/f_M^2$
30–100	137.3	$36.4/f_M$	10	$500\,000/f_M^2$
100–400	$47.3/f_M^{0.232}$	$0.125/f_M^{0.232}$	$5.93/f_M^{0.463}$	
400–2000	—	—	$5.93/f_M^{0.463}$	
2000–6000	—	—	200	

[a] Note 1 – Below 6 GHz, portable devices are typically tested for DRL compliance (SAR).
[b] Note 2 – S_E and S_H are plane-wave-equivalent power density values based on electric or magnetic field strength, respectively.
[c] Determined in air at the location of the body surface.
[d] Spatial and temporal peaks averaged over 6 min.
[e] f_M is the frequency in MHz.

This guideline is also equivalent to limiting the specific absorption (SA) (J kg^{-1}) below 6 GHz to one-fifth of what is allowed for a continuous field:[19]

$$\sum_{i=1}^{n} \left(\mathrm{SAR}_{i\ \mathrm{peak}} \times \tau_i\right) \leq \frac{\mathrm{DRL}_{\mathrm{local}} \times t_{\mathrm{avg}}}{5} \tag{9.42}$$

where:

- $\mathrm{SAR}_{i\,\mathrm{peak}}$ is the temporal peak SAR (W kg^{-1}) of the ith pulse;
- t_{avg} is 6 min; and
- $\mathrm{DRL}_{\mathrm{local}}$ is the SAR limit (see Table 9.1).

With the previous equations, it is possible to conclude that using pulsed signals with the intent of powering an implantable device would result in a smaller amount of energy being delivered to the device throughout the exposure rather than when a continuous wave is used. However, since the instantaneous power of a pulsed wave can be higher than that of a continuous one, a pulsed WPT can be used when the power delivered by the continuous wave is insufficient to, for example, polarize the diodes of the implant's RF-DC module. Nonetheless, special care must be given when designing such an application, as pulsed waves may elicit abnormal biological effects.[28]

For frequencies below 30 GHz, two effects of high peak energy deposition were reported: microwave heating and the stun reaction (loss of consciousness). Due to these effects, IEEE C.95.1-2019[19] states that the peak power density limits should be further limited for local exposures in both restricted and unrestricted environments for pulsed RF fields, as previously noted.

9.10 Wireless Power Transfer to Implantable and Ingestible Devices

The possibility of transferring power to implants *via* an EM wave has been extensively studied by researchers, since this technology can extend the lifetime of implants that require batteries to operate and reduce the number of surgeries a patient is submitted to by eliminating or reducing the need to replace said batteries. Additionally, the size of implantable devices is often a limiting factor in their applicability. Some locations of the human body have space available for larger devices, such as the chest, where pacemakers are implanted, and the digestive tract, where ingestible pills can travel. However, other locations such as the brain do not have this characteristic, requiring miniaturized implants. Providing energy to these devices is a challenge, since the traditional energy source of implants, *i.e.*, batteries, is ineffective. Even though battery technology has evolved significantly, the miniaturization of batteries still is not sufficient[2] as the energy densities that can be achieved are insufficient[29] and fabrication and integration with microsystems remains a challenge.[30]As such, WPT is a potential solution to enable miniaturized implants and maximize their lifetime.

Examples of works that report on implantable devices with wireless power transfer capabilities are presented in Table 9.6. Techniques such as inductive coupling (IC), midfield WPT (MF), and far-field WPT (FF) are presented, and they distinguish themselves by the distance at which the transmitting antenna is from the implant. In IC, a magnetic field generated in the transmitter coil induces an electrical current in the implant's coil. In MF, a focusing plate is used to concentrate EM waves on the predicted location of the implant. In FF, the antennas can be placed at the furthest distance from the implant, however the efficiency is typically the lowest of the three mentioned techniques, as seen in Table 9.6. Nevertheless, efforts

Table 9.6 Examples of implantable devices with wireless power transfer capabilities.

Ref.	Application	WPT technology	Frequency (MHz)	Power@ implant (μW)	Efficiency (%)
32	Oximetry	Inductive coupling	13.56	12 000	0.3
33	Pacemaker	Inductive coupling	13.56	—	0.03
34	Neuromodulator	Inductive coupling	1180	55.5	0.002
35	Glucose sensor	Inductive coupling	900	5	0.1
36	Neuromodulator	Inductive coupling	13.56	15 000	—
37	Neuromodulator	Inductive coupling	1500	10 650	0.33
38	Pacemaker	Midfield	1600	450	0.056
39	Pacemaker	Midfield	1600	195	0.04
40	—	Midfield	1200	123	0.03
41	Neuromodulator	Far-field	2340	100	0.005

towards the increase of the efficiency of FF WPT have been reported. For example, the interactions between co-polar coherent fields were explored to greatly increase the power transferred to implants while complying with exposure guidelines.[31]

9.11 Conclusion

The interaction of EM waves and biological tissue is a complex and demanding field that has been studied ever since the use of this radiation was proposed for therapeutic purposes. As seen in this chapter, there is not yet a consensus amongst academics of what the consequences of the interaction are for the human body. Some suggest EM waves can contribute to a plethora of health problems; others refute these claims.

In this chapter it was not possible to draw an unquestionable conclusion about the effects of EM radiation on our health. However, several independent organizations have thoroughly studied the available literature and developed standards based on scientific evidence which scientists and engineers must abide by when developing EM applications.

From the analysis presented in this chapter, some conclusions were drawn about WPT link parameters that can be manipulated to increase the power delivered to an implantable device safely. According to ref. 19 and eqn (9.41), the use of pulsed fields is not beneficial to the subject at hand, as the total energy delivered to the IMD over the averaging time is one-fifth of the energy provided by a continuous wave. Pulsed signals can, however, be of interest when high power is required to polarize the electronic components of the implant (*e.g.*, a rectifier's diodes). On the other hand, it is known that the SAR is calculated in small, predefined volumes of the propagation medium with the electric field intensity in that volume. Together with the knowledge that the EM wave attenuates as it propagates through the tissues, it is expected that the maximum SAR occurs near the surface of the medium. It becomes apparent that if the EM fields come from different locations around the medium, multiple SAR hotspots will occur, but it is likely that they won't be higher than the SAR on the surface. Finally, knowing that the constructive superposition of several coherent polarized fields can significantly increase the power in the medium, multiple EM waves should originate power hotspots inside the medium, which could be used to power an implantable device.

The optimization of the WPT link, together with the reduction of the energy requirements of miniaturized implantable and ingestible devices, may lead to the elimination of the need for batteries, allowing for even further device miniaturization. These efforts pave the way for wireless sensor networks inside the human body, enabling ubiquitous monitoring and control of biological and health parameters. However, as discussed in this chapter, the interaction of EM waves with biological tissue and the associated safety precautions impose limitations. As such, research efforts towards increasing the power transfer while keeping within safety regulations are expected to play a pivotal role in this field.

References

1. J. Lin, *Electromagnetic Fields in Biological Systems*, F. Barnes and B. Greenebaum, ed. CRC Press, Boca Raton, FL, 1st edn, 2012.
2. H. Dinis and P. M. Mendes, A comprehensive review of powering methods used in state-of-the-art miniaturized implantable electronic devices, *Biosens. Bioelectron.*, 2020, **2021**(172), 112781.
3. D. Pozar, *Microwave engineering*, John Wiley and Sons, Inc., Hoboken, NJ, 4th edn, 2011.
4. D. Fleisch, *A Student's Guide to Maxwell's Equations*, Cambridge University Press, New York, 1st edn, 2008.
5. F. Ulaby, E. Michielssen and U. Ravaioli, *Fundamentals of Applied Electromagnetics*, Prentice Hall, 6th edn, 2010.
6. R. W. Y. Habash Bioeffects and therapeutic applications of electromagnetic energy, *Bioeffects and Therapeutic Applications of Electromagnetic Energy*, CRC Press, Boca Raton, FL, 1st edn, 2007, pp. 1–379.
7. S. Baranski and P. Czerski, *Biological effects of microwaves*, Hutchinson & Ross, Dowden, 1986.
8. IT'IS Foundation. Dielectric Properties, 2020. Available from: https://itis.swiss/virtual-population/tissue-properties/database/dielectric-properties/.
9. J. Lin, *Electromagnetic Interaction with Biological Systems*, Plenum Press, New York, 1st edn, 1989.
10. C. A. Balanis, *Antenna Theory: Analysis and Design*, John Wiley and Sons, Inc., New Jersey, 3rd edn, 2005.
11. International Commission on Non-Ionizing Radiation Protection (ICNIRP), ICNIRP Guidelines for Limiting Exposure To Time-Varying Electric, Magnetic and Electromagnetic Field (up to 300 GHz), *Health Phys.*, 1998, **74**, 494–522.

12. D. J. Panagopoulos, O. Johansson and G. L. Carlo, Polarization: A Key Difference between Man-made and Natural Electromagnetic Fields, in regard to Biological Activity, *Sci. Rep.*, 2015, **5**(September), 1–10.
13. B. Greenebaum and F. Barnes, *Bioengineering and Biophysical Aspects of Electromagnetic Fields*, CRC Press, Boca Raton, FL, 4th edn, 2019.
14. K. R. Foster and R. Glaser, Thermal mechanisms of interaction of radiofrequency energy with biological systems with relevance to exposure guidelines, *Health Phys.*, 2007, **92**(6), 609–620.
15. E. R. Adair and R. C. Petersen, Biological effects of radio-frequency/microwave radiation, *IEEE Trans. Microwave Theory Tech.*, 2002, **50**(3), 953–962.
16. C. Sage, An Overview of Radiofrequency / Microwave Radiation Studies Relevant to Wireless Communications and Data. Proceeding of International Conference on Cell Tower Siting–Lingking Science & Public Health, 2000, pp. 513–521.
17. C. K. Chou, A need to provide explanations for observed biological effects of radiofrequency exposure, *Electromagn. Biol. Med.*, 2015, **34**(3), 175–179.
18. M. Kato, *Electromagnetics in Biology*, Springer, Japan, 1st edn, 2006.
19. Institute of Electrical and Electronics Engineers (IEEE), IEEE standard for safety levels with respect to human exposure to electric, magnetic, and electromagnetic fields, 0 Hz to 300 GHz, IEEE Std C95.1-2019. 2019, vol. 2019, pp. 1–312.
20. FCC, OET bulletin 65: Evaluating compliance with FCC guidelines for human exposure to radiofrequency electromagnetic fields, 1997, p. 65.
21. International Commission on Non-Ionizing Radiation Protection (ICNIRP), Guidelines for limiting exposure to time-varying electric and magnetic fields (1 Hz TO 100 kHz), *Health Phys.*, 2010, **99**(6), 818–836.
22. ICNIRP, ICNIRP Note - Critical Evaluation Of Two Radiofrequency Electromagnetic Fields Animal Carcinogenicity Studies Published In 2018, *Health Phys.*, 2019, **118**, 1.
23. International Telecommunications Union, Recommendation ITU-R SM.2151-0 - Guidance on frequency ranges for operation of wireless power transmission *via* radio frequency beam for mobile/portable devices and sensor networks, 2022.
24. International Telecommunications Union, Report ITU-R SM.2505-0 - Impact studies and human hazard issues for wireless power transmission *via* radio frequency beam, 2022.
25. IEEE, Standard for Assessment of Human Exposure to Electromagnetic Fields from Radiative Wireless Power Transfer Systems: Measurement and Computational Methods (Frequency Range of 30 MHz to 300 GHz), 2023. Available from: https://standards.ieee.org/ieee/63480/11319/.
26. D. J. Panagopoulos, O. Johansson and G. L. Carlo, Evaluation of Specific Absorption Rate as a Dosimetric Quantity for Electromagnetic Fields Bioeffects, *PLoS ONE*, 2013, **8**(6).
27. Z. Sienkiewicz, E. Van Rongen, R. Croft, G. Ziegelberger and B. Veyret, A closer look at the thresholds of thermal damage: workshop report by an ICNIRP task group, *Health Phys.*, 2016, **111**(3), 300–306.

28. S. Romanenko, R. Begley, A. R. Harvey, L. Hool and V. P. Wallace, The interaction between electromagnetic fields at megahertz, gigahertz and terahertz frequencies with cells, tissues and organisms: Risks and potential, *J. R. Soc., Interface*, 2017, **14**(137).
29. S. S. Karipott, B. D. Nelson, R. E. Guldberg and K. G. Ong, Clinical potential of implantable wireless sensors for orthopedic treatments, *Expert Rev. Med. Devices*, 2018, **15**(4), 255–264.
30. D. Karnaushenko, T. Kang, V. K. Bandari, F. Zhu and O. G. Schmidt, 3D Self-Assembled Microelectronic Devices: Concepts, Materials, Applications, *Adv. Mater.*, 2020, **32**(15), 1902994.
31. H. Dinis, I. Colmiais and P. M. Mendes, Smart wireless powering enabling IoT in inhomogeneous environments: a case study on biomedical applications, *IEEE Internet Things J.*, 2022, 1–1.
32. H. Zhang, P. Gutruf, K. Meacham, M. C. Montana, X. Zhao and A. M. Chiarelli, *et al.*, Wireless, battery-free optoelectronic systems as subdermal implants for local tissue oximetry, *Sci. Adv.*, 2019, **5**(3), eaaw0873.
33. H. Lyu, M. John, D. Burkland, B. Greet, A. Post and A. Babakhani, *et al.*, Synchronized Biventricular Heart Pacing in a Closed-chest Porcine Model based on Wirelessly Powered Leadless Pacemakers, *Sci. Rep.*, 2020, **10**(1), 2067.
34. A. Khalifa, Y. Liu, Y. Karimi, Q. Wang, A. Eisape and M. Stanaćević, *et al.*, The Microbead: A 0.009 mm3 Implantable Wireless Neural Stimulator, *IEEE Trans. Biomed. Circuits Syst.*, 2019, **13**(5), 971–985.
35. M. Mujeeb-U-Rahman, M. H. Nazari, M. Sencan and W. V. Antwerp, A Novel Needle-Injectable Millimeter scale Wireless Electrochemical Glucose Sensing Platform for Artificial Pancreas Applications, *Sci. Rep.*, 2019, **9**(1), 17421.
36. Y. Zhang, A. D. Mickle, P. Gutruf, L. A. McIlvried, H. Guo and Y. Wu, *et al.*, Battery-free, fully implantable optofluidic cuff system for wireless optogenetic and pharmacological neuromodulation of peripheral nerves, *Sci. Adv.*, 2019, **5**(7), eaaw5296.
37. K. L. Montgomery, A. J. Yeh, J. S. Ho, V. Tsao, S. Mohan Iyer and L. Grosenick, *et al.*, Wirelessly powered, fully internal optogenetics for brain, spinal and peripheral circuits in mice, *Nat. Methods*, 2015, **12**(10), 969–974.
38. D. R. Agrawal, Y. Tanabe, D. Weng, A. Ma, S. Hsu and S. Y. Liao, *et al.*, Conformal phased surfaces for wireless powering of bioelectronic microdevices, *Nat. Biomed. Eng.*, 2017, **1**(3), 0043.
39. J. S. Ho, A. J. Yeh, E. Neofytou, S. Kim, Y. Tanabe and B. Patlolla, *et al.*, Wireless power transfer to deep-tissue microimplants, *Proc. Natl. Acad. Sci. U. S. A.*, 2014, **111**(22), 7974–7979.
40. A. Abid, J. M. O'Brien, T. Bensel, C. Cleveland, L. Booth and B. R. Smith, *et al.*, Wireless Power Transfer to Millimeter-Sized Gastrointestinal Electronics Validated in a Swine Model, *Sci. Rep.*, 2017, **7**(1), 46745.
41. S. I. Park, D. S. Brenner, G. Shin, C. D. Morgan, B. A. Copits and H. U. Chung, *et al.*, Soft, stretchable, fully implantable miniaturized optoelectronic systems for wireless optogenetics, *Nat. Biotechnol.*, 2015, **33**(12), 1280–1286.

CHAPTER 10

Reconfigurable Wireless Power Transfer for Long-lasting Deep-body Bioelectronics

ICARO VELOSO SOARES,* ERDEM CIL AND DENYS NIKOLAYEV*

IETR – CNRS, Univ Rennes, Rennes, France
*Emails: icaro.soares@ieee.org; d@deniq.com

10.1 Introduction

Even though the development of electronic devices engineered to interact with the human body is not recent, the full potential of bioelectronics is yet to be unleashed. In fact, more than half a century has passed since the first pacemaker was clinically implanted in a human.[1] From then on, implantable biomedical devices have become smaller as time passes while the number of applications progressively increases. For instance, nowadays, bioelectronic devices have been applied to control or regulate the functioning of an organ or region of the body, such as cardiac,[2–5] neural,[6–9] spinal,[10,11] and cochlear stimulators.[12] In addition, they can also be employed to monitor biological processes, acquire and process vital sign data, and provide information for precise diagnostics, as in biotelemetry[13,14] and ingestible gastrointestinal endoscopy capsules.[15–17] Furthermore, groundbreaking bioelectronics applications have been recently developed for drug delivery,[18] prosthetic devices,[19] and brain–computer interfaces.[6] In this way, a technological paradigm is imposed on the development of

Detection Science Series No. 33
Energy Materials for Ingestible and Implantable Healthcare Devices
Edited by Santanu Patra and Anja Boisen

Published by the Royal Society of Chemistry, www.rsc.org

bioelectronics: developing ultra-miniaturized and low-power devices that can collect and transmit a large amount of data and simultaneously perform as many functions as possible.

In contrast with the groundbreaking potential of bioelectronics, the major constraint to their further development is that they still strongly rely on batteries to be powered. First, these batteries are bulky; thus, the bioelectronic device size is mainly dictated by the number of power cells and their dimensions. Apart from that, batteries require frequent maintenance to be replaced or recharged. In the context of implantable devices, these maintenances are carried out through invasive surgeries, increasing the patient's risk and making the treatment more expensive. Therefore, developing battery-free solutions like those summarized in Figure 10.1 would lead to significant miniaturization of implantable and ingestible bioelectronics, besides enhancing the range of their potential applications. One attempt towards battery-free bioelectronics is harvesting the energy from biological processes, such as biofuel cells,[20,21] in which the electric power is generated from glucose oxidation or even by converting kinetic energy from body movement[22,23] or thermal energy[24] into electric energy. However, the average power densities (10^{-1}–10^{1} $\mu W\,cm^{-2}$[25]) and, consequently, the harvested power levels are low and even more reduced due to losses in the converting process. Therefore, it is applied mainly to power ultra-low-power devices.

An alternative to achieve higher average power densities ($\sim 10^{2}$ $\mu W\,cm^{-2}$[26]) and deliver higher power levels to battery-free bioelectronics or to recharge implantable devices in a non-invasive manner is through wireless power

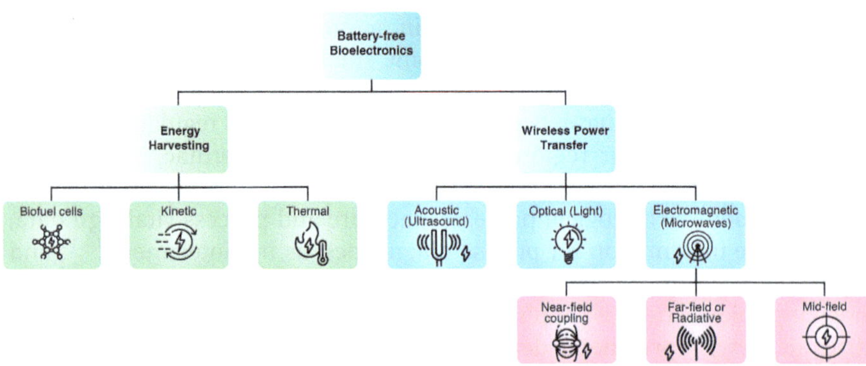

Figure 10.1 The most common approaches for powering battery-free bioelectronics. Energy-harvesting techniques generate power from ambient sources (*e.g.*, metabolic reactions, motion, and heating). Wireless power transfer (WPT) directly transmits energy from a source to a receiver without physical connections, thus being carried out by mechanical or electromagnetic waves. The three main microwave WPT techniques are near-field coupling, far-field radiation, and mid-field focusing.

transfer (WPT). The concept of wireless powering is not novel, with the first attempts being carried out by Nikola Tesla more than 100 years ago. Nevertheless, its application in biomedical engineering is still recent, with a growth in the development of wireless-powered bioelectronics being noticed in the last decade. In this context, an antenna or a radiating structure external to the body sends energy through electromagnetic waves in the microwave frequency range to the implanted receiver. The received power is converted to a continuous voltage that supplies the bioelectronic circuits or recharges its battery. Besides powering, this setup enables bio-telemetry applications once the implant can send valuable information and data back to the external antenna.

Most wireless-powered bioelectronics are based on conventional WPT techniques, and their choice is directly related to the nature of the application. For instance, near-field WPT is widely employed for subcutaneous applications in which the transmitter and receiver are close to each other. Once this method is based on the inductive or capacitive coupling between radiators, the delivered power is sharply reduced as the distance between them increases or in case of misalignment. On the other hand, radiative or far-field WPT can exhibit lower sensitivity to positional variations. Nevertheless, due to its operation at higher frequencies, the transmitted power experiences substantial attenuation in lossy body tissues, resulting in a notable decrease in the maximum achievable efficiency. Consequently, it is better suited for powering low-consumption implants. More recently, a new technique that combines the advantages of near-field coupling and far-field radiation has been developed, achieving considerable efficiency levels for deep-body implanted bioelectronics. This mid-field WPT approach allows the focusing of the electromagnetic field on a target, thus increasing the power delivered to the implant.

Therefore, the goal of this chapter is to present the main adaptive WPT techniques for achieving optimal wireless powering with a focus on deep-implanted bioelectronics as well as reconfiguration strategies to mitigate the effects of stochastic scattering and interindividual variability. For this purpose, this chapter starts by providing physical insights into the optimal energy exchange between an on-body radiation source and a miniaturized deep-implanted receiver, discussing the main loss mechanism that constrains the maximum achievable WPT efficiency. Then, the working principle of the WPT techniques that allows for electromagnetic radiation reconfiguration in lossy media, *i.e.*, radiative (far-field) and mid-field approaches, is detailed, and noteworthy implementations in the literature are presented. Subsequently, based on the analysis of the physical bounds and the state-of-the-art techniques, guidelines to achieve maximum WPT efficiency are proposed and discussed, with a focus on the importance of transmitter reconfigurability to establish an optimal wave path. Finally, the most novel reconfiguration approaches for electromagnetic focusing in lossy media, as well as WPT enhancement, are described.

10.2 Theory of Optimal Wireless Powering of Deep-body Bioelectronics

The optimal operating conditions that maximize the WPT from the on-body transmitter to a miniaturized implantable receiver are defined by the electromagnetic energy exchange between these two radiating structures across the heterogeneous, dispersive, and lossy medium, such as the body tissues. Hence, by investigating this problem, it is possible to establish the impact of the system's parameters on the WPT efficiency and electromagnetic exposure. For this intent, several numerical formulations have been proposed in the literature. For instance, a 2-D-axisymmetric full-wave approach is proposed[27] to evaluate the physical bounds on the maximum WPT efficiency as a function of the operating frequency, implantation depth, and electromagnetic nature of the receiver (electric or magnetic sources). This formulation has led to precise results compared with other formulations and experimental measurements, providing more physical insight than more realistic cases that require a costly full-wave analysis.

However, a 2D anatomical model that considers the complexities of the human body's shape, dimensions, and electromagnetic heterogeneity can better reproduce the conditions of physical applications, leading to a more accurate WPT efficiency prediction. Specifically, a case study comprising a miniature, leadless pacemaker deep-implanted within the human heart is shown in Figure 10.2. In this formulation, the transmitter is represented as an on-body conformal radiating electric-type source perfectly matched to the human tissues, while the receiver is modeled as a 1-mm radius circular region inside the heart. In this way, based on reciprocity and T-symmetry, the maximum achievable WPT efficiency is given by the ratio of the power flow

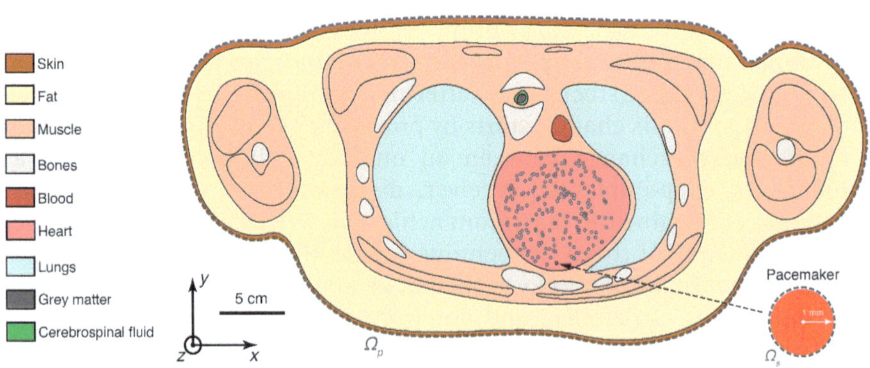

Figure 10.2 Cross-section of the human pectoral region comprising nine lossy tissues with dispersive electromagnetic properties.[109] In this two-dimensional formulation, the transmitter is an on-body source covering the entire body surface. Concurrently, the deep-implanted pacemaker (receiver) is at a depth of 37 mm from the skin, and it is filled by a lossless material matched to the wave impedance in the heart.

integrated across the body surface Ω_p by the same integral carried out along the receiver's encapsulation Ω_s:[27]

$$\eta_{\text{bounds}} = \frac{\oint_{\Omega_p} \left(\frac{1}{2}\mathbf{E} \times \mathbf{H}^*\right) \cdot \mathbf{ds}}{\oint_{\Omega_s} \left(\frac{1}{2}\mathbf{E} \times \mathbf{H}^*\right) \cdot \mathbf{ds}} \tag{10.1}$$

The results obtained with this proposed formulation are presented in Figure 10.3, in which $\eta_{\text{bounds}}^{\text{max}}$ and $\eta_{\text{bounds}}^{\text{min}}$ are the maximum efficiency in the co- and cross-polarization conditions, respectively. First, polarization plays a role in defining the field distribution inside the body. For instance, in the case of co-polarization, the direction of the propagating wave is parallel to the chest whose boundary defines the integration contour in eqn (10.1). Conversely, in the cross-polarization condition, the direction of maximum radiation is further away from that contour, which translates as a reduction on the average power delivered to the receiver, as can be noticed through the vertical offset between the $\eta_{\text{bounds}}^{\text{max}}$ and $\eta_{\text{bounds}}^{\text{min}}$ curves in Figure 10.3. However, as a miniature receiver is subjected to stochastic variations in its orientation due to the body movements, the maximum achievable efficiency is found in the shadowed region comprising these two curves.

However, the behavior of the maximum efficiency over the frequency is the same regardless of the polarization of the receiver. As shown in Figure 10.3, the efficiency increases with the frequency, reaching its maximum at frequencies between the ISM (industrial, scientific, and medical) bands of 915 MHz and 2.45 GHz. Then, it sharply drops at higher frequencies. This behavior can be explained by analyzing the different loss mechanisms that limit the achievable efficiency. First, at lower frequencies (<900 MHz), the reactive power is stored in the electric and magnetic fields confined in the lossy tissues between the transmitter and receiver. Therefore, most of the power is dissipated in these tissues, and only a small amount reaches the deep-implanted receiver. Conversely, as the loss tangent of the tissues increases with the frequency, at higher frequencies (>3 GHz), the power carried out by the propagating wave is mainly absorbed by the tissues before reaching the target receiver. Therefore, the trade-off between these two loss sources establishes an optimal frequency band, simultaneously minimizing the effects of reactive near-field coupling and propagating wave attenuation.

Apart from these two loss mechanisms defining the physical bounds on the WPT efficiency, other loss sources can also affect the power delivered to the receiver. First, the wave reflection between the on-body transmitter and the skin surface and the system-level losses (impedance mismatching, dissipation in the circuitry elements, *etc.*) are disregarded in this model since these devices are designed to mitigate these losses (the optimal engineering of the transmitter and receiver will be discussed in Section 10.5). Finally, since the human body configures a complex scattering media, the propagation path of the transmitted wave must be appropriately controlled to

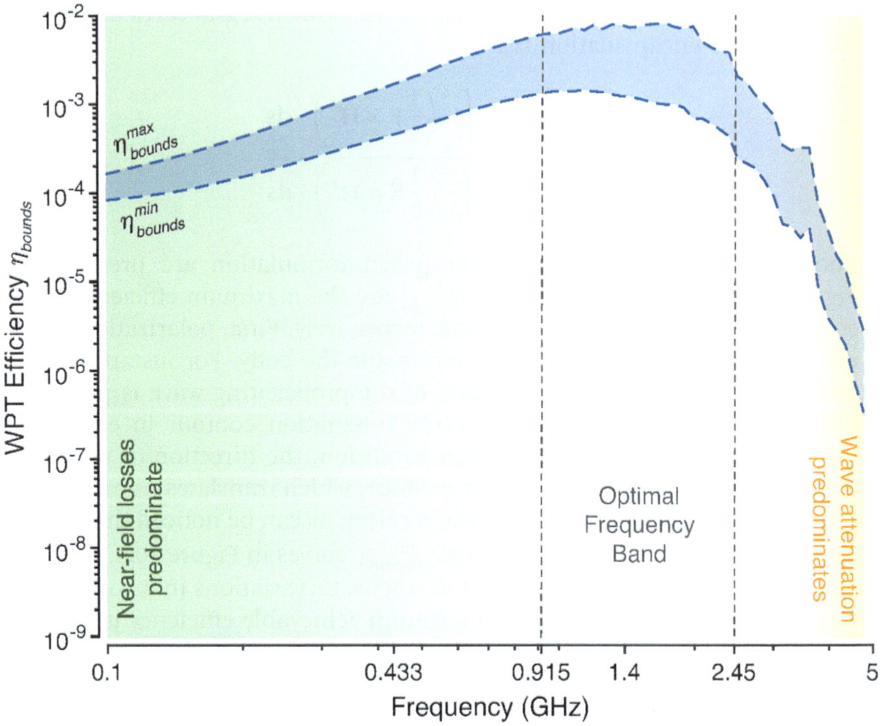

Figure 10.3 Maximum achievable WPT efficiency for an mm-size receiver deep-implanted in the human heart. The η^{max}_{bounds} and η^{min}_{bounds} curves are the efficiency in the case of co- and cross-polarization, respectively. The near-field losses are the most limiting factor to the maximum efficiency at low frequencies, whereas the propagating wave attenuation predominates at higher frequencies. Therefore, an optimal operation band for WPT is established between the ISM frequency bands of 915 MHz and 2.45 GHz.

focus the power density distribution toward the implanted receiver. This allows for mitigating the effects of wave scattering on the interface between different tissues and inner resonances in the organs. Therefore, the receiver and transmitter must be designed and reconfigured to establish this optimal wave path.

The following sections present the working principle of radiative and mid-field WPT in the context of implantable devices. Besides, different state-of-the-art implementations are described, highlighting the proposed strategies for enhancing efficiency.

10.3 Radiative Wireless Power Transfer Techniques

10.3.1 Working Principle

In some applications, the separation between the transmitter and receiver antennas on the human body exceeds the guided wavelength due to the high permittivity of biological tissues, which favors using radiative WPT techniques. Unlike near-field coupling approaches, where energy is stored within electromagnetic fields, in the radiative (far field) operation, energy is propagated through the medium *via* waves radiated from the transmitter antenna and absorbed by the implanted receiver. This method imposes minimal energy depletion from the source near the field, thus exerting less influence on the input impedance of the transmitter.[28] Moreover, the guided wavelengths in the centimeter range allow for reconfiguring the radiation patterns of transmitter antennas. Consequently, radiative WPT methods prove more suitable for effectively powering miniature devices implanted deep within the body.

A typical implementation of radiative WPT comprises one on-body transmitter antenna or antenna array connected to the power source and an implanted rectenna (antenna connected to a rectifier circuit), and their ideal characteristics depend on the nature of the application. For instance, a transmitter antenna with a broad beamwidth is best suited if the receiver moves through the body or is subjected to misalignments. However, as the power is radiated in a larger area, more energy is lost in the tissues, reducing the WPT efficiency. Therefore, for powering fixed implants, a transmitter with a directional radiation pattern can deliver energy to the receiver more efficiently. Alternatively, antenna arrays conformal to the body can be reconfigured based on beam-forming techniques[29–31] to maximize the power transfer towards the implant direction.

Regarding the receiver, as its gain and efficiency decrease as the antenna becomes smaller, the maximum received power is also intrinsically limited. However, one way to electrically increase its aperture is by using high-permittivity dielectrics and ceramic encapsulations,[32,33] thus enhancing the gain of the receiver antenna. As radiative WPT can deliver power to moving targets, it is mainly applied to powering ultra-low-power implants used in biotelemetry,[13,34] endoscopes, and leadless pacemakers,[4] as summarized in Table 10.1.

10.3.2 Applications

An innovative implementation of radiative WPT was proposed by Kaim *et al.*[13] based on a compact cubic rectenna MIMO system (CRMS) for implantable biotelemetry applications. The CRMS shown in Figure 10.4 is designed to receive incoming RF signals from the 360° environment with orthogonal circular polarizations in the 2.45 and 5.8 GHz ISM bands. This implantable antenna design allows the system to receive signals from multiple directions, regardless of the polarization of the incident wave – right hand circular polarization (RHCP) or left hand circular polarization (LHCP) – enhancing the WPT efficiency. The rectenna elements are interconnected in series and parallel configurations to improve the RF-DC conversion. The CRMS prototype achieved a high conversion efficiency of approximately 70% at low input powers in both frequency bands. The system was experimentally validated in a skin-fat-muscle phantom, ensuring human safety by reaching a maximum specific absorption rate (SAR) of 0.17 W kg^{-1} with a 30-dBm input power significantly below the standard limits.[35] Besides, with the

Table 10.1 Comparison between different implementations of radiative and mid-field WPT for in-body applications.

WPT	Ref.	Topology	Footprint	Implantation depth	Operating frequency	Efficiency	Max. SAR ($W\,kg^{-1}$)	Application	Comments
Radiative (far-field) WPT	13	Tx: Patch antenna Rx: Cubic MIMO	Tx: 50 × 50 mm Rx: 15 × 15 × 15 mm	50 mm	2.45 and 5.8 GHz	0.14–0.16%	0.17	Biotelemetry	The cubic topology enables collecting the power from both RHCP and LHCP in a 360° space
	36	Tx: Patch antenna Rx: Slot antenna	Tx: 120 × 120 mm Rx: 12 × 12 mm	2 mm (40 mm from the Tx)	2.45 GHz	0.07%	<1.6	Pacemakers and neural implants	Circularly polarized (CP) Tx and Rx antennas, a metamaterial slab is included
	110	Tx: Horn antenna Rx: Planar inverted F-antenna	Tx: off the shelf Rx: 8 × 4 mm	4 mm (500 mm from the Tx)	2.45 GHz	0.0074%	0.145	Implantable Medical Devices	An on-body parasitic patch is used to increase the gain of the implanted Rx
	111	Tx: Log-periodic antenna Rx: Planar inverted F-antenna	Tx: off the shelf Rx: 5 × 3.2 mm	40-60 mm (500 mm from the Tx)	673 MHz	0.00077% (rectified)	0.078	Pacemakers and neural implants	The performance of single-diode and doubler rectifiers are compared
	34	Tx: Log-periodic antenna Rx: Planar inverted F-antenna	Tx: off the shelf Rx: 16 × 14 mm	10 mm (500 mm from the Tx)	915 MHz	0.0102%	—	Biotelemetry	A 20-mm gel matching layer ($\varepsilon_{r=20}$, $\tan\delta = 0.003$) is used to improve WPT efficiency
	112	Tx: 2 × 4 Patch antenna array Rx: Fractal loop antenna	Tx: 130 × 90 mm Rx: 5 × 5 mm	8 mm (150 mm from the Tx)	2.4 GHz	0.5%	7.7×10^{-5}	Pacemaker and biotelemetry	A parasitic patch with the same Tx size and 80 mm behind the Tx is used to increase gain

	113	Tx: CP patch antenna Rx: Split resonant ring antenna	Tx: 39 × 40 mm Rx: 28.5 × 13 mm	55 mm (105 mm from the Tx)	1.4 GHz	0.12%	1.64	Endoscope	A conformal metasurface is used to control the wave incidence angle at the air–skin interface
Mid-field WPT	4	Tx: Slotted patch Rx: Meandered-line patch	Tx: 60 × 60 mm Rx: 7.5 × 18 mm	50 mm	1.470 GHz	0.67%	0.187	Endoscopes, pacemakers	Quad-band Tx (403, 915, 1470, 2400 MHz), matching layer ($\varepsilon_{r=78}$, $\tan\delta = 0.0004$)
	40	Tx: Loop antenna Rx: Meandered-line patch	Tx: 50 × 50 mm Rx: 5.6 × 6 mm	20 mm	1.9 GHz	0.26%	<1.6	Biotelemetry, neural implants	Data transfer at 915 MHz and WPT at 1.9 GHz, a slotted ground in Rx is used for miniaturization
	14	Tx: Slotted patch Rx: Slotted patch	Tx: 50 × 50 mm Rx: 18 × 10 mm	60 mm	1.470 GHz	1.07%	0.17	Biotelemetry	Data transfer at 915 MHz and WPT at 1.47 GHz, a matching layer of RO3010™ is used
	39	Tx: Loop antenna Rx: Slotted patch	Tx: 50 × 50 mm Rx: 18 × 10 mm	50 mm	915 MHz	0.06%	<1.6	Implantable medical devices	The self-phasing technique is used to achieve an optimum wave path focused toward the Rx
	42	Tx: Slotted patch with 4 ports Rx: Wire coil	Tx: 60 × 60 mm Rx: ø2 mm	55 mm	1.6 GHz	0.04% (rectified)	0.89	Leadless pacemakers	The slotted radiator is optimized to emulate the optimal current distribution
	41	Tx: Slotted patch with 2 ports Rx: Ground-slotted patch	Tx: 63 × 63 mm Rx: 20.5 × 10.25 mm	55 mm	1.5 GHz	0.45%	<1.6	Biotelemetry	A near-field plate is added close to the Tx to control power leakage

Table 10.1 (*Continued*)

WPT	Ref.	Topology	Footprint	Implantation depth	Operating frequency	Efficiency	Max. SAR ($W\,kg^{-1}$)	Application	Comments
	5	Tx: Bipolar spiral antenna Rx: Hooked-shaped dipole	Tx: 59 × 52 mm Rx: 12.5 × 12.5 mm	60 mm	1.87 GHz	0.61%	<2	Leadless pacemakers	The optimum current distribution is created by a bipolar spiral excited through an aperture
	16	Tx and Rx: Loop antenna	Tx and Rx: 6.8 × 6.8 mm	58 mm	1.2 GHz	0.02%	<2	Endoscope	An additional circuit is used to achieve resonance
	44	Tx: Concentric loaded loop Rx: Meandered-line dipole	Tx: ø30 mm Rx: 20 × 3 mm	10 mm	2.4 GHz	8%	<2	Neural interface	Ring-like surface currents coupled with lumped elements create a focused power pattern
	43	Tx: Phased surface radiator Rx: Wire coil	Tx: 60 × 60 mm Rx: ø1.5 × 3 mm	42 mm	1.6 GHz	0.06%	8.9	Leadless pacemakers	Capacitively loaded subwavelength resonant elements create a focused power pattern

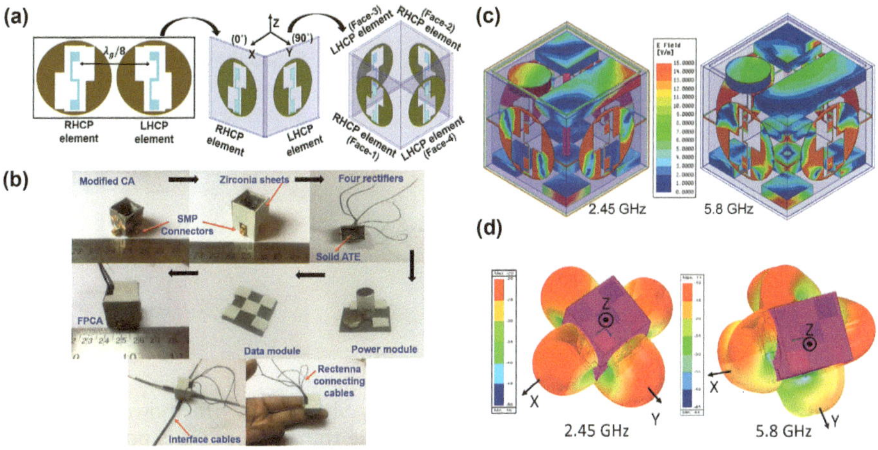

Figure 10.4 Implantable cubic rectenna proposed by Kaim *et al.*[13] for biotelemetry systems wireless powered through radiative WPT. (a) The cubic topology comprises two antenna elements with right hand circular polarization (RHCP) and left hand circular polarization (LHCP). (b) Step-by-step of the rectenna assembling. (c) Electric field distribution and (d) radiation pattern at both operating frequency bands. Adapted from ref. 13 with permission from IEEE, Copyright 2023.

transmitter and receiver antenna separated by 500 mm, the measured WPT efficiency was approximately 0.15% in both operating bands, delivering a DC power of 0.33 mW to the load.

A radiative WPT system for powering leadless pacemakers and neural implants based on circularly polarized (CP) antennas was proposed by Shaw *et al.*[36] The implementation shown in Figure 10.5 is designed to operate at 2.45 GHz. It employs a flexible CP slot antenna coated with a biocompatible alumina layer as the implanted receiver (Rx) and a CP patch antenna as the out-body transmitter (Tx). This system also incorporates a metamaterial (MTM) slab comprising 4×4 unit cells to enhance power transfer efficiency. The performance of the proposed approach is thoroughly explored in the human torso and head models, and its stability is analyzed in different scenarios for the Rx element, including rotation, bending, misalignment, and implantation depth variation. The results demonstrate that this system achieves an efficiency of 0.07% when the Tx and Rx are 40 mm apart, besides complying with the exposure level regulations.[35]

10.4 Mid-field Wireless Power Transfer Techniques

10.4.1 Working Principle

A clear distinction between the near-field coupling and radiative regions is less straightforward in complex media such as biological tissues than in free space. On the one hand, the high permittivity of the tissues reduces the

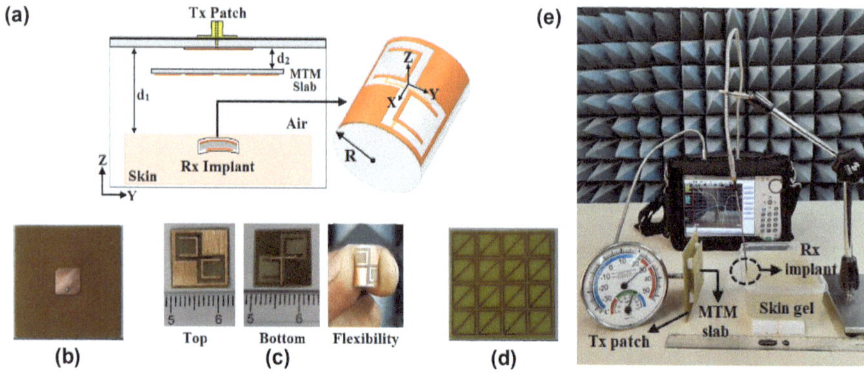

Figure 10.5 Radiative WPT system proposed by Shaw *et al.*[36] for powering deep-body implants. (a) System setup. The transmitter (Tx) is positioned at a distance $d_1 = 40$ mm from the skin, and a metamaterial (MTM) slab is inserted at $d_2 = 7$ mm from Tx to improve efficiency. (b) A circularly polarized (CP) patch antenna is used as a transmitter, while (c) the receiver comprises a flexible CP slot antenna. (d) The metamaterial slab comprises a 4×4 array of unit cells and presents zero effective refractive index at the operating frequency. (e) Experiment setup for measuring the WPT efficiency. Adapted from ref. 36 with permission from IEEE, Copyright 2021.

wavelength in the medium, thus decreasing the distance in which radiative effects predominate. On the other hand, the spatial distribution of the tissue is not uniform, which means that for the same implantation depth but at different positions on the body, a bioelectronic device can be more efficiently powered using near-field coupling or radiative WPT techniques. Finally, the wave reflections at the boundary between different organs induce surface currents that can spatially extend the region where near-field effects are observed. For instance, Figure 10.6 presents the wave impedance magnitude $|Z_{\mathrm{w}}|$ across several layers of body tissues. As can be seen, $|Z_{\mathrm{w}}|$ decreases as the wave propagates inside the body, converging to a constant value $|Z_{\mathrm{w}}|_{\text{far-field}}$. However, when there is a significant permittivity contrast, such as between the fat and muscle layers, the surface currents induced at the interface act as a radiation source, thus extending the near-field effects, which is verified by the wave impedance rising in this region.

Therefore, the WPT efficiency and distance range in implantable applications can be significantly improved if the advantages of near-field coupling and radiative WPT can be simultaneously exploited. For this reason, the mid-field WPT technique was proposed by combining the reactive (inductive coupling) and radiative effects.[37,38] Since the optimal frequency for WPT in implantable applications is around the sub- and low-GHz bands, the separation between the transmitter and receiver is comparable to the wavelength in the medium, and it is several times larger than the implant's size, which means that they are weakly coupled. For this reason, as shown in Figure 10.7, transmitters used in mid-field WPT strategies are designed to

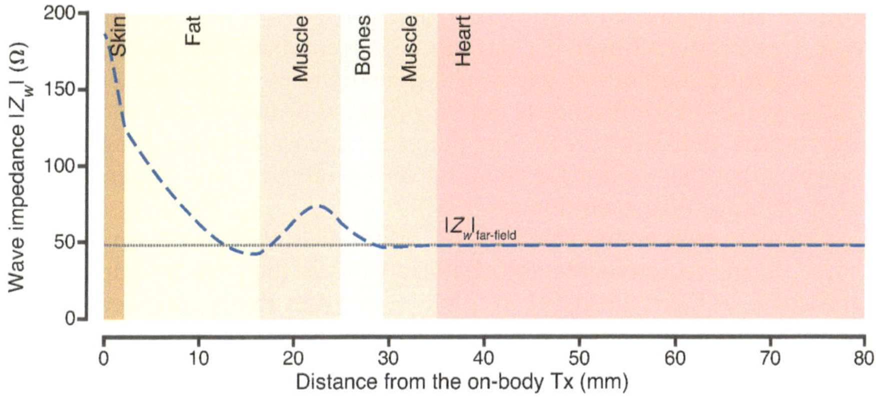

Figure 10.6 Wave impedance $|Z_\mathrm{w}|$ across a planar stratified human torso model, considering a normal plane wave incident upon the skin. $|Z_\mathrm{w}|$ decreases as the wave propagates inside the body, converging to a constant value $|Z_\mathrm{w}|_\text{far-field}$ in the far-field region. However, the surface currents induced at the boundary between two tissues with a significant permittivity contrast (*e.g.*, $\varepsilon_\text{fat} = 11.3$ and $\varepsilon_\text{muscle} = 54.8$ at 1 GHz[109]) act as a radiation source, locally extending the near-field and increasing $|Z_\mathrm{w}|$.

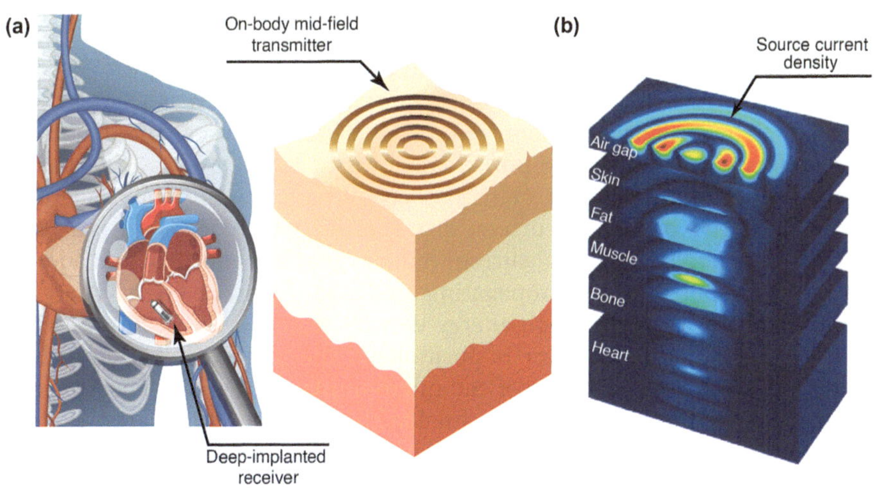

Figure 10.7 Mid-field wireless power transfer strategy for powering deep-implanted implantable devices. (a) An mm-sized receiver deeply implanted in the human body (*e.g.*, a leadless pacemaker) can be powered by an on-body transmitter tailored to enforce an optimal current density by alternating current paths in a periodicity of a half-wavelength.[38] (b) The power pattern generated by this optimal source current density is focused toward the implanted receiver, thus maximizing the WPT efficiency.[42] Adapted from ref. 42 with permission from PNAS, Copyright 2014.

enforce an optimal current density that generates a field distribution focused toward the receiver.[39] This optimal current distribution is obtained by reconfiguring the current paths in a periodicity of a half-wavelength.[38] In this way, the WPT efficiency is not only defined by the reactive coupling but also by the radiative power flow pattern, which allows reaching distance ranges larger than near-field coupling WPT strategies with higher efficiency than radiative WPT techniques.

Different physical implementations for mid-field WPT have been proposed based on the requirements and characteristics of the envisaged application, and some of them are listed in Table 10.1. From the transmitter side, the optimal current distribution can be achieved using capacitive-loaded concentric loops, slot antennas, or even reconfigurable topologies that act as an electric (dipole antenna) or magnetic source (loop antenna), depending on the operating frequency. As the transmitter creates the field focusing, its structure is generally more complex than the one used for the receiver. Therefore, simpler designs such as coils and meandered antennas are usually employed to absorb the transmitted power while maintaining a miniaturized form factor. Finally, as mid-field WPT can achieve greater efficiency at larger distance ranges, it is generally used for charging ingestible and deep-implanted biotelemetry devices,[14,40,41] endoscopes,[16] mm-sized leadless pacemakers,[5,42,43] and bioelectronics for peripheral nerve neuromodulation.[44]

10.4.2 Applications

In the seminal work that proposed the mid-field WPT strategy,[42] Ho *et al.* designed a WPT system for powering microimplants such as leadless pacemakers. The proposed system employs a patterned metal plate with slot structures fed through four ports, as shown in Figure 10.8a. The phase in each port is applied so that a circular current path flows on the transmitter, emulating the optimum current distribution, which leads to maximum power transfer. This optimal current induces spatially confined and adaptive energy transport through propagating modes in the tissue layers. Therefore, this approach creates a high-energy density region deep in tissue, within which the mm-sized receiver is positioned. The main results demonstrate successful wireless powering of the micro-implant for closed-chest wireless control of the heart, significantly smaller than conventional pacemakers, as seen in Figure 10.8b. Measurements in a porcine model showed that at a frequency of 1.6 GHz and with an input power of 500 mW, the proposed system delivers 195 μW to the receiver, significantly higher than the minimum power required by the pacemaker. Besides, exposure level assessment in an anthropomorphic phantom revealed that the maximum SAR level was 0.89 $\mathrm{W\,kg^{-1}}$, below human safety thresholds.[35]

A further development in mid-field WPT from Agrawal *et al.*[43] used conformal phased surfaces that can modulate phase with high resolution over non-planar body surfaces, enabling efficient wireless power transfer to microdevices implanted deep in the body. The phased surface is a metasurface

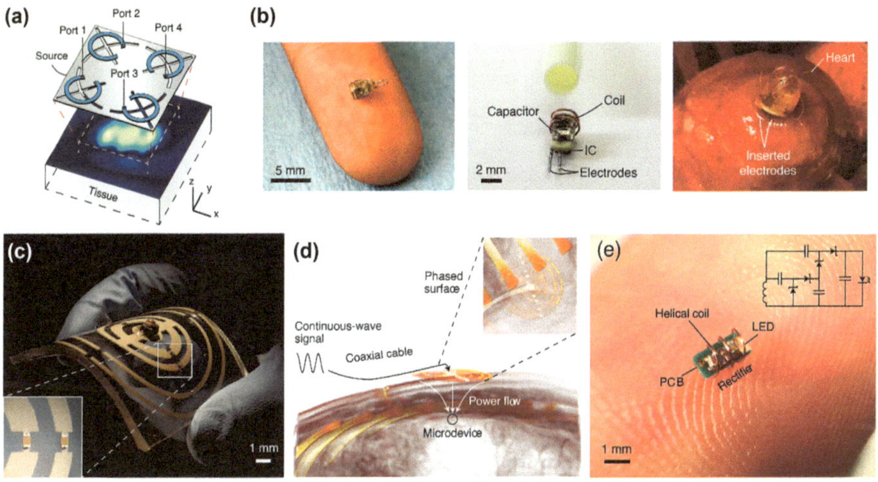

Figure 10.8 Mid-field wireless power transfer (WPT) systems for implantable medical devices. (a) and (b) A leadless pacemaker proposed by Ho *et al.*[42] is wirelessly powered through mid-field WPT. (a) The transmitter consists of a patterned metal plate excited by four ports with controlled phases, generating a focused power pattern on the skin surface, as shown by a time snapshot of the magnetic field. (b) The mm-sized receiver, implanted 35 mm below the skin in the lower epicardium of a rabbit, comprises a wire coil and an electrostimulation circuit. (c)–(e) A mid-field WPT system by Agrawal *et al.*,[43] featuring (c) a transmitter with concentric loops loaded with lumped capacitors to modulate the current density phase, emulates optimal current distribution. (d) The phased surface on the skin focuses power flow onto a deep-implanted microdevice. (e) The receiver, composed of a wire coil, rectifier circuit, and signalizing LED, demonstrates efficient energy transfer. (a) and (b) Adapted from ref. 42 with permission from PNAS, Copyright 2014. (c)–(e) Adapted from ref. 43 with permission from Springer Nature, Copyright 2017.

comprising capacitively loaded rings, as shown in Figure 10.8c, that generate an engineered phase response that focuses the radiated power pattern towards the receiver. Unlike conventional phased arrays, this structure does not require phase delay or control circuits, making it suitable for integration into conformal devices for use on non-planar body surfaces. *In vivo* experiments have proven that the proposed system can successfully power microdevices implanted in an adult pig's abdomen and neck, with implantation depths around 40 mm from the body surface. In this scenario, with an input power of 800 mW, the DC power delivered to the load was about 0.6 mW with a peak SAR of 8.9 $\mathrm{W\,kg^{-1}}$, under the safety threshold in which no adverse effects are verified.[35] Finally, this system was applied to wirelessly power a miniaturized leadless pacemaker, achieving target heart rates in different stimulation sites.

Das and Yoo proposed a biotelemetry implant powered by mid-field WPT.[41] This system, presented in Figure 10.9, comprises a multiband

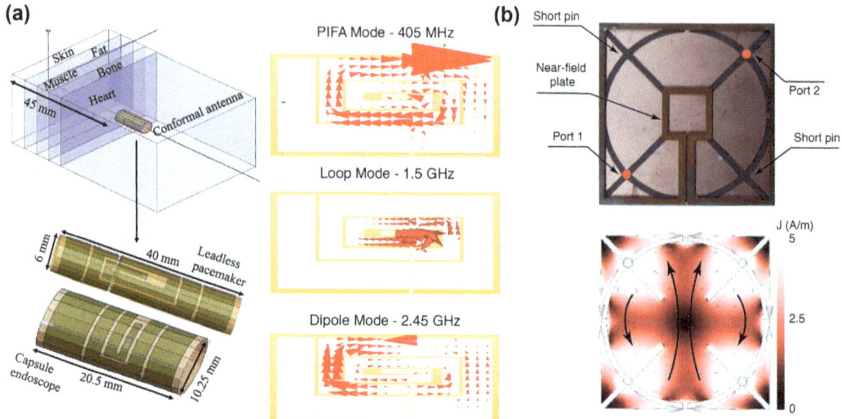

Figure 10.9 A multiband mid-field WPT strategy was proposed by Das and Yoo[41] for powering leadless pacemakers and endoscopes. (a) The proposed receiver antenna operates at 405 MHz as a Planar Inverted-F Antenna (PIFA), at 1.5 GHz as a loop antenna, and at 2.45 GHz as a dipole antenna. A T-shaped ground slot is used to tune the desired operating mode. (b) The mid-field transmitter comprised a patterned metal plate with slots, two ports, and two short pins. The phases applied to the ports induce a circular current path that emulates the optimal current density, focusing the power pattern toward the receiver. In addition, a near-field plate composed of an external closed loop and an inner open loop is positioned in front of the mid-field transmitter to reduce power leakage. Adapted from ref. 41 with permission from IEEE, Copyright 2017.

receiver antenna that can be easily tuned by adjusting a T-shaped slot in its ground plane. In this way, it operates as a Planar Inverted-F Antenna (PIFA) at 405 MHz, a loop antenna at 1.5 GHz, and a dipole at 2.45 GHz. In addition, they proposed a redesigned mid-field transmitter with only two ports, providing an alternative to conventional mid-field WPT radiators. Besides, a near-field plate comprising a large square loop and an inner open loop is added to the transmitter to control power leakage. The current distribution in these loops reduces the backward propagating magnetic field and focuses the power pattern into the receiver, acting as a magnetic screen.[45,46] *Ex vivo* experiments in a porcine heart revealed that the WPT efficiency achieved using the near-field plate was around 0.45%, five times higher than the efficiency measured without the focusing plate. Besides, the proposed mid-field scheme complies with the electromagnetic exposure level regulations.[35]

10.5 Guidelines for the Optimal Design of Deep-body Wireless Power Transfer Systems

Based on the investigation of the loss mechanisms that affect the in-body WPT efficiency and the conditions to define the physical bounds, it is possible to establish fundamental guidelines to achieve near-optimum performance in wireless powering implantable bioelectronics. These insights, drawn from the findings discussed in Section 10.2, are related to the choice of the WPT technique and its operating frequency, the electromagnetic nature and polarization of the sources, as well as the on- and in-body receiver design. Therefore, the optimum WPT conditions can be defined given the characteristics of the application.

10.5.1 Choice of WPT Approach

Firstly, the choice of WPT technique and operating frequency depends on the implantation depth of the receiver. Near-field coupling techniques like inductive and capacitive WPT excel for subcutaneous implants. In contrast, mid-field and radiative techniques perform better for deep-implanted devices within the frequency range between the 915 MHz to the 2.45 GHz ISM bands.

While both radiative and mid-field approaches are best suited for wireless powering ultraminiature and deep-implanted bioelectronic devices than near-field coupling techniques, radiative WPT typically has a broader radiation pattern from the external antenna, leading to significant energy loss through absorption in surrounding tissues before reaching the target area, thus reducing the power delivered to the implant and limiting overall efficiency. In contrast, the mid-field strength lies in its ability to be adaptively reconfigured, focusing the transmitted power towards the deep implant, regardless of stochastic variations due to human body dynamics. In this way, in mid-field implementations, the electromagnetic field distribution is shaped to create a region of higher power density at the implant's location, minimizing energy dissipation in neighboring tissues, thereby achieving significantly higher power transfer efficiency compared to conventional unfocused radiative methods. This focused transfer allows mid-field WPT to reach implementation depths beyond near-field techniques while often delivering more power to deep targets than simple radiative methods.

To achieve this reconfigurability, mid-field implementations are inherently more complex, requiring the transmitter to incorporate advanced on-body conformal radiating structures such as phased arrays or metasurfaces.[47–49] These elements enable spatial control of the transmitted electromagnetic fields inside the body. Furthermore, this real-time reconfigurability requires dedicated transceiver circuits capable of dynamically computing and applying the optimal excitation to the radiating elements based on the implant's position and surrounding anatomical variations. However, a radiative WPT approach may be more practical in scenarios where the bioelectronic implant is fixed and less affected by stochastic tissue movement or physiological variability. In such cases, a single conformal antenna can suffice for power transmission, allowing for simpler and more cost-effective system design. To enhance efficiency in these setups, a buffer layer can be inserted between the antenna and the skin to confine the near-field close to the body surface, thereby reducing wave impedance mismatch and minimizing power dissipation in the lossy tissue.[27] This not only increases the power delivered to the implant but also contributes to a significant reduction in local SAR.

10.5.2 Choice of the Operating Frequency

Beyond the fundamental choice between these different WPT strategies, the operating frequency is another crucial design aspect. At lower frequencies, reactive power stored in the fields in lossy tissues limits efficiency, while propagating wave attenuation predominates at higher frequencies. Therefore, the optimal range, typically between 900 MHz and 3 GHz, balances these loss mechanisms. Inside this range, three frequencies are standardized for medical applications: 915 MHz, 1.4 GHz, and 2.45 GHz.

10.5.3 Choice of Antenna Type

A conscientious choice of transmitter and receiver antenna types is essential to achieve WPT efficiency as close as possible to the physical bounds. First, the receiver antenna type and its polarization play a crucial role, and the constraints and characteristics of a given application define its choice. Even though magnetic loop antennas may theoretically exhibit higher efficiency at frequencies below 300 MHz,[27,50] their practical implementation leads to high ohmic losses[51,52] that are disregarded in problem formulation to define the physical bounds. Conversely, the dielectric loading caused by the high permittivity of human tissues benefits the operation of electric-type antennas at the optimal frequency band.[53] Apart from that, the polarization of these antennas significantly affects the WPT efficiency. Therefore, reconfigurable topologies of implantable antennas[33,54–56] are best suited to mitigate the impact of stochastic variations on the antenna position and orientation caused by body movements.[57] Finally, the receiver miniaturization intrinsically limits the maximum achievable values of antenna efficiency. However, engineering techniques, such as high-permittivity encapsulations and ceramic shells,[59] can be applied to approach these maximum values as much as possible.

The on-body transmitter is the component of the WPT system that can be most tailored to enhance efficiency. First, the design transmitter should minimize system-level losses, such as impedance mismatches and losses in its electronic circuitry. Additionally, using low-loss, high-permittivity superstrates as a buffer interfacing the on-body transmitter and the skin layer is another strategy to enhance the WPT efficiency. Besides acting as impedance-matching layers, these superstrates can be optimized to confine the near-field in a low-loss region, thus mitigating power dissipation in the lossy tissues.[27,34,58] Another strategy for implementing a matching layer for on-body transmitters in the literature is using metasurfaces with their effective guided wave impedance tailored to match the one in the skin, thus reducing the losses due to the reflection at this interface.[59–61]

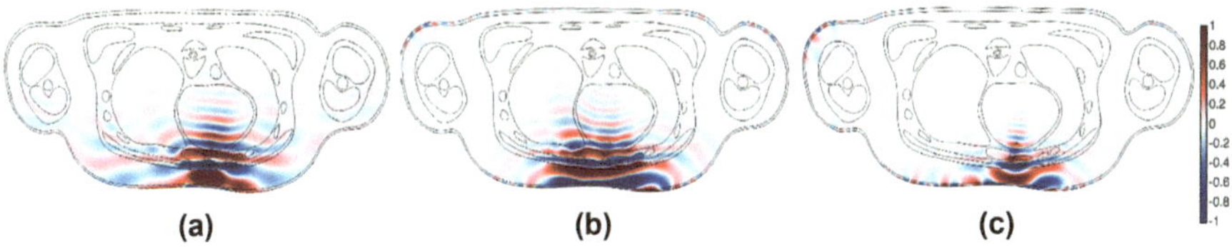

Figure 10.10 Magnetic field (H_z) distribution inside the human pectoral cross-section considering an on-body radiation source comprising (a) a single $\lambda/2$ dipole antenna, (b) an antenna array with eight elements with uniform excitation (same amplitude and phase in all antennas), and (c) an antenna array with eight elements optimally reconfigured by wavefront shaping techniques.

Finally, when the efficiency values achieved by the works presented in Sections 10.3 and 10.4 and listed in Table 10.1 are compared with the maximum efficiency in Figure 10.3, these solutions currently found in the literature are proven to be suboptimal. This can be explained by the fact that most of these approaches rely on single-antenna transmitters, whereas the physical bounds are established by considering a radiating structure conformal to the body. Generally, the field distribution of a single-antenna transmitter shown in Figure 10.10a resembles that of a dipole or a loop antenna, depending on whether it is an electric-type or magnetic-type antenna. Although it can transmit some power to the implantable receiver, most of the power is dissipated in the neighboring tissues, thus leading to a WPT efficiency about three orders of magnitude lower than the theoretical maximum, as seen in Figure 10.11.

Conversely, using a conformal antenna array as an on-body transmitter provides more degrees of freedom for maximizing efficiency. However, an antenna array by itself is not enough. Figure 10.10b reveals that an array with uniform excitation (same amplitude and phase in all elements) increases efficiency, but it is still lower than the expected maximum. Therefore, the transmitter array excitation must be reconfigured so that a field distribution is focused on the deep-implanted receiver. Apart from that, the transmitter reconfiguration ensures the robustness of the WPT efficiency against stochastic variations on the receiver orientation and the inherent inter-individual variability.

The problem formulation used to establish the physical bounds can be recalled to engineer an optimal transmitter reconfiguration scheme. Based on this analysis of the propagation and scattering of the electromagnetic waves inside the human body, the power injected in each element of the on-body transmitter array can be optimized in order to control the impinging wave. This strategy, known as **wavefront shaping**, leads to a perfectly focused field distribution between the on-body transmitter and the implanted receiver, as shown in Figure 10.10c. Therefore, the transmitted power is more efficiently delivered to the implant, thus increasing the efficiency, as seen in Figure 10.11. Moreover, the optimal wave path established allows for

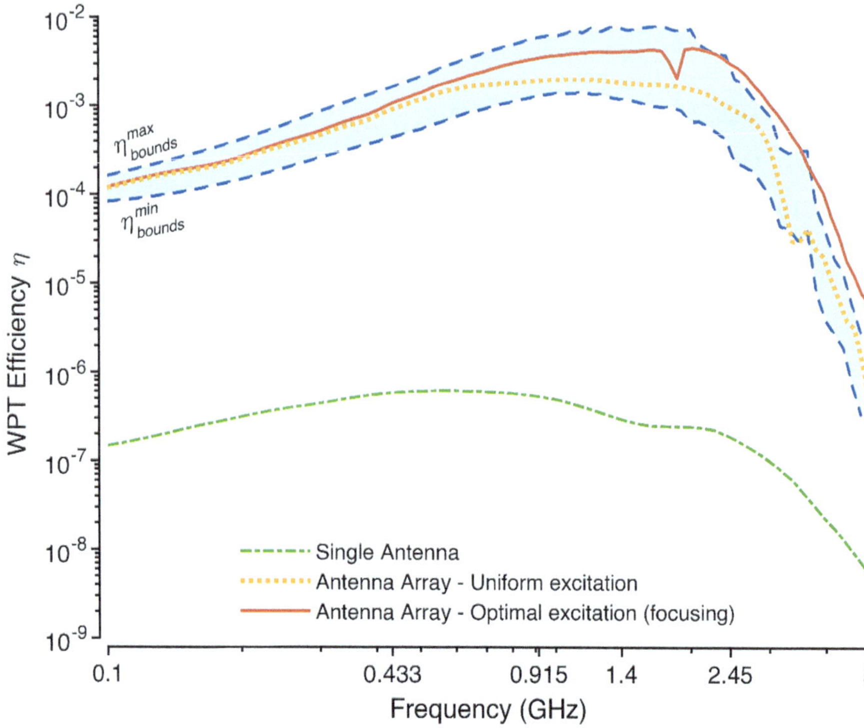

Figure 10.11 WPT efficiency for an mm-size receiver deep-implanted in the human heart as a function of the frequency. The dot-dashed green line indicates the efficiency of considering a single ideal electric-type antenna ($\lambda/2$ dipole) as the on-body transmitter. The dashed yellow and continuous orange lines are the efficiency values considering a source comprising a conformal antenna array with eight elements with uniform and optimal excitation, respectively.

mitigating the effect of resonant modes, especially at higher frequencies, which affect the efficiency values expected from the physical bounds formulation. For this reason, at the frequencies above the optimal band, the optimal excitation can slightly surpass the maximum theoretical values. Finally, since less power is dissipated in the neighboring tissues, the electromagnetic exposure level is reduced, increasing user safety.

In summary, selecting a WPT strategy for bioelectronic implants is primarily guided by the implantation depth and the nature of the target application. While near-field is still suitable for powering subcutaneous devices, radiative and mid-field techniques must be employed for deeper implants. However, mid-field WPT offers a distinct advantage through its ability to focus energy towards the target, achieving higher power transfer efficiency in lossy tissue compared to unfocused radiative methods. Furthermore, optimal system performance for deep implants relies on careful frequency selection, enhancing receiver efficiency through design, and employing reconfigurable external transmitter arrays capable of wave-front shaping.

10.6 Reconfigurability Techniques for On-body Transmitters

Other approaches have been proposed to reconfigure the excitation of the on-body transmitter arrays. For instance, the mid-field WPT implementation proposed by Agrawal *et al.*[62] and shown in Figure 10.8 employs a conformal phased surface as a transmitter. In this case, the concentric loops and the reactive elements were optimized to emulate the amplitude and phase that solve the matched filtering problem. For this, a fixed receiver is assumed to be implanted in a homogeneous planar phantom with muscle-like permittivity. Then, the mathematical formulation proposed in ref. 63 is applied to obtain optimal focusing.

Several advancements have been made regarding the design of on-body conformal antenna arrays, as summarized in ref. 64. However, beam-forming in attenuating media remains a challenge.[65] One alternative is the application of techniques developed for microwave biomedical imaging. In ref. 66, for instance, an inverse problem is solved to determine the input amplitude and phase to be applied to each transmitter antenna.[67] This work also discusses using the singular vectors of the matrix of scattered fields as the array excitation.[68,69] Apart from that, an algorithm is proposed in ref. 70 to determine the optimal amplitude of each element in a magnetic dipole array conformal to the human body. These amplitudes are determined by minimizing the deviation between the desired electric field pattern and the one produced by the array, considering several layers of biological tissues. Besides, some techniques employed for ultrasonic power transfer[71–75] can inspire alternatives for beamforming in the microwave range.

Different near-field shaping techniques have been employed for microwave focusing in lossy and heterogeneous media. For instance, ref. 76 presents several antenna array design techniques and near-field focusing algorithms proposed for medical imaging, hyperthermia, and tissue stimulation, which could be directly translated for WPT applications. Another option for near-field focusing is the use of resonant Bessel beam launchers. These launchers comprise reactively loaded surfaces in which the excited current distribution generates a radiation beam that follows a Bessel function pattern. It is mainly used in free-space applications as it produces a radiation beam that does not spread out as it propagates.[77–79] Therefore, it has also been applied to increase the wireless power transfer range for radiative near-field approaches.[80] Even though the lossy and heterogeneous nature of human body tissues imposes challenges to achieving such behavior, implantable WPT solutions can be developed based on the concept of non-diffracting beams. Alternatively, Iliopoulos *et al.*[81] propose a focusing strategy based on convex optimization to maximize the electric field inside a

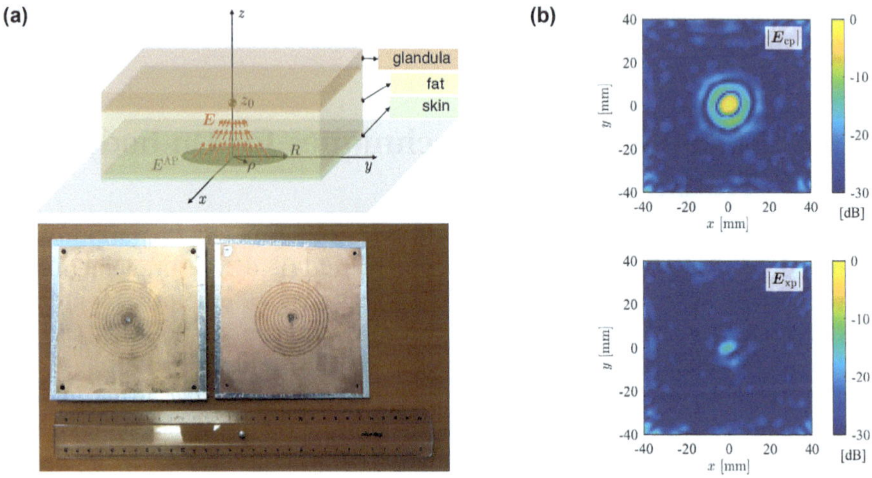

Figure 10.12 Focusing strategy for breast cancer detection proposed by Iliopoulos *et al.*[81] (a) A radial line slot array implements the focusing technique based on near-field convex optimization to increase the mm-wave penetration in a planar layered phantom representative of the human breast. (b) Measured near-field mapping of the co- $|E_{cp}|$ and cross-polar $|E_{xp}|$ component of the designed antenna (normalized to the maximum $|E_{cp}|$). Adapted from ref. 81 with permission from IEEE, Copyright 2021.

breast phantom and increase the penetration of mm waves in biomedical imaging. In this case, as shown in Figure 10.12a, a radial line slot array operating at 30 GHz is used as a transmitter and fed to focus at a distance of 40 mm. The measured 2-D near-field map in Figure 10.12b shows that a focused beam is achieved, with a 15 dB increase in the field level compared with uniform excitation.

Apart from wavefront shaping strategies, other implementations can be explored to implement optimal focusing. For instance, non-linear backscattering has been used to control the wireless powering of implantable bioelectronics.[82] As shown in Figure 10.13, the harmonic components $n\omega$ generated by a non-linear load on the receiver side (the diodes on the rectifier, for example) affect the reflection coefficient spectrum $\Gamma^{(n)}$ on the transmitter side. Based on the power of the spectral components $P_H^{(n)}$, the transmitted power level P_T is adjusted to achieve maximum efficiency. This approach has been tested for powering a miniature device implanted in a rat's brain, as shown in Figure 10.13b.

10.7 Technological Challenges and Perspectives on Wireless Powered Bioelectronics

Wireless powering implantable and ingestible bioelectronic devices remains a technological challenge, primarily due to the inherent complexities of the

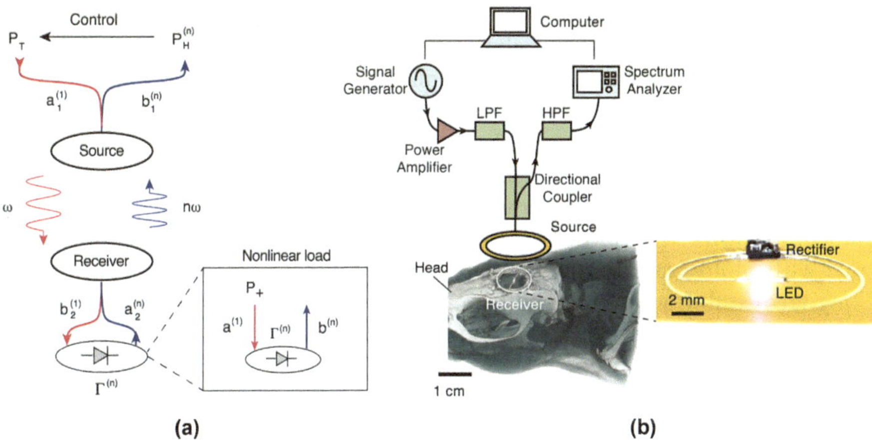

Figure 10.13 Implementation of the harmonic feedback control of wireless-powered implantable bioelectronics proposed by Tian *et al.*[82] (a) A nonlinear element in the receiver side generates harmonic components $n\omega$ that affect the reflection coefficient spectrum $\Gamma^{(n)}$ of the source. The transmitted power P_T is then controlled based on the knowledge of the harmonic component power $P_H{}^{(n)}$. (b) Implementation setup for powering a receiver implanted in the rat brain (LHP and HPF indicate, respectively, a low- and a high-pass filter). Adapted from ref. 82, https://doi.org/10.1063/1.5030600, under the terms of the CC BY 4.0 license, http://creativecommons.org/licenses/by/4.0/.

human body as an electromagnetic medium and the demanding requirements of miniaturized, long-lasting implants. While WPT offers a compelling solution to the limitations of battery-powered implants, such as bulkiness and the need for maintenance through invasive medical procedures, achieving efficient, reliable, and safe power delivery deep within the body requires overcoming significant technical obstacles.

One of the primary challenges is the interaction of electromagnetic fields with heterogeneous, dispersive, and lossy biological tissues. As the waves propagate through the body, they undergo substantial attenuation due to absorption by tissues, which significantly reduces the power reaching the implanted device, particularly for deep implants. Furthermore, the complex structure of the body, with different organs, muscles, fat layers, and bones, acts as a complex scattering medium, causing wave scattering and reflection at tissue interfaces, further reducing the power delivered to the implant and limiting efficiency.[27] Reflection losses, particularly at the outer body surface (skin–air interface), can be substantial, sometimes accounting for over 80% of the total transmission loss in clinically relevant scenarios.[83]

Beyond fundamental propagation issues, practical integration challenges include maintaining reliable power transfer despite the dynamic nature of the biological environment. For instance, body movements can cause stochastic variations in the implant's position and orientation. Besides, inter-individual variability in tissue properties also affects wave propagation, requiring WPT systems designed to robustly operate regardless of such uncertainties. Another technical challenge is implant miniaturization. While it is crucial for minimally invasive implants, reducing the implant size inherently limits the efficiency of the receiver antenna.[53,84] Moreover, strict safety standards for human EM exposure impose limitations on the maximum power that can be transmitted into the body,[35] requiring highly efficient WPT systems that minimize energy dissipation in non-target tissues. Biocompatibility is another essential consideration, as all materials in contact with biological tissues must avoid inflammatory responses and ensure long-term stability without compromising device performance. Additionally, the increasing interest in distributed implantable networks, where multiple devices operate simultaneously within the body, poses new challenges in managing power allocation and avoiding interference. Such scenarios demand WPT systems capable of dynamically reconfiguring the transmitted fields to efficiently and safely power multiple implants in real time. The current gap between the theoretical maximum achievable WPT efficiency and that of physically realized systems reported in the literature highlights the extent of these challenges.

To address these challenges, several technological solutions have been developed. For dynamic and uncertain biological environments, reconfigurable transmitter structures comprising phased arrays or metasurfaces enable implementing beamforming algorithms[85] in radiative WPT and adaptive wavefront shaping in mid-field approaches,[83] allowing the system to steer and focus the transmitted power toward implants. Besides, advanced control algorithms, often based on harmonic feedback,[82] can further enhance real-time reconfiguration. To overcome receiver miniaturization constraints, researchers have explored high-permittivity dielectric substrates,[86] non-linear and active matching networks,[87,88] and novel antenna geometries to enhance the effective aperture and sensitivity of mm-sized receivers.[33] Safety constraints related to EM exposure are being mitigated through the use of field-confining buffer structures and frequency optimization to operate within medical ISM bands where tissue absorption is lower.[27] Regarding biocompatibility, encapsulation techniques using materials such as parylene, silicone, or ceramic prevent tissue inflammation and preserve device performance in physiological conditions over extended periods.[89] Finally, for powering networks of implants,[90,91] multiplexed WPT systems have shown promise in addressing simultaneous multi-device power delivery while minimizing cross-device interference and maintaining overall system efficiency.[92]

Furthermore, emerging engineering strategies and novel physical concepts offer new perspectives on designing efficient and patient-adaptive WPT solutions to unlock the full potential of battery-free miniaturized bioelectronics for deep-implanted applications. For instance, the latest advancements on mesoscopic scattering and non-Hermitian physics,[83] such as coherent perfect absorption (CPA),[93] parity-time symmetry,[94] bound states in the continuum (BIC),[95] and anapoles,[96] are promising avenues for introducing novel ways to control wave behavior in complex media, potentially leading to more efficient and robust WPT systems. These concepts could offer new methods for enhancing wavefront shaping, mitigating losses (*e.g.*, radiation losses *via* BICs or anapoles), and enabling adaptive WPT tailored to individual variability and dynamic conditions. In addition, developing hybrid WPT systems that combine multiple modalities (*e.g.*, ultrasound, biochemical, thermal) is another alternative to enhance power delivery levels.[97,98]

On the hardware front, innovations in metamaterials,[49] stretchable electronics,[99,100] and ultra-thin biocompatible coatings[89] can lead to further miniaturization while improving integration with soft tissues. In parallel, the rise of bioresorbable electronics[101,102] opens new possibilities for temporary monitoring and therapeutic applications, eliminating the need for surgical removal. Finally, machine learning and AI-driven control algorithms[103–105] are another research path explored to dynamically manage power delivery across implant networks and mitigate the impact that fibrosis and endothelialization[106–108] can have on the WPT efficiency over time by learning from tissue responses and user activity to maximize efficiency and safety.

10.8 Conclusions

Wireless power transfer techniques based on optimal reconfigurability are essential for designing battery-free miniature bioelectronics for deep-implanted applications. The challenges of efficiently transferring power through the lossy and heterogeneous tissues of the human body are significant, but advancements in establishing tailored WPT techniques and transmitter reconfiguration strategies have achieved significant progress toward overcoming these obstacles.

First, the optimal design of WPT systems is deeply influenced by the depth of the implanted receiver, the choice of operating frequency, and the electromagnetic properties of both the transmitter and receiver. For subcutaneous implants, near-field coupling techniques such as inductive or capacitive WPT are effective, while for deeper implants, mid-field and radiative WPT techniques operating between 915 MHz and 2.45 GHz have been shown to provide higher efficiency. Apart from that, a conscious choice of the receiver antenna is crucial in achieving high efficiency. Even though the characteristics of the receiver depend on the application, reconfigurable topologies are ideal for mitigating the impact of body movement and ensuring efficient power transfer despite variations in receiver orientation.

However, the on-body transmitters are the WPT component that can be mostly tailored to enhance power transfer efficiency. First, impedance-matching structures based on low-loss superstrates or metasurfaces can reduce power dissipation in tissues and increase the power delivery to the implant. However, single-antenna transmitters, primarily used in the literature, often result in significant power loss, as most of the transmitted energy is absorbed by surrounding tissues. Therefore, introducing reconfigurable conformal antenna arrays, coupled with optimal excitation achieved with wavefront shaping techniques, offers a promising solution by focusing the field distribution on the implanted device, thereby maximizing efficiency and reducing electromagnetic exposure.

Overall, by combining reconfigurable transmitter arrays with optimized receiver designs and advanced focusing techniques, WPT systems can achieve near-optimal performance, ensuring long-lasting, reliable power delivery for deep-body implantable bioelectronics. This not only enhances device longevity but also improves user safety by minimizing power dissipation in non-target tissues.

Acknowledgements

This study was supported in part by the French Agence Nationale de la Recherche (ANR) through the Project MedWave under Grant ANR-21-CE19-0045.

References

1. U. F. Tesler, *A history of cardiac surgery: an adventurous voyage from antiquity to the artificial heart*, Cambridge Scholars Publishing, Newcastle upon Tyne, UK, 1st edn, 2020.
2. U. Anwar, Z. Liu and D. Markovic, A 6.78-MHz burst-mode controlled inductive wireless power transfer system for biomedical implants with back-channel communication eliminated using transmitter q-factor detection, *IEEE J. Emerging Sel. Top. Power Electron.*, 2022, **10**, 6383–6395.
3. J. Wang, E. G. Lim, M. P. Leach, Z. Wang, R. Pei, Z. Jiang and Y. Huang, A 403 MHz wireless power transfer system with tuned split-ring loops for implantable medical devices, *IEEE Trans. Antennas Propag.*, 2022, **70**, 1355–1366.
4. A. Basir and H. Yoo, Efficient wireless power transfer system with a miniaturized quad-band implantable antenna for deep-body multitasking implants, *IEEE Trans. Microwave Theory Tech.*, 2020, **68**, 1943–1953.
5. H. Le-Huu and C. Seo, Bipolar spiral midfield wireless power transfer for cardiac implants application, *IEEE Antennas Wirel. Propag. Lett.*, 2021, 1.
6. W. Ouyang, W. Lu, Y. Zhang, Y. Liu, J. U. Kim, H. Shen, Y. Wu, H. Luan, K. Kilner, S. P. Lee, Y. Lu, Y. Yang, J. Wang, Y. Yu, A. J. Wegener, J. A. Moreno, Z. Xie, Y. Wu, S. M. Won, K. Kwon, C. Wu, W. Bai, H. Guo, T. Liu, H. Bai, G. Monti, J. Zhu, S. R. Madhvapathy, J. Trueb, M. Stanslaski, E. M. Higbee-Dempsey, I. Stepien, N. Ghoreishi-Haack, C. R. Haney, T. Kim, Y. Huang, R. Ghaffari, A. R. Banks, T. C. Jhou, C. H. Good and J. A. Rogers, A wireless and battery-less implant for

multimodal closed-loop neuromodulation in small animals, *Nat. Biomed. Eng.*, 2023, 1–18.

7. G. Gaugain, L. Quéguiner, M. Bikson, R. Sauleau, M. Zhadobov, J. Modolo and D. Nikolayev, Quasi-static approximation error of electric field analysis for transcranial current stimulation, *J. Neural Eng.*, 2023, **20**, 016027.
8. G. L. Barbruni, P. M. Ros, D. Demarchi, S. Carrara and D. Ghezzi, Miniaturised wireless power transfer systems for neurostimulation: a review, *IEEE Trans. Biomed. Circuits Syst.*, 2020, **14**, 1160–1178.
9. K. L. Kozielski, A. Jahanshahi, H. B. Gilbert, Y. Yu, Ö. Erin, D. Francisco, F. Alosaimi, Y. Temel and M. Sitti, Nonresonant powering of injectable nanoelectrodes enables wireless deep brain stimulation in freely moving mice, *Sci. Adv.*, 2021, **7**, eabc4189.
10. C. Rong, B. Zhang, Z. Wei, L. Wu and X. Shu, A wireless power transfer system for spinal cord stimulation based on generalized parity-time symmetry condition, *IEEE Trans. Ind. Appl.*, 2021, 1.
11. G. Mao, Z. Zhou, H. Su, Y. Chen, J. Zhang, C. Zhang, Z. Wang and X. Lü, A fully implantable and programmable epidural spinal cord stimulation system for rats with spinal cord injury, *IEEE Trans. Neural Syst. Rehabil. Eng.*, 2023, **31**, 818–828.
12. S. Abolpour Moshizi, C. J. Pastras, R. Sharma, M. A. Parvez Mahmud, R. Ryan, A. Razmjou and M. Asadnia, Recent advancements in bioelectronic devices to interface with the peripheral vestibular system, *Biosens. Bioelectron.*, 2022, **214**, 114521.
13. V. Kaim, N. Singh, B. K. Kanaujia, L. Matekovits, K. P. Esselle and K. Rambabu, Multi-channel implantable cubic rectenna MIMO system with CP diversity in orthogonal space for enhanced wireless power transfer in biotelemetry, *IEEE Trans. Antennas Propag.*, 2023, **71**, 200–214.
14. A. Iqbal, M. Al-Hasan, I. B. Mabrouk, A. Basir, M. Nedil and H. Yoo, Biotelemetry and wireless powering of biomedical implants using a rectifier integrated self-diplexing implantable antenna, *IEEE Trans. Microwave Theory Tech.*, 2021, **69**, 3438–3451.
15. E. Cil, I. V. Soares, D. Renaudeau, R. Lucas, S. Dumanli, R. Sauleau and D. Nikolayev, On the use of impedance detuning for gastrointestinal segment tracking of ingestible capsules, *IEEE Trans. Antennas Propag.*, 2023, **71**, 1977–1981.
16. A. Abid, J. M. O'Brien, T. Bensel, C. Cleveland, L. Booth, B. R. Smith, R. Langer and G. Traverso, Wireless power transfer to millimeter-sized gastrointestinal electronics validated in a swine model, *Sci. Rep.*, 2017, **7**, 46745.
17. S. Sharma, K. B. Ramadi, N. H. Poole, S. S. Srinivasan, K. Ishida, J. Kuosmanen, J. Jenkins, F. Aghlmand, M. B. Swift, M. G. Shapiro, G. Traverso and A. Emami, Location-aware ingestible microdevices for wireless monitoring of gastrointestinal dynamics, *Nat. Electron.*, 2023, **6**, 242–256.
18. Y. Zhang, F. Liu, Y. Zhang, J. Wang, D. D'Andrea, J. B. Walters, S. Li, H.-J. Yoon, M. Wu, S. Li, Z. Hu, T. Wang, J. Choi, K. Bailey, E. Dempsey,

K. Zhao, A. Lantsova, Y. Bouricha, I. Huang, H. Guo, X. Ni, Y. Wu, G. Lee, F. Jiang, Y. Huang, C. K. Franz and J. A. Rogers, Self-powered, light-controlled, bioresorbable platforms for programmed drug delivery, *Proc. Natl. Acad. Sci. U. S. A.*, 2023, **120**, e2217734120.

19. J. C. Yang, J. Mun, S. Y. Kwon, S. Park, Z. Bao and S. Park, Electronic skin: recent progress and future prospects for skin-attachable devices for health monitoring, robotics, and prosthetics, *Adv. Mater.*, 2019, **31**, 1904765.
20. Y. Yu, H. Y. Y. Nyein, W. Gao and A. Javey, Flexible electrochemical bioelectronics: the rise of *in situ* bioanalysis, *Adv. Mater.*, 2020, **32**, 1902083.
21. Y. Zhou, X. Jia, D. Pang, S. Jiang, M. Zhu, G. Lu, Y. Tian, C. Wang, D. Chao and G. Wallace, An integrated Mg battery-powered iontophoresis patch for efficient and controllable transdermal drug delivery, *Nat. Commun.*, 2023, **14**, 297.
22. H. Sheng, X. Zhang, J. Liang, M. Shao, E. Xie, C. Yu and W. Lan, Recent advances of energy solutions for implantable bioelectronics, *Adv. Healthcare Mater.*, 2021, **10**, 2100199.
23. H. Zhou, Y. Zhang, Y. Qiu, H. Wu, W. Qin, Y. Liao, Q. Yu and H. Cheng, Stretchable piezoelectric energy harvesters and self-powered sensors for wearable and implantable devices, *Biosens. Bioelectron.*, 2020, **168**, 112569.
24. X. Huang, L. Wang, H. Wang, B. Zhang, X. Wang, R. Y. Z. Stening, X. Sheng and L. Yin, Materials Strategies and Device Architectures of Emerging Power Supply Devices for Implantable Bioelectronics, *Small*, 2020, **16**, 1902827.
25. M. Hyland, H. Hunter, J. Liu, E. Veety and D. Vashaee, Wearable thermoelectric generators for human body heat harvesting, *Appl. Energy*, 2016, **182**, 518–524.
26. X. L. Chang, P. S. Chee and E. H. Lim, Compact conformal tattoo-polymer antenna for on-body wireless power transfer, *Sci. Rep.*, 2023, **13**, 9678.
27. I. V. Soares, M. Gao, E. Cil, Z. Šipuš, A. K. Skrivervik, J. S. Ho and D. Nikolayev, Wireless powering efficiency of deep-body implantable devices, *IEEE Trans. Microwave Theory Tech.*, 2023, **71**, 2680–2692.
28. C. A. Balanis, *Antenna theory: analysis and design*, Wiley, Hoboken, New Jersey, 4th edn, 2016.
29. I. Soares, P. Vadher, A. Skrivervik, G. Sacco and D. Nikolayev, in *2023 17th European Conference on Antennas and Propagation (EuCAP)*, Florence, Italy, 2023.
30. D. Nikolayev, I. V. Soares, A. Mazzinghi and A. K. Skrivervik, Conformal beam-scanning arrays: survey, analysis, and benchmarking, *IEEE Antennas Propag. Mag*, 2025, **67**(4), 6–17.
31. Y. Su, I. V. Soares, S. Balon, J. Cao, D. Nikolayev and A. K. Skrivervik, Compact irregular conformal phased array: high-gain wide-scan on-board system for enhancing UAV communication, *IEEE Trans. Veh. Technol.*, 2024, 1–12.

32. D. Nikolayev, W. Joseph, A. Skrivervik, M. Zhadobov, L. Martens and R. Sauleau, Dielectric-loaded conformal microstrip antennas for versatile in-body applications, *IEEE Antennas Wirel. Propag. Lett.*, 2019, **18**, 2686–2690.
33. D. Nikolayev, A. K. Skrivervik, J. S. Ho, M. Zhadobov and R. Sauleau, Reconfigurable dual-band capsule-conformal antenna array for in-body bioelectronics, *IEEE Trans. Antennas Propag.*, 2022, **70**, 3749–3761.
34. S. Bakogianni and S. Koulouridis, A dual-band implantable rectenna for wireless data and power support at sub-GHz region, *IEEE Trans. Antennas Propag.*, 2019, **67**, 6800–6810.
35. IEEE, IEEE standard for safety levels with respect to human exposure to electric, magnetic, and electromagnetic fields, 0 Hz to 300 GHz, IEEE Std C951-2019 Revis. IEEE Std C951-2005 Inc. IEEE Std C951-2019Cor 1-*2019*, 2019, pp. 1–312.
36. T. Shaw, G. Samanta and D. Mitra, Efficient wireless power transfer system for implantable medical devices using circular polarized antennas, *IEEE Trans. Antennas Propag.*, 2021, **69**, 4109–4122.
37. S. Kim, J. S. Ho and A. S. Y. Poon, Wireless power transfer to miniature implants: transmitter optimization, *IEEE Trans. Antennas Propag.*, 2012, **60**, 4838–4845.
38. J. S. Ho, S. Kim and A. S. Y. Poon, Midfield wireless powering for implantable systems, *Proc. IEEE*, 2013, **101**, 1369–1378.
39. K. Zhang, C. Liu, Z. H. Jiang, Y. Zhang, X. Liu, H. Guo and X. Yang, Near-field wireless power transfer to deep-tissue implants for biomedical applications, *IEEE Trans. Antennas Propag.*, 2020, **68**, 1098–1106.
40. S. A. A. Shah and H. Yoo, Radiative near-field wireless power transfer to scalp-implantable biotelemetric device, *IEEE Trans. Microwave Theory Tech.*, 2020, **68**, 2944–2953.
41. R. Das and H. Yoo, A multiband antenna associating wireless monitoring and nonleaky wireless power transfer system for biomedical implants, *IEEE Trans. Microwave Theory Tech.*, 2017, **65**, 2485–2495.
42. J. S. Ho, A. J. Yeh, E. Neofytou, S. Kim, Y. Tanabe, B. Patlolla, R. E. Beygui and A. S. Y. Poon, Wireless power transfer to deep-tissue microimplants, *Proc. Natl. Acad. Sci. U. S. A.*, 2014, **111**, 7974–7979.
43. D. R. Agrawal, Y. Tanabe, D. Weng, A. Ma, S. Hsu, S.-Y. Liao, Z. Zhen, Z.-Y. Zhu, C. Sun, Z. Dong, F. Yang, H. F. Tse, A. S. Y. Poon and J. S. Ho, Conformal phased surfaces for wireless powering of bioelectronic microdevices, *Nat. Biomed. Eng.*, 2017, **1**, 1–9.
44. Y. Tanabe, J. S. Ho, J. Liu, S.-Y. Liao, Z. Zhen, S. Hsu, C. Shuto, Z.-Y. Zhu, A. Ma, C. Vassos, P. Chen, H. F. Tse and A. S. Y. Poon, High-performance wireless powering for peripheral nerve neuromodulation systems, *PLoS One*, 2017, **12**, e0186698.
45. M. F. Imani and A. Grbic, A unidirectional subwavelength focusing near-field plate, *J. Appl. Phys.*, 2014, **115**, 044904.
46. F. Gao, F. Zhang, M. Huang and D. F. Sievenpiper, Programmable Screen for Patterning Magnetic Fields, *IEEE Trans. Microwave Theory Tech.*, 2014, **62**, 481–490.

47. G. Ren, M. Wang, H. Zheng and E. Li, Wireless Power Transmission Efficiency Improved by Conformal Phase Gradient Metasurface for Implanted Devices, *IEEE J. Electromagn. RF Microwaves Med. Biol.*, 2024, **8**, 325–331.
48. I. A. Shah, M. Zada, S. A. A. Shah, A. Basir and H. Yoo, Flexible Metasurface-Coupled Efficient Wireless Power Transfer System for Implantable Devices, *IEEE Trans. Microwave Theory Tech.*, 2024, **72**, 2534–2547.
49. S. Mahmud, A. Nezaratizadeh, A. B. Satriya, Y.-K. Yoon, J. S. Ho and A. Khalifa, Harnessing metamaterials for efficient wireless power transfer for implantable medical devices, *Bioelectron. Med.*, 2024, **10**, 7.
50. I. V. Soares, M. Gao, A. K. Skrivervik, Z. Sipus, M. Zhadobov, R. Sauleau and D. Nikolayev, in *2021 IEEE Wireless Power Transfer Conference (WPTC)*, San Diego, USA, 2021.
51. J. Liska, M. Gao, L. Jelinek, M. Capek and A. Skrivervik, in *2023* 17th European Conference on Antennas and Propagation (EuCAP), 2023, pp. 1–5.
52. J. Liska, M. Gao, L. Jelinek, E. R. Algarp, A. K. Skrivervik and M. Capek, Maximum radiation efficiency of arbitrarily shaped implantable antennas, *IEEE Trans. Antennas Propag.*, 2024, **72**, 3507–3516.
53. D. Nikolayev, W. Joseph, M. Zhadobov, R. Sauleau and L. Martens, Optimal radiation of body-implanted capsules, *Phys. Rev. Lett.*, 2019, **122**, 108101.
54. H. Wong, W. Lin, L. Huitema and E. Arnaud, Multi-polarization reconfigurable antenna for wireless biomedical system, *IEEE Trans. Biomed. Circuits Syst.*, 2017, **11**, 652–660.
55. M. Chang, J. Han, Y. Li, X. Ma, H. Liu and L. Li, Self-powered polarization-reconfigurable rectenna for wireless power transfer system, *IEEE Trans. Antennas Propag.*, 2023, **71**, 6297–6307.
56. M. Karthikeyan, J. Pradeep, M. Harikrishnan, R. Sitharthan, N. Mishra, M. Rajesh and T. Sivanesan, in *Advances in Flexible and Printed Electronics: Materials, fabrication, and applications*, IOP Publishing, 2023.
57. I. V. Soares and D. Nikolayev, in *2022 Wireless Power Week (WPW)*, IEEE, Bordeaux, France, 2022, pp. 284–288.
58. I. V. Soares, A. K. Skrivervik and D. Nikolayev, in 20212 16th European Conference on Antennas and Propagation (EuCAP), Madrid, Spain, 2022.
59. F. Yang, P. M. Lee, Z. Dong, X. Tian and J. S. Ho, Enhancing wireless transmission from the body with wearable diffractive patterns, *Phys. Rev. Appl.*, 2019, **12**, 054020.
60. F. Yang, D. T. Nguyen, B. O. Raeker, A. Grbic and J. S. Ho, Near-reflectionless wireless transmission into the body with cascaded metasurfaces, *IEEE Trans. Antennas Propag.*, 2022, **70**, 8379–8388.
61. L.-J. Xu, L. Zhu and M. Liang, Impedance matching at skin–air interface for transmission-range improvement of implantable antennas, *IEEE Trans. Antennas Propag.*, 2023, **71**, 2235–2246.
62. D. R. Agrawal, Y. Tanabe, D. Weng, A. Ma, S. Hsu, S.-Y. Liao, Z. Zhen, Z.-Y. Zhu, C. Sun, Z. Dong, F. Yang, H. F. Tse, A. S. Y. Poon and J. S. Ho,

Conformal phased surfaces for wireless powering of bioelectronic microdevices, *Nat. Biomed. Eng.*, 2017, **1**, 1–9.
63. J. S. Ho, B. Qiu, Y. Tanabe, A. J. Yeh, S. Fan and A. S. Y. Poon, Planar immersion lens with metasurfaces, *Phys. Rev. B: Condens. Matter Mater. Phys.*, 2015, **91**, 125145.
64. H. Herssens, W. Joseph and A. Thielens, A Survey of On-Body Antenna Arrays: Future improvements, new designs, and lessons learned [Bioelectromagnetics], *IEEE Antennas Propag. Mag.*, 2023, **65**, 86–96.
65. G. Buonanno and S. Costanzo, Fresnel-Zone Focused Antenna Arrays: Tolerance Analysis for Biomedical Applications, *IEEE Trans. Antennas Propag.*, 2023, **71**, 7261–7272.
66. M. Converse, E. J. Bond, S. C. Hagness and B. D. Van Veen, Ultrawide-band microwave space-time beamforming for hyperthermia treatment of breast cancer: a computational feasibility study, *IEEE Trans. Microwave Theory Tech.*, 2004, **52**, 1876–1889.
67. X. Li, S. K. Davis, S. C. Hagness, D. W. vanderWeide and B. D. VanVeen, Microwave Imaging *via* Space–Time Beamforming: Experimental Investigation of Tumor Detection in Multilayer Breast Phantoms, *IEEE Trans. Microwave Theory Tech.*, 2004, **52**, 1856–1865.
68. R. F. Remis and P. M. van den Berg, On the equivalence of the Newton-Kantorovich and distorted Born methods, *Inverse Probl.*, 2000, **16**, L1.
69. R. Chandra, H. Zhou, I. Balasingham and R. M. Narayanan, On the Opportunities and Challenges in Microwave Medical Sensing and Imaging, *IEEE Trans. Biomed. Eng.*, 2015, **62**, 1667–1682.
70. V. Harid, H. Kim, B.-Z. Li and T. Lei, A method for non-destructive microwave focusing for deep brain and tissue stimulation, *PLoS One*, 2023, **2**, e0278765.
71. B. C. Benedict, M. M. Ghanbari and R. Muller, Phased Array Beamforming Methods for Powering Biomedical Ultrasonic Implants, *IEEE Trans. Ultrason. Ferroelectr. Freq. Control*, 2022, **69**, 2756.
72. Z. Kashani and M. Kiani, A Study on Ultrasonic Wireless Power Transfer With Phased Array for Biomedical Implants, *IEEE Trans. Biomed. Circuits Syst.*, 2023, 1–11.
73. H. Firew and M. T. Barros, An Adaptable Lateral Resolution Acoustic Beamforming for the Internet of Bio-Nano Things in the Brain, *IEEE Trans. Mol. Biol. Multi-Scale Commun.*, 2023, **9**, 217–221.
74. H. Rivandi and T. L. Costa, A 2D Ultrasound Phased-Array Transmitter ASIC for High-Frequency US Stimulation and Powering, *IEEE Trans. Biomed. Circuits Syst.*, 2023, **17**, 701–712.
75. J. Yang, E. Chérin, J. Yin, P. A. Dayton, F. S. Foster and C. E. M. Demore, Superharmonic and Micro-Ultrasound Imaging with Plane Wave Beamforming Techniques, *IEEE Trans. Ultrason. Ferroelectr. Freq. Control*, 2023, 1–1.
76. P. Nepa and A. Buffi, Near-field-focused microwave antennas: near-field shaping and implementation, *IEEE Antennas Propag. Mag.*, 2017, **59**, 42–53.

77. S. C. Pavone, M. Ettorre and M. Albani, Analysis and design of bessel beam launchers: longitudinal polarization, *IEEE Trans. Antennas Propag.*, 2016, **64**, 2311–2318.
78. E. S. Gamez Rodriguez, M. Machnoor and G. Lazzi, On the generation of nondiffracting beams in extremely subwavelength applications, *IEEE Trans. Antennas Propag.*, 2017, **65**, 5228–5237.
79. S. C. Pavone, M. Ettorre, M. Casaletti and M. Albani, Analysis and design of bessel beam launchers: transverse polarization, *IEEE Trans. Antennas Propag.*, 2021, **69**, 5175–5180.
80. E. Negri, W. Fuscaldo, P. Burghignoli and A. Galli, Leaky-wave analysis of TM-, TE-, and hybrid-polarized aperture-fed bessel-beam launchers for wireless-power-transfer links, *IEEE Trans. Antennas Propag.*, 2023, **71**, 1424–1436.
81. I. Iliopoulos, S. D. Meo, M. Pasian, M. Zhadobov, P. Pouliguen, P. Potier, L. Perregrini, R. Sauleau and M. Ettorre, Enhancement of penetration of millimeter waves by field focusing applied to breast cancer detection, *IEEE Trans. Biomed. Eng.*, 2021, **68**, 959–966.
82. X. Tian, P. M. Lee and J. S. Ho, Control of wireless power transfer to a bioelectronic device by harmonic feedback, *AIP Adv.*, 2018, **8**, 095308.
83. I. V. Soares, Z. Sipus, A. K. Skrivervik, R. Sauleau, A. Alu, J. S. Ho and D. Nikolayev, Wireless Powering for Long-Lasting Deep-Body Bioelectronic Devices, *IEEE Antennas Propag. Mag.*, 2025, **67**(3), 74–86.
84. A. K. Skrivervik, M. Bosiljevac and Z. Sipus, Fundamental limits for implanted antennas: maximum power density reaching free space, *IEEE Trans. Antennas Propag.*, 2019, **67**, 4978–4988.
85. D. Nikolayev, I. V. Soares, A. Mazzinghi and A. K. Skrivervik, Conformal Beam-Scanning Arrays: Survey, analysis, and benchmarking, *IEEE Antennas Propag. Mag.*, 2024, 2–14.
86. D. Nikolayev, M. Zhadobov, L. Le Coq, P. Karban and R. Sauleau, Robust ultraminiature capsule antenna for ingestible and implantable applications, *IEEE Trans. Antennas Propag.*, 2017, **65**, 6107–6119.
87. T.-Y. Shih and N. Behdad, Wideband, Non-Foster Impedance Matching of Electrically Small Transmitting Antennas, *IEEE Trans. Antennas Propag.*, 2018, **66**, 5687–5697.
88. X. Yang, Z. Zhang, M. Xu, S. Li, Y. Zhang, X.-F. Zhu, X. Ouyang and A. Alù, Digital non-Foster-inspired electronics for broadband impedance matching, *Nat. Commun.*, 2024, **15**, 4346.
89. Y. Zhou, G. H. Burgoyne Morris and M. Nair, Current and emerging strategies for biocompatible materials for implantable electronics, *Cell Rep. Phys. Sci.*, 2024, **5**, 101852.
90. J. E. Woods, F. Alrashdan, E. C. Chen, W. Tan, M. John, L. Jaworski, D. Bernard, A. Post, A. Moctezuma-Ramirez, A. Elgalad, A. G. Steele, S. M. Barber, P. J. Horner, A. H. Faraji, D. G. Sayenko, M. Razavi and J. T. Robinson, *Nat. Biomed. Eng.*, 2026, **10**, 532–544.
91. X. Tian, Q. Zeng, S. A. Kurt, R. R. Li, D. T. Nguyen, Z. Xiong, Z. Li, X. Yang, X. Xiao, C. Wu, B. C. K. Tee, D. Nikolayev, C. J. Charles and

J. S. Ho, Implant-to-implant wireless networking with metamaterial textiles, *Nat. Commun.*, 2023, **14**, 4335.

92. W. Geyi, The Method of Maximum Power Transmission Efficiency for the Design of Antenna Arrays, *IEEE Open J. Antennas Propag.*, 2021, **2**, 412–430.
93. A. Krasnok, D. G. Baranov, A. Generalov, S. Li and A. Alù, Coherently enhanced wireless power transfer, *Phys. Rev. Lett.*, 2018, **120**, 143901.
94. A. Krasnok, N. Nefedkin and A. Alu, Parity-time symmetry and exceptional points [electromagnetic perspectives], *IEEE Antennas Propag. Mag.*, 2021, **63**, 110–121.
95. C. W. Hsu, B. Zhen, A. D. Stone, J. D. Joannopoulos and M. Soljačić, Bound states in the continuum, *Nat. Rev. Mater.*, 2016, **1**, 1–13.
96. M. Song, P. Jayathurathnage, E. Zanganeh, M. Krasikova, P. Smirnov, P. Belov, P. Kapitanova, K. Simovski, S. Tretyakov and A. Krasnok, Wireless power transfer based on novel physical concepts, *Nat. Electron.*, 2021, **4**, 707–716.
97. N. Dutta, S. Mukherjee and K. Mandal, A Highly Efficient Hybrid WPT System Possessing Beam-Steering Beam-Splitting and Near-field Focusing Capability at ISM Band, *IEEE Trans. Antennas Propag.*, 2025, 1.
98. M. Meng and M. Kiani, A Hybrid Inductive-Ultrasonic Link for Wireless Power Transmission to Millimeter-Sized Biomedical Implants, *IEEE Trans. Circuits Syst. II Express Briefs*, 2017, **64**, 1137–1141.
99. M. S. Rahman, A. Shon, R. Joseph, A. Pavlov, A. Stefanov, M. Namkoong, H. Guo, D. Bui, R. Master, A. Sharma, J. Lee, M. Rivas, A. Elati, Y. Jones-Hall, F. Zhao, H. Park, M. A. Hook and L. Tian, Soft, stretchable conductive hydrogels for high-performance electronic implants, *Sci. Adv.*, 2025, **11**, eads4415.
100. L. Wang, K. Jiang and G. Shen, Wearable, Implantable, and Interventional Medical Devices Based on Smart Electronic Skins, *Adv. Mater. Technol.*, 2021, **6**, 2100107.
101. Y. Zhang, G. Lee, S. Li, Z. Hu, K. Zhao and J. A. Rogers, Advances in Bioresorbable Materials and Electronics, *Chem. Rev.*, 2023, **123**, 11722–11773.
102. J. Hwangbo, H. Seo, G. Sim, R. Avila, M. Nair, B. Kim and Y. Choi, Bioresorbable polymers for electronic medicine, *Cell Rep. Phys. Sci.*, 2024, **5**, 102099.
103. M. Jafari, G. Marquez, H. Dechiraju, M. Gomez and M. Rolandi, Merging machine learning and bioelectronics for closed-loop control of biological systems and homeostasis, *Cell Rep. Phys. Sci.*, 2023, **4**(8), 101535.
104. M. Jafari, G. Marquez, J. Selberg, M. Jia, H. Dechiraju, P. Pansodtee, M. Teodorescu, M. Rolandi and M. Gomez, Feedback Control of Bioelectronic Devices Using Machine Learning, *IEEE Control Syst. Lett.*, 2021, **5**, 1133–1138.
105. S. Tang, P. Cheng, H. Liang, D. Sha, L. Zhao, D. Ding, G. Liu, C. Liu, Y. Liu, Y. Ma and Y. Yuan, Deep-Learning Integrated Bioelectronic-

Tissue Interface for Cardiovascular Diagnosis and Prognosis, *Adv. Funct. Mater.*, 2025, 2423264.

106. A. Suwardi, F. Wang, K. Xue, M.-Y. Han, P. Teo, P. Wang, S. Wang, Y. Liu, E. Ye, Z. Li and X. J. Loh, Machine Learning-Driven Biomaterials Evolution, *Adv. Mater.*, 2022, **34**, 2102703.
107. D. Wei, D. Deng, S. Gui, W. You, J. Feng, X. Meng, X. Chen, J. Lv, Y. Tang, T. Chen and P. Liu, Machine learning to predict in-stent stenosis after Pipeline embolization device placement, *Front. Neurol.*, 2022, **13**, 912984.
108. C. Krittanawong, A. J. Rogers, K. W. Johnson, Z. Wang, M. P. Turakhia, J. L. Halperin and S. M. Narayan, Integration of novel monitoring devices with machine learning technology for scalable cardiovascular management, *Nat. Rev. Cardiol.*, 2021, **18**, 75–91.
109. S. Gabriel, R. W. Lau and C. Gabriel, The dielectric properties of biological tissues: III. Parametric models for the dielectric spectrum of tissues, *Phys. Med. Biol.*, 1996, **41**, 2271–2293.
110. C. Liu, Y.-X. Guo, H. Sun and S. Xiao, Design and safety considerations of an implantable rectenna for far-field wireless power transfer, *IEEE Trans. Antennas Propag.*, 2014, **62**, 5798–5806.
111. A. Abdi and H. Aliakbarian, A Miniaturized UHF-Band Rectenna for Power Transmission to Deep-Body Implantable Devices, *IEEE J. Transl. Eng. Health Med.*, 2019, **7**, 1–11.
112. D. Sharma, S. Kumar, N. Singh, B. K. Kanaujia, S. P. Singh and Aimé Lay-Ekuakille, Far-field wireless power transmission and measurement for a leadless transcatheter pacing system, *IEEE Trans. Instrum. Meas.*, 2023, **72**, 1–12.
113. G. Ren, M. Wang, H. Zheng and E. Li, Wireless power transmission efficiency improved by conformal phase gradient metasurface for implanted devices, *IEEE J. Electromagn. RF Microw. Med. Biol.*, 2024, 1–7.

CHAPTER 11

Trends and Challenges of Energy Material Integration with Biomedical Devices for In Vivo *Applications*

KARTIKEYA SHARMA,*[a] MORTEN BO SØNDERGAARD SVENDSEN,[a] DURVA A. NAIK[b] AND ANJA BOISEN*[a]

[a] Drug Delivery and Sensing (IDUN), Department of Health Technology, Technical University of Denmark, Ørsted Plads, 2800, Kgs. Lyngby, Denmark; [b] Materials Science and Engineering, Carnegie Mellon University, 5000 Forbes Avenue, Wean Hall 3325, Pittsburgh, PA 15213, USA
*Emails: karsha@dtu.dk; aboi@dtu.dk

11.1 Introduction

Implantable and ingestible electronic devices are establishing a new standard for non-invasive or minimally invasive healthcare towards monitoring, diagnostic, and therapeutic functions.[1] Advancements in materials science, microfabrication techniques, microelectronics, and biomedical engineering have resulted in miniature electronic devices which can be implanted or ingested into the human body for real-time activities. Most implantable and ingestible electronic devices comprise a power unit, functional units in the form of sensors and actuators, embedded electronics, and a wireless communication unit.[2] The power unit forms an essential part of the device, supplying electrical energy to the whole system. Since their application in a pacemaker in 1958, batteries have been the main power unit

Detection Science Series No. 33
Energy Materials for Ingestible and Implantable Healthcare Devices
Edited by Santanu Patra and Anja Boisen

Published by the Royal Society of Chemistry, www.rsc.org

for biomedical electronic devices. Despite their widespread utility, conventional batteries have certain drawbacks associated with their use in implantable and ingestible devices, owing mainly to their rigidness and bulkiness, and risk of toxicity.[2] Consequently, scientific advancements have yielded improved alternatives like transient batteries, flexible batteries, solid-state batteries, supercapacitors, wireless power transfer (WPT), and energy harvesting.

As the technologies around implantable and ingestible devices are maturing, so are the needs of more sustainable power supply systems. Current devices require the power sources to be small, lightweight, durable, flexible, biocompatible/bioresorbable, and energy dense.[3] In the context of these requirements, energy materials are becoming increasingly valuable. Most devices, implantable or ingestible, are in contact with certain organs of the human body where physiological parameters like temperature, mechanical forces, pH, and bodily fluids are omnipresent. This provides an opportunity to harvest energy from these different parameters and power the implantable or ingestible devices. Such methods have already been demonstrated by harvesting mechanical energy,[4] thermal energy,[5] gastric and intestinal fluid,[6] and enzymes,[7] among others. Another benefit of using energy materials as power sources is their multifunctionality, which allows them to be used for pressure sensing,[8] temperature sensing,[9] drug delivery,[10] and electrostimulation[11] while harvesting energy simultaneously. Although the technology around these energy materials is advancing rapidly, a major challenge is presented in the integration of such technologies in devices. Devices based on energy materials often comprise power management electronics before they can be used as primary or secondary power sources,[12] which adds significantly to the complexity of the system in terms of the device size and biocompatibility.

Translating discoveries in energy materials research into practical biomedical devices for *in vivo* applications presents a unique set of challenges. These challenges arise mainly due to the application-specific requirements that translate into challenges in fabrication and miniaturization of integrated biomedical devices and risk mitigation for *in vivo* use.[13] Various challenges arise from the anatomy and physiology of the different body organs, outlining different constraints and requirements. In the context of energy materials, these challenges can be further reduced to (i) matching the choice of energy material to end-use application, (ii) matching of the energy density to the required performance, and (iii) matching the choice of energy material to the risk/benefit in relation to device performance. In this chapter, we address the challenges that arise in the integration of energy materials into biomedical electronic devices, specifically in terms of the integrated device fabrication, application-specific requirements, and risk mitigation (as shown in Figure 11.1) during their use highlighting the interplay between clinical needs, fabrication challenges, and risk–performance considerations.

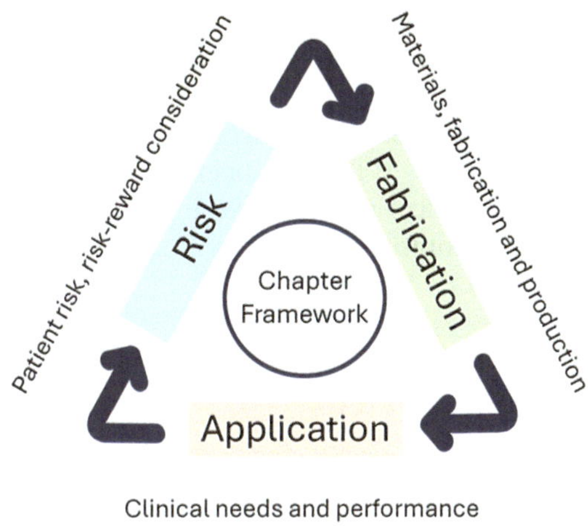

Figure 11.1 The framework devised to address the challenges of energy material integrated miniature biomedical devices.

11.2 State of the Art

The human body is a repository of energy forms, existing in the forms of thermal, biochemical, and kinetic energy. Depending on the function and location of biomedical devices, different types of energies can be harvested and translated into electrical energy as power sources. The translation of energy can be achieved using thermoelectric devices, biofuel cells, piezoelectric devices, triboelectric devices, and supercapacitors.[14] Alternatively, energy can be transferred non-invasively and wirelessly into the body using ultrasound or radio waves.[15,16] Such technologies can be implemented for implantable as well as ingestible devices.

Mechanical energy from the intrinsic movements of various organs has been used to power devices for sensing, drug delivery, and electrostimulation using triboelectric (TENGs) and piezoelectric nanogenerators (PENGs). The mechanical motion of the heartbeat has been used to power cardiac pacemakers and their pacing systems.[17] TENG-powered nonplanar devices have also been implanted into cardiac ventricles to harvest energy and monitor cardiac contractions remotely and wirelessly.[18] Motion caused by respiration has been used to harvest energy and power implanted pacemakers.[19] Electrostimulation has been achieved using TENG-driven devices for muscular rehabilitation[20] and weight control.[11] TENGs have also been applied to drive drug delivery devices *in vivo*.[21] Alternatively, PENGs have also been employed for powering functional implantable devices. PENGs have been shown to harvest energy from the motions of the heart, the lungs, and the diaphragm.[22] Ceramic material-based PENGs in forms of thin films and fibers have been used for cardiac stimulation and physiological sensing.[17] Biological fluids like blood, sweat, glucose, and gastric and intestinal fluid have been harvested into biochemical energy through biofuel cells

(BFCs).[23] Hybrid devices, combining BFCs and supercapacitors, have been demonstrated to enhance the energy output and exhibit the potential of energy harvesting inside the body using fluids.[24] Similarly, BFCs have been used in developing a self-powered biosensing ingestible capsule for the gastrointestinal (GI) tract.[25] WPT has been achieved to power deep implanted miniature biomedical devices, like ingestible electronics,[26] cardiac pacemakers,[27] and implantable electrochemical sensors.[28]

11.3 Challenges of Integrating Energy Materials into Biomedical Devices

11.3.1 Miniaturization and Fabrication of Energy Material Integrated Devices

Miniaturization of implantable and ingestible electronic devices is majorly driven by enhanced patient comfort, less invasive surgical procedures, lower infection risks, and shorter hospitalization time.[29] However, integration of energy materials to replace the conventional power sources becomes a bigger challenge as a result of this miniaturization. For energy materials in general, the electrical properties are directly proportional to the surface area of the active materials. Therefore, determining a method for obtaining high energy density while minimizing the weight and size of the devices is essential for energy material-powered devices. Despite their dependency on the surface area, ambient energy-powered devices rely less on bulky conventional batteries, providing greater scope for miniaturization.[30]

11.3.1.1 Miniaturization

Miniaturization of energy materials-powered *in vivo* devices results not only in lower surface area, but also in increased internal resistance, reduced power output, and higher effects of fluid dynamics on operational sites.[30] Hence, owing to the reduced available volume within devices, it is necessary to implement technologies that result in increased energy density and higher energy conversion efficiencies. Some approaches to achieve this include the use of organic/inorganic composites,[31] nanomaterial composites,[32] fiber-based devices,[33] surface modification,[34] ultrathin and flexible encapsulation materials,[35] enhanced tissue-device interface,[36] and flexible integrated electronics.[37]

To overcome the limitation of reduced surface for triboelectric energy-harvesting devices, alternative strategies are used to modulate the contact surface area. By introducing surface roughness on multiple scales, *e.g.*, micro and nano structures on the triboelectric layers, the current output can be significantly enhanced. To improve the charge density, nanomaterials like graphene, carbon nanotubes, graphite micro-flakes, silver nanowires, and metal fillers are also used to dope the triboelectric layers, which leads to an improved dielectric constant and higher capacitance of the layers.[38] Similarly, piezoelectric nanogenerators are fabricated with inorganic–organic or nanomaterial composites to enhance energy output. Nanomaterials are also being used to facilitate electron transfer at the electrode–biofluid interface in microbial and enzymatic fuel cells and to modify the electrode architecture to yield a higher surface area and reduce the impedance of electrodes at the surface.[39] The electrodes are being modified to achieve better bioelectrocatalysis and charge storage properties, which result in higher power output.[24]

Another challenge with miniaturization of energy materials integrated devices is power management electronics. Output from energy-harvesting devices is not only a consequence of the material properties, but also energy management and energy storage components. Energy output of analog nature requires rectification before it can be stored for further use. Devices having digital output usually require boost circuits to amplify the output to meet the voltage or current requirements.[40] For example, an enzymatic fuel cell which was used to power a pacemaker utilized a standard electronic DC–DC converter as a charge pump to increase the output voltage of a single cell.[41] These components themselves pose a size challenge in their integration into devices but also further add to the size requirement for the energy material. However, with advancements in microelectronics, these components are being made available with more advanced technologies and in smaller form factors, allowing for less components and more compact circuit designs. The scientific growth in the field of organic transistors has also shown promise in microfabricated and ink-based electronic circuits for energy harvesting.[42] Recently, an innovative approach towards rectification of energy output from triboelectric nanogenerators was reported, where conventional diodes were replaced with short-circuited thin-film transistors to fabricate a full-bridge rectification circuit.[37] The study also reported a soft and stretchable supercapacitor for storing the rectified energy obtained from the triboelectric nanogenerator and the rectification circuit, as shown in Figure 11.2.

Overall miniaturization of implantable and ingestible devices for real-time monitoring also requires miniaturization of wireless communication modules on-board. Antennas form an essential part of wireless communication, where the size of the antenna dictates the communication frequency, bandwidth, and gain. A reduction in the size of antennas leads to deterioration of the gain and the bandwidth, limiting the use of devices *in vivo*. Hence, to achieve an optimal tradeoff between the size and characteristics of an antenna, micro- and nanofabrication techniques are utilized. Most biomedical devices operate in a frequency range of 300 MHz to 3 GHz, providing the possibility of reducing the size of transmitting and receiving coils and antennas.[43] A capsule-based device was recently reported where a flexible thin-film inductive coil (coil antenna) was integrated onto the outer surface of the capsule for continuous fully wireless and passive pressure monitoring in the GI tract.[44] Similarly, various other strategies have been tested for developing miniature planar or conformal antennas for capsule-based

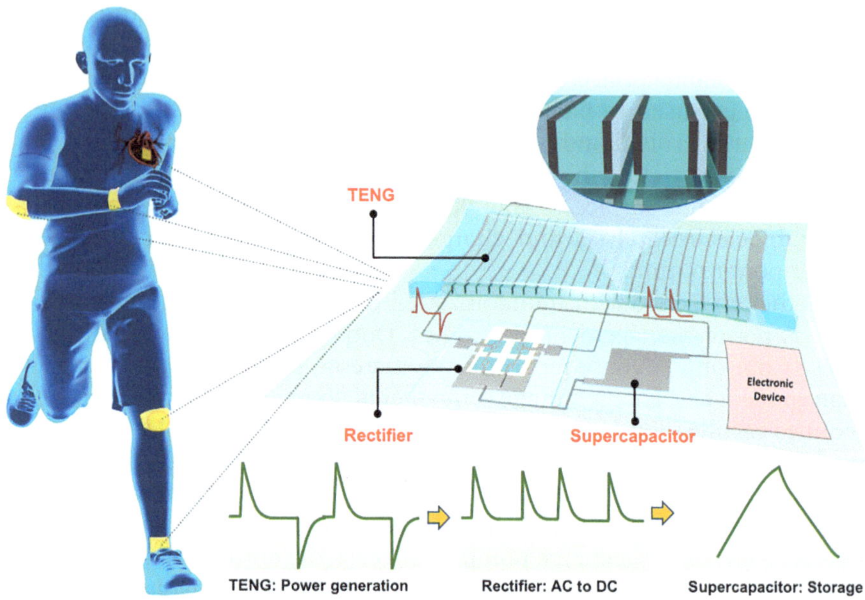

Figure 11.2 An intrinsically stretchable implantable TENG as a power source for implantable devices consisting of flexible electronics for energy management. Reproduced from ref. 37 with permission from Elsevier, Copyright 2023.

endoscopy devices operating in different frequency ranges.[45,46] 3D printing, inkjet printing, and microfabrication have been utilized to demonstrate miniature antenna designs for implantable electronic devices.

11.3.1.2 Fabrication of Miniaturized Devices

A direct consequence of miniaturization of biomedical devices is the challenge in their fabrication. The fabrication of energy material powered devices primarily involves separate micro- or nanofabrication of power supply component, sensing/actuation component, and integrated electronic component, followed by their compilation. The complexity of the devices often defines the choice of fabrication technique.[30] The shape, size, and structure of biomedical devices for *in vivo* applications are determined majorly by the dimensional constraints imposed by different body organs, or sites of application. These constraints also determine the choice of materials for the fabrication and mechanical properties of the devices. Owing to the softness of organ tissues and muscles, it is also highly desirable that these devices conform to the tissue walls without causing any damage or restrictions in their function.[36] As an example, self-powered pressure-sensing devices wrapped around blood vessels should be able to modulate their mechanics based on the blood flow in order to not interfere with the vessel's ability to compress or dilate the blood flow.[47]

Owing to these constraints, soft materials like polymers and their composites are mainly used for the fabrication of biomedical devices powered by energy materials. The use of such materials allows for flexible and stretchable devices which enhance the mechanical interface between the device and the biological tissue.[48] Additionally, biocompatible metals are also used as conductive materials for electrodes. Advances in micro- and nanofabrication have led to techniques that can provide from atomically to micro-scale thin devices, aiding the process of miniaturization. The most common techniques employed for fabrication of state-of-the-art devices include inkjet and screen printing,[49] high-resolution additive manufacturing,[50] two-photon polymerization,[51] photolithography,[52] and electrospinning,[53] among others.

Fabrication of integrated devices requires an electrical connection across the independently fabricated components which would traditionally be achieved using wires. However, recent advancements have led to the development of conductive inks, conductive fibers and textiles, and conductive filaments which, despite being soft and pliable, are highly conductive and can be used for establishing electrical connections.[54] The advancements in ink and printing technologies have provided printed electronic components which can replace the conventional electronic components to fabricate a completely soft device. These have also led to the development of flexible, stretchable, and biodegradable biomedical electronic devices. Transient implantable electronics devices have been demonstrated to develop customized lifetime devices using biocompatible and bioresorbable materials.[55] Fiber-based devices have been demonstrated for energy harvesting, energy storage, and as electrodes.[14]

11.3.2 Application-specific Requirements

The integration of energy materials into implantable and ingestible biomedical devices presents further challenges than those relating to technology itself. These system-level challenges must be addressed to ensure optimal performance and patient safety.

11.3.2.1 Conformity to Tissues

One of the primary requirements for functioning of implantable and ingestible devices is their ability to conform to soft tissues and organs of the body.[56] Further, the energy transfer for effective mechanical energy harvesting is a direct result of good conformity to the surrounding tissues, requiring the devices to be soft, flexible, and stretchable to match the mechanical properties of the tissues.[57] This also prevents any organ damage that might occur during the operation of the device *in vivo*. Mechanical properties of biological tissues vary from macro- to nanoscale,[58] which makes it difficult to exactly mimic the mechanics to enhance the compatibility and functionality. Mechanical conformation also allows for better contact between the device and the target soft tissue, allowing for better sensing, stimulation, and energy transfer.[59] Conformable devices are often achieved by the use of soft materials like hydrogels,[60] silicones,[59] and responsive materials,[61] which are inherently flexible or stretchable. Even though many applications require the use of inorganic materials as active energy material, like lead zirconate titanate (PZT) as a piezoelectric material, the fabrication process is such that the materials are used in a thin-layer form which does not compromise the ability of the device to conform with the body organs to a very high degree.[22] Other approaches to achieve conformation to the tissue include mechanical interlocking with the microstructures on the tissue surface,[62] stimuli-responsive mucoadhesion,[63] and the use of bioadhesives,[64] among others.

11.3.2.2 Functioning of the Device

Considering the previous section, the intended function of implantable or ingestible devices significantly influences their energy requirements and utilization strategies. Yang *et al.* explain a framework that categorizes these devices into three main functional types: closed-loop systems, diagnostic systems, and therapeutic systems, each with increasing levels of complexity and energy demands.[65] This systems approach is highly relevant when considering energy utilization of energy-harvesting or energy-converting systems, particularly in relation to charge-release cycles.

Closed-loop systems represent the most energy-efficient category, operating autonomously without external intervention. These devices typically monitor physiological parameters continuously or provide immediate feedback or stimulus. In these systems, energy harvesting or conversion can be directly coupled with immediate use, optimizing energy utilization.[66]

Diagnostic systems require more energy than closed-loop systems due to their data collection and transmission functions. These devices could be monitoring physiological parameters over time and store or transmit data for analysis. Ingestible sensors for measuring medication adherence and wireless motility capsules for assessing gut transit time are examples of diagnostic systems. For these devices, short-term energy storage may be necessary to power data transmission or processing functions. Therapeutic systems are the most demanding, as they actively intervene to treat medical conditions.[65] These devices deliver therapies such as electrical stimulation or drug release and would require control or timing of energy release. Therapeutic systems typically require long-term energy storage and sophisticated control mechanisms for energy utilization.[65]

The energy utilization strategy for a device depends on its function and energy source. Immediate use is ideal for closed-loop systems with differing consistency in energy harvest. Short-term storage, suitable for systems with intermittent data transmission, might involve supercapacitors or small batteries to accumulate harvested energy for burst transmissions. Long-term storage is necessary for therapeutic systems or long-term implants, typically involving batteries with high energy densities.

The function of implantable and ingestible devices directly impacts their energy requirements and utilization strategies. As devices progress from closed-loop to diagnostic to therapeutic functions, they require increasingly complex energy management systems. Developers must carefully consider the device's intended function when selecting energy materials and developing power management strategies to ensure optimal performance and longevity in the challenging *in vivo* environment. This approach aligns with the model introduced in the chapter, highlighting the interplay between clinical needs, production challenges, and risk–benefit considerations in the development of these advanced biomedical devices.

11.3.2.3 In Vitro *to* In Vivo *Performance Translation*

Performance in the lab *versus* performance *in vivo* presents significant challenges for biomedical devices, particularly when it comes to energy materials.[67] The discrepancy between controlled laboratory conditions *versus* the complex, dynamic environment of the human body is referred to as *in vitro*/*in vivo* correlation (IVIVC).[68]

IVIVC is a critical aspect of pharmaceutical development, aiming to predict *in vivo* performance based on *in vitro* data. However, establishing good correlations for implantable and ingestible devices is particularly challenging due to several factors.[69] The human body presents a highly variable environment both physiologically and chemically. There is also a significant variation between humans, with varying tissue properties, metabolism, and immune responses among different individuals.[68] The dynamic nature of the human body is unlike static laboratory settings—the *in vivo* environment is constantly changing due to factors such as body movement, temperature fluctuations, and varying pH levels. The operating conditions are not fixed other than what human homeostasis offers. Likewise, the *in vivo* performance is observed over longer periods of time, whereas *in vitro* tests may provide short-term data which often fail to capture the long-term effects of device implantation or ingestion, including tissue remodeling, inflammatory responses, and so on. Finally, the device–tissue interactions, including the tissue properties, and biological environment, are difficult to replicate *in vitro*. Efforts are best spent in isolated *in vitro* tests, making relevant parts realistic and compromising on the whole test-system realism. For example, the intestinal wall involves anatomical features varying in several orders of magnitude, from mucus pores and microvilli to tissue folds and muscles, which are generally not replicated in intestinal models. Similarly, lab-on-chip models, revealing information on organ and organoid interactions with drugs, isolate the cells from their biological environment, compromising the realism of the organ physiology during the experimentation.

Dividing *in vitro* tests into smaller, simpler, and more realistic units enables a much easier identification of device behavior in different scenarios, along with better understanding of the changes needed in the system to meet more realistic conditions. Even in the case of clinical trials, literature reporting real-world data, generated by daily delivery and administration of healthcare, can show different results compared to those of controlled experimental settings in clinical trials.[70] Transitioning from an *in vitro* to an *in vivo* environment, the operation becomes more uncontrolled with varying conditions at all scales, often leading to non-ideal device performance, which diminishes the actual effect of the technology.

11.3.3 Risk Mitigation for *In Vivo* Use

Energy materials with high volumetric energy density are a key design consideration for *in vivo* medical devices owing to the strict size limitations, while weight restrictions are generally less of a concern.[71] However, these high energy densities are achieved by employing materials that may not necessarily be biocompatible. Hence, owing to the delicate balance between high performance and biocompatibility of the devices, either of the parameters may be compromised during the fabrication and application of energy material integrated devices for *in vivo* use. Consequently, it is important to explore the biological risks and safety challenges associated with the deployment of miniature devices *in vivo*. Figure 11.3 shows the important characteristics to consider while designing miniature biomedical devices powered by energy materials.

Despite advances in materials science and energy-harvesting technologies, batteries remain as the primary power source for electronic biomedical devices.[72,73] Drawbacks associated with the use of batteries have led to the development of batteries based on alternate technologies like solid-state electrolytes (SSEs)[74] and biodegradable batteries,[75] among others. Polymer- or inorganic-based SSEs are emerging as safer and more efficient alternatives to conventional batteries, due to their design flexibility and mechanical flexibility.[74] Although low ionic conductivity and poor interfacial properties limit their use in devices, various efforts are being made to enhance the conductivity and to develop better energy storage systems to advance the use of SSE-based batteries.[76] Similarly, biodegradable batteries are being explored as solutions to mitigate the long-term risks associated with implanted

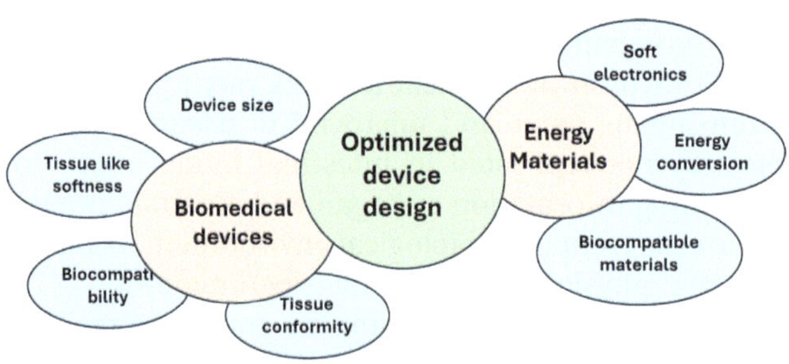

Figure 11.3 Strategies for the design of optimized energy material integrated miniature biomedical devices for *in vivo* use.

batteries *in vivo*. Magnesium-based batteries,[77] zinc-based batteries,[78] and sodium ion-based batteries[79] have shown potential in powering biomedical devices *in vivo*. The lifespan of biodegradable batteries can be regulated by adjusting the electrode fabrication,[80] by using biofluids as electrolytes (*e.g.*, blood, saliva, urine),[80] and by employing biopolymers.[81]

Although devices harvesting energy *in vivo* provide better biocompatibility compared to batteries, a major drawback is the low output current resulting in lower energy densities. Microbial and enzymatic fuel cells have shown potential in harvesting energy from abundant biological compounds, however the possibility of bioelectrode instability and biological contamination and biofouling have prevented their widespread use. Moreover, all biofuel cells are constrained by their dependence on local oxygen concentrations at the anode.[82] Oxygen diffusion limitations can reduce current density, leading to local oxygen depletion, which may worsen tissue fibrosis or induce hypoxia in the surrounding area. For biofuels to become mainstream in miniature self-powered biomedical devices these issues must be addressed to ensure their long-term viability and minimize risks *in vivo*.

For PENG-powered devices, the composition of the PENG significantly impacts both their energy output and biocompatibility. Prominent inorganic piezoelectric materials, such as lead magnesium niobate-lead titanate (PMN-PT),[83] lead zirconate titanate (PZT),[22] and barium titanate ($BaTiO_3$),[84] and organic materials like PVDF,[85] offer high piezoelectric coefficients but pose significant health risks due to their toxicity and rigidity. To overcome these drawbacks without compromising on the energy output, one strategy is to encapsulate PENGs with biocompatible materials like polydimethylsiloxane (PDMS), polytetrafluoroethylene (PTFE), polyimide, and SU-8 epoxy.[86] Alternatively, sodium-, potassium-, and lithium-based niobates are emerging as highly piezoelectric biocompatible alternatives. Zinc oxide (ZnO)-based piezoelectric structures have also been shown to power devices *in vivo* in their various nano-structural forms.[87] For applications where energy-harvesting performance can be compromised for biocompatibility, biological macromolecules, such as poly(lactic acid) (PLA) and poly(3-hydroxybutyrate) (P3HB), and naturally occurring polymers and proteins like elastin, silk fibroin, amylose, gelatin, and chitin can be used.[88] These materials offer a promising approach for creating on-demand electronic power sources in biomedical applications.

A significant challenge in the practical deployment of TENGs is ensuring their continuous operation under harsh environmental conditions. To address this, scientists have used encapsulation methods using robust materials for enhanced protection.[82,89,90] For implantable and *in vivo* applications, additional challenges must be addressed, including ensuring that the TENG's durability, flexibility, and biocompatibility complement the specific needs of the area where the device will be implanted. Furthermore, the immune system's foreign body response is a significant concern, as it can encapsulate the device with fibroblasts, potentially inhibiting the TENG's ability to receive biomechanical stimuli and impairing its long-term functionality.

11.3.4 Risk Management and Safety Standards

In the development and deployment of miniature biomedical devices for *in vivo* use, effective risk management is critical to ensure their durability and reliability over time. Risk assessment should be carefully integrated throughout the lifecycle of technology—from early-stage research and experimentation to clinical trials and post-market surveillance. This proactive approach helps identify potential hazards early while also safeguarding the opportunity to maximize patient benefits, ensuring that innovations do not get sidelined by overly cautious risk assessments. Risk avoidance may often lead to avoiding potential benefits, where opportunities to improve patient outcomes are missed due to overly cautious risk assessment.

As such, adherence to established standards—such as ISO 14971, which provides a framework for identifying, assessing, and mitigating risks associated with medical devices, and ISO 14708-1:2014, which ensures compliance of active implantable medical devices through specific requirements and testing for electrically powered devices or those utilizing alternative power sources—is essential. These guidelines are particularly relevant when integrating novel energy materials, as they help mitigate the risks associated with potential new materials and technologies, while ensuring their safe use. Novel materials may introduce unforeseen biocompatibility issues, unknown long-term safety concerns, or failure modes that have not yet been encountered in clinical settings. By carefully considering and following these ISO standards, researchers and manufacturers can balance the need for innovation with the imperative of patient safety, reducing the likelihood of overlooking critical risks while still embracing opportunities to enhance medical outcomes. Table 11.1 lists some relevant standards and guidelines

Table 11.1 Relevant standards encompassing the development of energy material integrated miniature biomedical devices for *in vivo* use.

Requirement	Standard/guideline	Focus
Biocompatibility	ISO10993 FDA Biocompatibility	Biocompatibility chemical test, biological evaluation, material safety, cytotoxicity
Electromagnetic compatibility (EMC)	IEC60601-1 ISO14117	Electromagnetic compatibility of active implantable devices, performance limits of EM emitters, EM shielding
Battery performance	ISO14708 ISO62133	Guidelines for active implantable medical devices, energy densities, alternative energy sources, safety tests for batteries
Ingestible device safety	FDA 21 CFR §876.1300	Specific requirements, identification/classification of ingestible devices
QMS|post-market surveillance	EU MDR, FD&C Act (21 U.S.C. §301 et seq.) ISO14971	Risk management of active devices, biocompatibility, electricity, moving parts, radiation, and usability

to be considered during the development of energy material integrated miniature biomedical devices for *in vivo* use. However, the list is not exhaustive and hence should only be considered as a starting point towards research and development of energy material-powered devices.

11.4 Conclusion

As it stands, the research around different materials aimed at alternative sources of energy *in vivo* shows potential in powering low-power devices and sensor systems. Despite the drawbacks, such devices powered *via* energy materials other than batteries exhibit biocompatibility and reduced risk in biological applications, albeit with a compromise on the performance of the device. In order to overcome the challenges associated with the miniaturization of the devices, various techniques are being explored to fabricate the energy materials and devices with low surface area but high energy densities. To optimize the energy output and ensure conformity of the devices with the biological tissues, various materials are being explored to develop soft, flexible, and stretchable devices with soft electronics and a better tissue–device interface. Despite the efforts, a compromise must be achieved between the energy material performance and the risk associated with the device use, owing to the drawbacks of high energy density materials and the biocompatibility of low energy density materials. Hence, it is important to consider the use case of the device during the development stage to make appropriate choices regarding the material selection, the energy technology, the fabrication process, and the most suitable method of minimizing the risk posed by the device during its operation. For this purpose, this chapter introduces a framework that highlights and correlates the major challenges in the fabrication and integration of energy materials with biomedical devices. The framework connects the fabrication of integrated devices, their application-specific requirements, and the steps taken to mitigate risks during their use *in vivo*, thereby serving as a helpful tool for understanding the challenges involved in developing miniature biomedical devices powered by energy materials.

References

1. A. Sohail, A. Ali, H. Shaukat, F. M. Bhatti, S. Ali and S. A. Kouritem, *et al.*, Integrating self-powered medical devices with advanced energy harvesting: A review, *Energy Strategy Rev.*, 2024, **52**, 101328.
2. S. Yang, V. Sencadas, S. S. You, N. Z. Jia, S. S. Srinivasan and H. Huang, *et al.*, Powering Implantable and Ingestible Electronics, *Adv. Funct. Mater.*, 2021, **31**(44), 2009289.
3. M. Xu, Y. Liu, K. Yang, S. Li, M. Wang and J. Wang, *et al.*, Minimally invasive power sources for implantable electronics, *Exploration*, 2024, **4**(1), 20220106.

4. H. J. Yoon and S. W. Kim, Nanogenerators to Power Implantable Medical Systems, *Joule*, 2020, **4**(7), 1398–1407.
5. T. Janes, S. Petrosky, T. Buhr, A. I. Karsilayan, J. Silva-Martinez, D. Genzer, *et al.* Thermoelectric Energy Harvesting for Implantable Medical Devices, in 2021 43rd Annual International Conference of the IEEE Engineering in Medicine & Biology Society (EMBC), IEEE, 2021, pp. 1547–1550.
6. Y. Han, C. Yu and H. Liu, A microbial fuel cell as power supply for implantable medical devices, *Biosens. Bioelectron.*, 2010, **25**(9), 2156–2160.
7. A. Berezovska, P. H. M. Buzzetti, Y. Nedellec, C. Gondran, F. Giroud and A. J. Gross, *et al.*, A hollow microcavity enzymatic fuel cell for in vivo energy harvesting, *Cell Rep. Phys. Sci.*, 2024, **5**(9), 102203.
8. L. Du, H. Hao, Y. Ding, A. Gabros, T. C. E. Mier and J. Van der Spiegel, *et al.*, An implantable, wireless, battery-free system for tactile pressure sensing, *Microsyst. Nanoeng.*, 2023, **9**(1), 130.
9. G. Clementi, I. Neri, F. Cottone, A. Di Michele, M. Mattarelli and L. Sforna, *et al.*, Self-powered temperature sensors harnessing membrane potential of living cells, *Nano Energy*, 2024, **121**, 109211.
10. M. Bok, Y. Lee, D. Park, S. Shin, Z. J. Zhao and B. Hwang, *et al.*, Microneedles integrated with a triboelectric nanogenerator: an electrically active drug delivery system, *Nanoscale*, 2018, **10**(28), 13502–13510.
11. G. Yao, L. Kang, J. Li, Y. Long, H. Wei and C. A. Ferreira, *et al.*, Effective weight control *via* an implanted self-powered vagus nerve stimulation device, *Nat. Commun.*, 2018, **9**(1), 5349.
12. V. Raghunathan and P. H. Chou, Design and power management of energy harvesting embedded systems, in Proceedings of the 2006 international symposium on Low power electronics and design - ISLPED '06, ACM Press, New York, New York, USA, 2006, pp. 369.
13. A. Pfenniger, M. Jonsson, A. Zurbuchen, V. M. Koch and R. Vogel, Energy Harvesting from the Cardiovascular System, or How to Get a Little Help from Yourself, *Ann. Biomed. Eng.*, 2013, **41**(11), 2248–2263.
14. M. Xu, Y. Liu, K. Yang, S. Li, M. Wang and J. Wang, *et al.*, Minimally invasive power sources for implantable electronics, *Exploration*, 2024, **4**(1), 20220106.
15. G. Hwang, H. Park, J. Lee, S. Oh, K. Park and M. Byun, *et al.*, Self-Powered Cardiac Pacemaker Enabled by Flexible Single Crystalline PMN-PT Piezoelectric Energy Harvester, *Adv. Mater.*, 2014, **26**(28), 4880–4887.
16. Z. Xie, R. Avila, Y. Huang and J. A. Rogers, Flexible and Stretchable Antennas for Biointegrated Electronics, *Adv. Mater.*, 2020, **32**(15), 1902767.
17. G. Hwang, H. Park, J. Lee, S. Oh, K. Park and M. Byun, *et al.*, Self-Powered Cardiac Pacemaker Enabled by Flexible Single Crystalline PMN-PT Piezoelectric Energy Harvester, *Adv. Mater.*, 2014, **26**(28), 4880–4887.
18. X. Qu, S. Cheng, Y. Liu, Y. Hu, Y. Shan and R. Luo, *et al.*, Bias-Free Cardiac Monitoring Capsule, *Adv. Mater.*, 2024, **36**(33), 2402457.
19. Q. Zheng, B. Shi, F. Fan, X. Wang, L. Yan and W. Yuan, *et al.*, In Vivo Powering of Pacemaker by Breathing-Driven Implanted Triboelectric Nanogenerator, *Adv. Mater.*, 2014, **26**(33), 5851–5856.

20. J. Wang, H. Wang, N. V. Thakor and C. Lee, Self-Powered Direct Muscle Stimulation Using a Triboelectric Nanogenerator (TENG) Integrated with a Flexible Multiple-Channel Intramuscular Electrode, *ACS Nano*, 2019, **13**(3), 3589–3599.
21. P. Song, S. Kuang, N. Panwar, G. Yang, D. J. H. Tng and S. C. Tjin, *et al.*, A Self-Powered Implantable Drug-Delivery System Using Biokinetic Energy, *Adv. Mater.*, 2017, **29**(11), 1605668.
22. C. Dagdeviren, B. D. Yang, Y. Su, P. L. Tran, P. Joe and E. Anderson, *et al.*, Conformal piezoelectric energy harvesting and storage from motions of the heart, lung, and diaphragm, *Proc. Natl. Acad. Sci. U. S. A.*, 2014, **111**(5), 1927–1932.
23. J. Olivo, S. Carrara and G. De Micheli, Biofuel Cells and Inductive Powering as Energy Harvesting Techniques for Implantable Sensors, *Sci. Adv. Mater.*, 2011, **3**(3), 420–425.
24. G. Pankratova, P. Bollella, D. Pankratov and L. Gorton, Supercapacitive biofuel cells, *Curr. Opin. Biotechnol*, 2022, **73**, 179–187.
25. E. De la Paz, N. H. Maganti, A. Trifonov, I. Jeerapan, K. Mahato and L. Yin, *et al.*, A self-powered ingestible wireless biosensing system for real-time *in situ* monitoring of gastrointestinal tract metabolites, *Nat. Commun.*, 2022, **13**(1), 7405.
26. A. Iqbal, P. R. Sura, M. Al-Hasan, I. B. Mabrouk and T. A. Denidni, Wireless power transfer system for deep-implanted biomedical devices, *Sci. Rep.*, 2022, **12**(1), 13689.
27. M. Zada, I. A. Shah, A. Basir and H. Yoo, Ultra-Compact Implantable Antenna With Enhanced Performance for Leadless Cardiac Pacemaker System, *IEEE Trans. Antennas Propag.*, 2021, **69**(2), 1152–1157.
28. T. Karacolak, A. Z. Hood and E. Topsakal, Design of a Dual-Band Implantable Antenna and Development of Skin Mimicking Gels for Continuous Glucose Monitoring, *IEEE Trans. Microwave Theory Tech.*, 2008, **56**(4), 1001–1008.
29. C. Kim, Evolution of Advanced Miniaturization for Active Implantable Medical Devices, in *Nano-Bio- Electronic, Photonic and MEMS Packaging*, Cham, Springer International Publishing; 2021, pp. 407–415.
30. B. Shi, Z. Li and Y. Fan, Implantable Energy-Harvesting Devices, *Adv. Mater.*, 2018, **30**(44), 1801511.
31. A. Mahmud, A. A. Khan, S. Islam, P. Voss and D. Ban, Integration of organic/inorganic nanostructured materials in a hybrid nanogenerator enables efficacious energy harvesting *via* mutual performance enhancement, *Nano Energy*, 2019, **58**, 112–120.
32. T. Huang, M. Lu, H. Yu, Q. Zhang, H. Wang and M. Zhu, Enhanced Power Output of a Triboelectric Nanogenerator Composed of Electrospun Nanofiber Mats Doped with Graphene Oxide, *Sci. Rep.*, 2015, **5**(1), 13942.
33. H. Kim, D. C. Nguyen, T. T. Luu, Z. Ding, Z. H. Lin and D. Choi, Recent Advances in Functional Fiber-Based Wearable Triboelectric Nanogenerators, *Nanomaterials*, 2023, **13**(19), 2718.

34. Y. Zou, J. Xu, K. Chen and J. Chen, Advances in Nanostructures for High-Performance Triboelectric Nanogenerators, *Adv. Mater. Technol.*, 2021, **6**(3), 2000916.
35. N. Mintz Hemed, A. Pham, E. T. Zhao, P. Wang and N. A. Melosh, Enhanced Thin-Film Encapsulation Through Micron-Scale Anchors, *Adv. Funct. Mater.*, 2024, **34**(38), 2402661.
36. R. Feiner and T. Dvir, Tissue–electronics interfaces: from implantable devices to engineered tissues, *Nat. Rev. Mater.*, 2017, **3**(1), 17076.
37. P. Cheng, S. Dai, Y. Liu, Y. Li, H. Hayashi and R. Papani, *et al.*, An intrinsically stretchable power-source system for bioelectronics, *Device*, 2024, **2**(1), 100216.
38. C. Cao, Z. Li, F. Shen, Q. Zhang, Y. Gong and H. Guo, *et al.*, Progress in techniques for improving the output performance of triboelectric nanogenerators, *Energy Environ. Sci.*, 2024, **17**(3), 885–924.
39. S. Choi, Biofuel Cells and Biobatteries: Misconceptions, Opportunities, and Challenges, *Batteries*, 2023, **9**(2), 119.
40. N. Almarri, J. Chang, W. Song, D. Jiang and A. Demosthenous, Piezoelectric energy harvesting and ultra-low-power management circuits for medical devices, *Nano Energy*, 2024, **131**, 110196.
41. X. Zhang, H. Ren, S. Pyo, J. I. Lee, J. Kim and J. Chae, A High-Efficiency DC–DC Boost Converter for a Miniaturized Microbial Fuel Cell, *IEEE Trans. Power Electron.*, 2015, **30**(4), 2041–2049.
42. A. Petritz, E. Karner-Petritz, T. Uemura, P. Schäffner, T. Araki and B. Stadlober, *et al.*, Imperceptible energy harvesting device and biomedical sensor based on ultraflexible ferroelectric transducers and organic diodes, *Nat. Commun.*, 2021, **12**(1), 2399.
43. D. K. Freeman and S. J. Byrnes, Optimal Frequency for Wireless Power Transmission Into the Body: Efficiency Versus Received Power, *IEEE Trans. Antennas Propag.*, 2019, **67**(6), 4073–4083.
44. V. Mitrakos, G. Cummins, F. J. Tauber, B. F. Cox, S. K. Pavuluri and G. S. Wood, *et al.*, PressureCap: An endoscopic sensor capsule for real-time gastrointestinal pressure monitoring, *Device*, 2024, **2**(5), 100325.
45. Z. Duan, L. J. Xu, S. Gao and W. Geyi, Integrated Design of Wideband Omnidirectional Antenna and Electronic Components for Wireless Capsule Endoscopy Systems, *IEEE Access*, 2018, **6**, 29626–29636.
46. N. Abbas, A. Basir, A. Iqbal, M. Yousaf, A. Akram and H. Yoo, Ultra-Miniaturized Antenna for Deeply Implanted Biomedical Devices, *IEEE Access*, 2022, **10**, 54563–54571.
47. R. Herbert, H. R. Lim, B. Rigo and W. H. Yeo, Fully implantable wireless batteryless vascular electronics with printed soft sensors for multiplex sensing of hemodynamics, *Sci. Adv.*, 2022, **8**(19), eabm1175.
48. J. H. Lee, H. Kim, J. H. Kim and S. H. Lee, Soft implantable microelectrodes for future medicine: prosthetics, neural signal recording and neuromodulation, *Lab Chip*, 2016, **16**(6), 959–976.
49. T. Huang and W. Wu, Inkjet-Printed Wearable Nanosystems for Self-Powered Technologies, *Adv. Mater. Interfaces*, 2020, **7**(12), 2000015.

50. Q. Li, T. Wu, W. Zhao, Y. Li, J. Ji and G. Wang, 3D printing stretchable core-shell laser scribed graphene conductive network for self-powered wearable devices, *Composites, Part B*, 2022, **240**, 110000.
51. Z. Huang, G. Chi-Pong Tsui, Y. Deng and C. Y. Tang, Two-photon polymerization nanolithography technology for fabrication of stimulus-responsive micro/nano-structures for biomedical applications, *Nanotechnol. Rev.*, 2020, **9**(1), 1118–1136.
52. M. J. Bathaei, R. Singh, H. Mirzajani, E. Istif, M. J. Akhtar and T. Abbasiasl, *et al.*, Photolithography-Based Microfabrication of Biodegradable Flexible and Stretchable Sensors, *Adv. Mater.*, 2023, **35**(6), 2207081.
53. E. J. Curry, T. T. Le, R. Das, K. Ke, E. M. Santorella and D. Paul, *et al.*, Biodegradable nanofiber-based piezoelectric transducer, *Proc. Natl. Acad. Sci. U. S. A.*, 2020, **117**(1), 214–220.
54. E. Dimitriou and N. Michailidis, Printable conductive inks used for the fabrication of electronics: an overview, *Nanotechnology*, 2021, **32**(50), 502009.
55. C. Dagdeviren, S. Hwang, Y. Su, S. Kim, H. Cheng and O. Gur, *et al.*, Transient, Biocompatible Electronics and Energy Harvesters Based on ZnO, *Small*, 2013, **9**(20), 3398–3404.
56. R. Herbert, J. W. Jeong and W. H. Yeo, Soft Material-Enabled Electronics for Medicine, Healthcare, and Human-Machine Interfaces, *Materials*, 2020, **13**(3), 517.
57. A. Amar, A. Kouki and H. Cao, Power Approaches for Implantable Medical Devices, *Sensors*, 2015, **15**(11), 28889–28914.
58. C. F. Guimarães, L. Gasperini, A. P. Marques and R. L. Reis, The stiffness of living tissues and its implications for tissue engineering, *Nat. Rev. Mater.*, 2020, **5**(5), 351–370.
59. G. Schiavone, F. Fallegger, X. Kang, B. Barra, N. Vachicouras and E. Roussinova, *et al.*, Soft, Implantable Bioelectronic Interfaces for Translational Research, *Adv. Mater.*, 2020, **32**(17), 1906512.
60. S. Oh, S. Lee, S. W. Kim, C. Y. Kim, E. Y. Jeong and J. Lee, *et al.*, Softening implantable bioelectronics: Material designs, applications, and future directions, *Biosens. Bioelectron*, 2024, **258**, 116328.
61. J. H. Koo, J. K. Song, D. H. Kim and D. Son, Soft Implantable Bioelectronics, *ACS Mater. Lett.*, 2021, **3**(11), 1528–1540.
62. D. Naik, G. Balakrishnan, M. Rajagopalan, X. Huang, N. Trivedi and A. Bhat, *et al.*, Villi Inspired Mechanical Interlocking for Intestinal Retentive Devices, *Adv. Sci.*, 2023, **10**(30), 2301084.
63. J. Lee, D. I. Kim, S. Bang and S. Park, Drug-Loaded Mucoadhesive Patch with Active Delivery and Controlled Releasing Ability, *Adv. Intell. Syst.*, 2022, **4**(4), 2100203.
64. J. Park, H. W. Kim, S. Lim, H. Yi, Z. Wu and I. G. Kwon, *et al.*, Conformal Fixation Strategies and Bioadhesives for Soft Bioelectronics, *Adv. Funct. Mater.*, 2024, **34**(22), 2313728.
65. A. D. Mickle, S. M. Won, K. N. Noh, J. Yoon, K. W. Meacham and Y. Xue, *et al.*, A wireless closed-loop system for optogenetic peripheral neuromodulation, *Nature*, 2019, **565**(7739), 361–365.

66. Y. Yang, L. Xu, D. Jiang, B. Z. Chen, R. Luo and Z. Liu, *et al.*, Self-Powered Controllable Transdermal Drug Delivery System, *Adv. Funct. Mater.*, 2021, **31**(36), 2104092.
67. A. J. Petty, R. L. Keate, B. Jiang, G. A. Ameer and J. Rivnay, Conducting Polymers for Tissue Regeneration *in Vivo*, *Chem. Mater.*, 2020, **32**(10), 4095–4115.
68. G. Hulsart-Billström, J. Dawson, S. Hofmann, R. Müller, M. Stoddart and M. Alini, *et al.*, A surprisingly poor correlation between *in vitro* and *in vivo* testing of biomaterials for bone regeneration: results of a multicentre analysis, *Eur. Cells Mater.*, 2016, **31**, 312–322.
69. J. Bassi da Silva, S. B. Ferreira, S. de, O. de Freitas and M. L. Bruschi, A critical review about methodologies for the analysis of mucoadhesive properties of drug delivery systems, *Drug Dev. Ind. Pharm.*, 2017, **43**(7), 1053–1070.
70. S. Dagenais, L. Russo, A. Madsen, J. Webster and L. Becnel, Use of Real-World Evidence to Drive Drug Development Strategy and Inform Clinical Trial Design, *Clin. Pharmacol. Ther.*, 2022, **111**(1), 77–89.
71. N. A. Kyeremateng and R. Hahn, Attainable Energy Density of Microbatteries, *ACS Energy Lett.*, 2018, **3**(5), 1172–1175.
72. D. C. Bock, A. C. Marschilok, K. J. Takeuchi and E. S. Takeuchi, Batteries used to power implantable biomedical devices, *Electrochim. Acta*, 2012, **84**, 155–164.
73. S. Yang, V. Sencadas, S. S. You, N. Z. Jia, S. S. Srinivasan and H. Huang, *et al.*, Powering Implantable and Ingestible Electronics, *Adv. Funct. Mater.*, 2021, **31**(44), 2009289.
74. J. Feng, L. Wang, Y. Chen, P. Wang, H. Zhang and X. He, PEO based polymer-ceramic hybrid solid electrolytes: a review, *Nano Convergence*, 2021, **8**(1), 2.
75. L. Zhai, J. Duan, T. Lin and H. Shao, Recent advances in implantable batteries: Development and challenge, *J. Alloys Compd.*, 2024, **979**, 173551.
76. Y. Wang and W. Zhong, Development of Electrolytes towards Achieving Safe and High-Performance Energy-Storage Devices: A Review, *ChemElectroChem*, 2015, **2**(1), 22–36.
77. O. Yue, X. Wang, L. Xie, Z. Bai, X. Zou and X. Liu, Biomimetic Exogenous "Tissue Batteries" as Artificial Power Sources for Implantable Bioelectronic Devices Manufacturing, *Adv. Sci.*, 2024, **11**(11), 2307369.
78. X. Huang, H. Hou, B. Yu, J. Bai, Y. Guan and L. Wang, *et al.*, Fully Biodegradable and Long-Term Operational Primary Zinc Batteries as Power Sources for Electronic Medicine, *ACS Nano*, 2023, **17**(6), 5727–5739.
79. Y. J. Kim, W. Wu, S. E. Chun, J. F. Whitacre and C. J. Bettinger, Biologically derived melanin electrodes in aqueous sodium-ion energy storage devices, *Proc. Natl. Acad. Sci. U. S. A.*, 2013, **110**(52), 20912–20917.
80. N. Mittal, A. Ojanguren, M. Niederberger and E. Lizundia, Degradation Behavior, Biocompatibility, Electrochemical Performance, and Circularity Potential of Transient Batteries, *Adv. Sci.*, 2021, **8**(12), 2004814.

81. N. Patra, P. Ramesh, V. Donthu and A. Ahmad, Biopolymer-based composites for sustainable energy storage: recent developments and future outlook, *J. Mater. Sci.: Mater. Eng.*, 2024, **19**(1), 34.
82. Q. M. Saqib, I. Ahmad, A. Mannan, J. Mahmood, S. Ameen and C. S. Patil, *et al.*, Triboelectric Energy Harvesting from Highly Conjugated Fused Aromatic Ladder Structure Under Extreme Environmental Conditions, *Adv. Mater.*, 2024, **36**(19), 2311029.
83. G. Hwang, H. Park, J. Lee, S. Oh, K. Park and M. Byun, *et al.*, Self-Powered Cardiac Pacemaker Enabled by Flexible Single Crystalline PMN-PT Piezoelectric Energy Harvester, *Adv. Mater.*, 2014, **26**(28), 4880–4887.
84. D. Berlincourt and H. Jaffe, Elastic and Piezoelectric Coefficients of Single-Crystal Barium Titanate, *Phys. Rev.*, 1958, **111**(1), 143–148.
85. J. Gomes, J. Serrado Nunes, V. Sencadas and S. Lanceros-Mendez, Influence of the β-phase content and degree of crystallinity on the piezo- and ferroelectric properties of poly(vinylidene fluoride), *Smart Mater. Struct.*, 2010, **19**(6), 065010.
86. H. G. Yeo, J. Jung, M. Sim, J. E. Jang and H. Choi, Integrated Piezoelectric AlN Thin Film with SU-8/PDMS Supporting Layer for Flexible Sensor Array, *Sensors*, 2020, **20**(1), 315.
87. S. Goel and B. Kumar, A review on piezo-/ferro-electric properties of morphologically diverse ZnO nanostructures, *J. Alloys Compd.*, 2020, **816**, 152491.
88. M. Ali, M. J. Bathaei, E. Istif, S. N. H. Karimi and L. Beker, Biodegradable Piezoelectric Polymers: Recent Advancements in Materials and Applications, *Adv. Healthcare Mater.*, 2023, **12**(23), 2300318.
89. C. Zhao, H. Feng, L. Zhang, Z. Li, Y. Zou and P. Tan, *et al.*, Cancer Therapy: Highly Efficient In Vivo Cancer Therapy by an Implantable Magnet Triboelectric Nanogenerator (Adv. Funct. Mater. 41/2019), *Adv. Funct. Mater.*, 2019, **29**(41), 1808640.
90. X. Xia, Q. Liu, Y. Zhu and Y. Zi, Recent advances of triboelectric nanogenerator based applications in biomedical systems, *EcoMat*, 2020, **2**(4), e12049.

CHAPTER 12

Energy Materials for Ingestible and Implantable Healthcare Devices: Conclusion and Future Perspective for Autonomous and Long-lasting Solutions

SANTANU PATRA* AND ANJA BOISEN

Drug Delivery and Sensing (IDUN), Department of Health Technology, Technical University of Denmark, Ørsted Plads, 2800, Kgs. Lyngby, Denmark
*Email: sanpa@dtu.dk

12.1 Introduction

Ingestible and implantable devices are gaining popularity in the healthcare sector due to their ability to provide real-time monitoring, diagnostics, and therapeutic interventions for a range of health issues.[1,2] These miniaturized biomedical devices allow less invasive procedures, which can be integrated with continuous data collection from and within the body.[3] One of the key components for these devices is the power supply, as it enables active monitoring, triggers overall systems, and transfers data.[4,5] However, conventional power sources like batteries have limitations like short lifetime, and their physical characteristics like rigidity, bulkiness, and associated toxicity limit their use and potential in real-life applications.[6] Also, they require surgical replacements as one of the mandatory

Detection Science Series No. 33
Energy Materials for Ingestible and Implantable Healthcare Devices
Edited by Santanu Patra and Anja Boisen

Published by the Royal Society of Chemistry, www.rsc.org

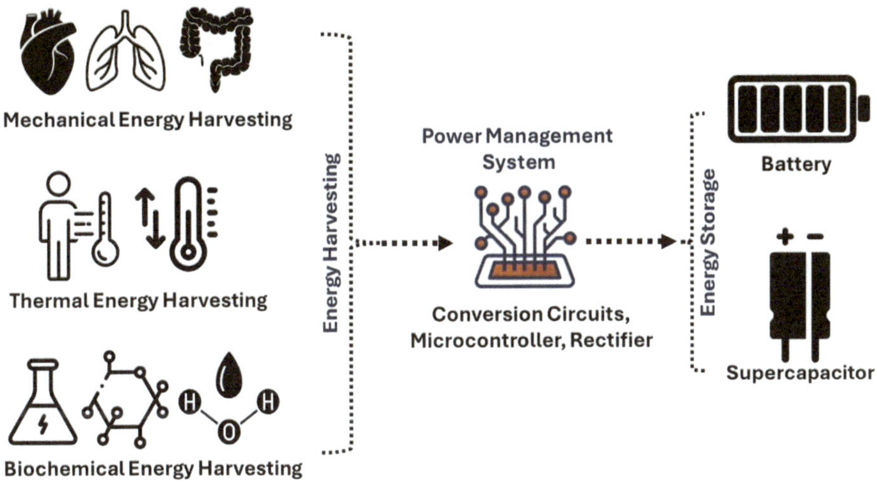

Figure 12.1 Hybrid system for powering implantable and ingestible devices.

steps. These substantial challenges have generated an increasing demand for advanced energy materials and innovative power systems appropriate for prolonged usage within the human body, facilitating autonomous and sustainable healthcare solutions. Thus, current research is extensively focused on developing innovative energy harvesting and storage systems to meet increasing energy demands. Depending on the application, researchers are targeting different internal stimuli to test the various energy harvesting systems and attempting to design different energy storage devices, such as batteries and supercapacitors.[1,4] Hybrid devices are gaining popularity to overcome the challenges in the existing system, which comprises an energy harvesting and storage system along with power management (Figure 12.1).

12.2 Powering Ingestible and Implantable Devices: Essential Criteria

To qualify as an appropriate internal power supply system, it must adhere to some essential criteria, such as:[4]

- **Biocompatibility:** biocompatibility is one of the key criteria, as the device must not cause any toxic side effects to the surrounding environment and tissues.[4,6,7] Specifically, biocompatible materials employed in biomedical energy transfers must not release or decompose into any toxic substances. Biomedical energy device materials also possibly need to be biodegradable and chemically stable. For this reason, it is crucial that these materials undergo regulated degradation so that the device can continue to function normally throughout its lifetime and no harmful consequences in biological applications can develop over time.

- **Miniaturization:** as the main target is applying them inside the body, the ingestible and implantable devices always need to be sufficiently small.[8] Accordingly, the power system must be miniaturized to accommodate the space limitations. Miniaturizing the device requires smaller components, additional materials handling, processing, and innovative design. This domain is of significant interest to researchers seeking to investigate novel materials and methods for device miniaturization.
- **Long-term stability:** power sources must function continuously in extreme biological conditions (*e.g.*, acidic stomach, oxidative tissue). Thus, the materials should be designed in such a way that we can avoid any change or deformation with time. In biomedical energy devices, materials go through repeated mechanical strains during loading and unloading processes. In addition, body movement and deformations suggest designing flexible biomedical devices for their long-term operation.
- **Sufficient energy density and efficiency:** for long-term operation, the power-supplying devices must have sufficient energy capacity to sustain the function for days, weeks, or months. Additionally, certain *in vivo* devices must be multifunctional, capable of supplying power for sensing or operating a pacemaker while simultaneously facilitating wireless data transmission. Consequently, they require high energy density and efficiency for prolonged autonomous functionality.

12.3 Possible Energy Harvesting Technologies Along with Their Challenges and Future Perspectives

Considering the internal stimuli, there are three principal types of energy harvesters, which are mechanical, thermal, and biochemical, that can be utilized inside the body.[4,9,10] Mechanical energy is produced by the movement of muscles and joints, resulting from the heart's contraction and expansion during blood circulation, as well as the motion of muscles in daily activities.[11] In the case of ingestible devices, the mechanical movement of the gastrointestinal tract, such as peristalsis, can produce mechanical energy.[12] Thermal energy is generated as a by-product of different metabolic activities and the regulation of body temperature.[13] The thermal energy generated by the human body constitutes a significant source of energy that can be harvested. Biochemical energy is generated when nutrients are transformed into a usable form through processes such as cellular respiration. For example, glucose is broken down to create adenosine triphosphate (ATP) inside the body.[13] Below, we discuss these energy harvesting approaches and summarize their potential challenges and opportunities for future research.[1,4,10,13] The same has been outlined in Table 12.1.

Table 12.1 Challenges and future perspectives of the energy harvesting, storage, and power management systems. Data collected from Chapters 1–11 and ref. 1, 4, 6, 16, 26, 27–31, 39–42, 52–57, 59, 60, 63, 72, and 78–80.

Category	Type	Key challenges	Future perspectives
Mechanical energy harvesting	Triboelectric nanogenerators (TENGs) and piezoelectric nanogenerators (PENGs)	• Often produces intermittent to low current output and lack of optimum force *in vivo* • Sensitivity to moisture and issues with materials degradation in biological environments • Stability issues like mechanical deformation after cycles • Issues with biocompatibility • Rigidity and fragility of certain piezoelectric materials • Miniaturization of devices • Poling of piezoelectric materials is compulsory for the operation of the device • Complex power management circuits are required due to high impedance	• Selection of highly efficient piezo- and triboelectric materials • Enhance surface charge density through various surface engineering technologies, including nano-structuring and micropatterning • By changing the material composition to increase the current output • Increase materials stability to withstand load and overcome degradation issues • Selecting biocompatible materials with high efficiency • Designing a hybrid system, so the harvested energy can be stored for desired applications
Thermal energy harvesting	Thermo- and pyroelectric nanogenerators (TEGs and PEGs)	• Requires a thermal gradient for operation • Low output performance with respect to the temperature gradient • Materials limitations and biocompatibility	• Enhancing the thermal contact between materials and the body to increase power output • Designing a hybrid system by incorporating thermal energy harvesting with a storage system • Emerging nanostructured, biocompatible, and flexible thermoelectric materials can significantly enhance the effectiveness of energy harvesting

Table 12.1 (*Continued*)

Category	Type	Key challenges	Future perspectives
Biochemical energy harvesting	Biofuel cells (BFCs)	• The output power from biochemical energy harvesting is typically very low, mostly a few microwatts • Enzyme degradation and low metabolic rates of microbes • Immobilization of enzymes and growth of microbes • Lack of oxygen *in vivo* • Reliance on the amount of chemicals like glucose in body fluid can provide an irregular power supply • Issues with biocompatibility and biofouling • Miniaturization of electrodes and overall BFC setup	• Research on enhancing the electrochemical activities of the materials, efficient microbe/enzyme selection, and system design for improved efficiency • Designing biocompatible and edible materials • Focus on developing pH-sensitive components for on-demand power generation • Implementing hybrid systems by integrating BFCs with energy storage to improve output power • Using low-powered electronics in the power management system
Energy storage	Battery	• Issues with large size and limited battery life • Demand regular surgical replacement, thus increasing patient risks and healthcare costs • Issues with biocompatibility and toxic to the human body • Suitable electrode materials and their integration into biomedical devices • Slow charging and discharging sometimes limit their ingestible applications	• Development of biocompatible and biodegradable batteries. Zn- and Mg-based batteries are frontrunners in this regard • Miniaturization to reduce the size constraints. Utilize efficient electrode materials to overcome the challenges associated with a low physical surface area • Choice of solid and non-toxic electrolytes. Possibilities of using gastric fluid as an electrolyte to increase effectiveness • Hybrid devices to recharge the battery using harvested energy

Energy storage	Supercapacitors	• Low energy density • Fast/self-discharge issues • Lack of a flat voltage plateau • Miniaturization issues, as reducing their size often leads to a significant decrease in energy output	• Designing a hybrid supercapacitor or supercapattery system to enhance the energy density, wherein one electrode functions as a battery and the other as a supercapacitor • Overcoming self-discharge or voltage leakage by employing a gel electrolyte with pseudocapacitive materials • Hybrid setup, integrating a supercapacitor with energy harvesting and the use of low-powered electronics • Efficient electrode materials to overcome the issues with miniaturization
Power management	Power management system (PMS)	• The miniaturization of PMS design is a very complex task • When PMS interacts with biological tissues, the performance is somehow impacted, and there is always an issue with biocompatibility • In the case of wireless transfer, the interaction between biological tissues and electromagnetic waves is complex, and greater understanding is required for safety and optimization • Low-frequency and high-power wireless transmission can impact human tissues • Underperformance in deep-implanted devices and antenna/transmitter structure is generally more complex • The power consumption in the PMS must be managed efficiently between all the operations. The PMS should have the ability to regulate voltage, protect against leakage, and withstand the biological environment in the long term	• Advanced hybrid system by integrating energy storage and harvesting • Research on low-power electronics and AI/ML-based control systems • Focus on biocompatible or biodegradable power units • Development of reconfigurability techniques for on-body transmitters to establish an optimal wave path and mitigate variability • Advancements in reconfigurable antennas • Alternatives for mitigating heating concerns in the future • PMS must integrate a security mechanism for operation and data handling without consuming extra energy

12.3.1 Mechanical Energy Harvesting

Mechanical energy harvesting utilizes mechanical movement inside the body, with the goal of providing self-sustaining and long-lasting solutions.[14] Using this method, the mechanical energy inside the body could be converted into useful electrical energy. Different body movements, such as heart/lung motions, or peristaltic movements, are considered the internal stimuli for mechanical energy harvesting.[15] Piezoelectric nanogenerators (PENGs) and triboelectric nanogenerators (TENGs) are two common terms in this case.[16] PENGs utilize the piezoelectric effect, whereby piezoelectric crystalline materials generate an electric charge when subjected to mechanical deformation, such as compression, bending, and stretching.[17] On the other hand, TENGs operate *via* the sliding or contact–separation of two materials with different electrical affinities, resulting in the generation of current and charge transfer.[18] The efficient materials in TENG are selected *via* the established triboelectric series.[19] For example, polytetrafluoroethylene (PTFE) and polydimethylsiloxane (PDMS) are two widely used triboelectric materials known for their strong charge affinity and reported biocompatibility.[20,21] Different natural polymers like cellulose, starch, proteins, gelatin, chitin, and silk fibroin, along with synthetic biodegradable polymers such as polylactic acid (PLA), PLGA, PPDO, PCL, and PBS, have also been employed.[22,23] In the case of PENG, polyvinylidene fluoride (PVDF)-based materials have been widely employed.[24] Certain piezoelectric ceramics, such as lead zirconate titanate (PZT), provide significant energy conversion but may pose cytotoxicity hazards.[25] Thus, various biocompatible materials, such as PDMS, PTFE, polyimide, and SU-8 epoxy, are utilized in this regard.

Although mechanical energy harvesters like PENG/TENG offer great promise for their use in biomedical devices, they have several limitations,[16,26,27] including the following:

- **Current output and efficiency:** in the case of both PENG/TENG, harvesting sufficient current to function the device and its components is a significant problem. The output current should be adjusted according to the needs of the particular application and device.
- **Materials selection, stability, and degradation:** both piezoelectric and triboelectric materials have stability issues and degradation after prolonged operation. When it comes to PENGs, the use of ceramic materials suffers from their issue with fragility and thus restricts their applications. In TENGs, mechanical deformation caused by persistent friction can affect stability and output voltage due to the generation of heat.
- **Optimum force for energy generation:** inside the body, the force is relatively small, which also impacts the output power. With less available force, the output power will also be less and might not be sufficient for the intended applications.

- **Miniaturization:** the miniaturization of energy materials for *in vivo* devices leads to a decrease in surface area, an increase in internal resistance, reduced power production, and increased influence of fluid dynamics on operating sites.
- **Biocompatibility:** biocompatibility is another issue for each of the energy harvesting and storage devices. Some of the high-performance materials used in either PENGs/TENGs are not biocompatible, thus posing cytotoxic effects.

To overcome these challenges, future directions include:[27–31]

- **Hybrid energy harvesting and storage systems:** due to the limited output power of TENGs/PENGs, it would be beneficial to develop hybrid devices by integrating a storage system to identify suitable applications. The harvested energy is thus stored in the storage system *via* power management, thus increasing the efficiency.
- **Material innovation:** to address the limitations of reduced surface area in mechanical energy harvesting devices, several solutions are employed to adjust the contact surface area. Introducing surface roughness at numerous scales, such as micro- and nanostructures on the triboelectric layers, can substantially boost the current output. The current output can be increased by adding micro- and nanoscale surface roughness to the surface layers. As such, biocompatibility and mechanical stability phenomena should be taken into consideration. To generate power from a low force, the materials should have a high efficiency to convert that force into useful power.

12.3.2 Thermal Energy Harvesting

Thermal energy harvesting converts waste heat or temperature differences in the human body into electrical energy.[32–34] Thermoelectric generators (TEGs) and pyroelectric generators (PEGs) are the two primary types of energy harvesters used to harvest thermal energy within the body by collecting the body's natural heat and converting it into useful energy.[33] PEGs use the difference in temperature over time to convert thermal energy into electrical energy, while TEGs use spatial temperature differences to do the same.[35,36] Thermoelectric materials rely on the Seebeck effect to transform heat into electricity directly.[37] Although thermoelectric nanogenerators often provide low voltage, they possess the advantage of generating direct current (DC) voltage, hence eliminating the necessity for energy rectification. There are primarily three categories of thermal materials: inorganic (such as Sb_2Te_3 and Bi_2Te_3), organic (predominantly sourced from carbon and conductive polymers), and hybrid (a synthesis of both).[4] Pyroelectric materials are a subclass of piezoelectric materials that respond to alterations in polarization resulting from temperature fluctuations.[38] The examples of pyroelectric materials include triglycine sulfate, polyvinylidene fluoride, lithium tantalate, gallium nitride, zinc oxide, *etc.*[4] The principal disadvantage of these materials is their dependence on continuous temperature variations.

Although it has potential, there are only a few works reported on thermal energy harvesting for implantable and ingestible applications. The challenges include:[39,40]

- **Low output**: the output from thermal energy harvesting is very low, which restricts its *in vivo* applications.
- **Minimal temperature gradient**: the temperature difference is directly proportional to the power output. However, within the human body, the core temperature remains around 37 °C, with minimal change. Thus, it is difficult to maintain the temperature gradient inside the body.
- **Material limitations**: further optimization of thermoelectric materials is required in terms of biocompatibility, mechanical stability, and cytotoxicity.

Considering the challenges, future research in thermal energy harvesting should focus on:[41,42]

- **Thermal interface engineering**: research should focus on enhancing the thermal contact between materials and the body to increase power output.
- **Hybrid systems**: as the output power is both inadequate and inconsistent, it would be advantageous to incorporate thermal energy harvesting with a storage system. Consequently, the harvested energy will be stored in the energy storage system *via* the power management system and can subsequently be employed as needed.
- **Material innovation**: materials play a crucial role in enhancing efficiency. Emerging nanostructured and flexible thermoelectric materials can significantly enhance the effectiveness of energy harvesting. Also, the materials should be biocompatible and non-toxic.

12.3.3 Biochemical Energy Harvesting

Biochemical energy harvesting is a promising area of research having the potential to power both implantable and ingestible medical devices by converting chemical energy from biological processes or body fluids into electrical energy, enabling self-powered, autonomous operation over extended periods.[43] Biofuel cells (BFCs) and galvanic cells are two significant domains that utilize biochemical energy harvesting to harvest energy in the gastrointestinal system. BFCs convert chemical energy into electrical energy *via* enzymatic (enzymatic fuel cell) or microbiological extracellular electron transfer reactions (microbial fuel cell).[44,45] BFCs utilize biocatalysts, such as enzymes or microorganisms, to oxidize chemicals found in body fluids, like glucose, lactate, or sweat, while simultaneously reducing an electron acceptor, typically oxygen, to produce an electrical current.[45] Typically, in the anode, carbon-based conductive materials such as graphene, graphite, or carbon cloth, are used for the growth of microbes and immobilization of the enzyme, whereas in the cathode, a state-of-the-art platinum-based oxygen reduction reaction catalyst, oxygen-rich metal oxides, conductive polymers, or carbon materials are utilized.[46–48] Galvanic cells also use the biochemical energy harvesting approach, where the natural fluids in the body can act as electrolytes to activate the system and generate electricity.[49,50] In implantable devices, the surrounding body fluid acts as an electrolyte, while in ingestible devices, gastric juice serves as both the electrolyte and a source of ions.[49] The galvanic cells are generally housed within a capsule system that activates upon entering the stomach. Mostly, Zn–Cu and Zn–Pd couples are used as typical galvanic cells inside the body.[49,51]

Although biochemical energy harvesting technologies have a huge potential to be applied in biomedical devices, they still face several challenges:[52,53]

- **Low power output**: the output power from biochemical energy harvesting is typically very low, mostly a few micro-watts, thus limiting their applications to low-power electronics.
- **Irregular power supply**: biochemical energy harvesting relies on several factors, starting from the amount of glucose in the gastric fluid to the pH. Additionally, the activity of microbes and enzymes decreases significantly after operation for several hours, which makes continuous power supply difficult.
- **Biocompatibility and biofouling**: biocompatibility of the electrode materials is another concern. Also, there is a risk of biofouling and instability of the electrodes after biological contamination for a long time.
- **Lack of oxygen**: the cathode in BFCs follows the oxygen reduction reaction and presents maximum performance in aerobic conditions. However, inside the body, there is a lack of oxygen, which lowers the efficiency of BFCs.
- **Miniaturization**: as the intended application is inside the body, the system should be sufficiently small. As we know, the miniaturization of the surface area of the electrodes decreases, thereby decreasing the power output.

Future advancements in biochemical energy harvesting for their applications in ingestible and implantable devices should focus on:[54–57]

- **Harvesting–storage hybrid systems**: since the output power is sufficiently low and irregular, it would be helpful to integrate biochemical energy harvesting with a storage system. Thus, the harvested energy will be stored in the energy storage through the power management system and can then be utilized whenever necessary.

- **Material innovation**: materials in biochemical energy harvesting play an essential role. Firstly, the electrode materials should be biocompatible and non-toxic. Additionally, to enhance power output and overcome the issues associated with miniaturization, innovative materials with a higher electroactive surface area will be helpful. So, although the size of the electrodes will be smaller, they will still contribute efficiently to the output power. Also, to overcome the lack of oxygen inside the body, the cathode materials in BFCs should have oxygen-rich/storage properties.
- **Low-power electronics**: developing low-power bioelectronic devices can significantly enhance their practical applications.

12.4 Energy Storage Technologies Along with Their Challenges and Future Perspectives

Despite the energy harvesting approaches providing a viable method for powering implantable and ingestible biomedical devices, their output is mostly very low and significantly influenced by physiological conditions.[58] These somehow limit their applications within the body. Energy storage devices, such as batteries and supercapacitors, are advantageous since they provide a steady, reliable, and fast power source, ensuring continuous and consistent operation despite environmental fluctuations.[59] Therefore, energy storage is crucial for these devices to work properly; in fact, it is often the main power source in hybrid systems. Below we discuss these energy storage approaches and summarize their potential challenges and opportunities for future research. The same has been outlined in Table 12.1.

12.4.1 Batteries

Batteries are important for implantable and ingestible medical devices that convert chemical energy into electrical energy through electrochemical reactions.[60] They are suitable for powering these devices, having a higher energy density, and can deliver a consistent and stable electrical power supply for continuous operation. A battery comprises two components: an anode that releases electrons *via* oxidation and a cathode that receives electrons through reduction, therefore completing a redox reaction. An electrolyte in gel, liquid, or solid states enables ion flow within the system, and a separator is also integrated to avoid the electrodes' contact. In terms of powering biomedical devices, lithium batteries are the first choice owing to their various advantages, like high energy density and long cycle life.[61] However, the associated safety concerns, degradation issues, and toxicity prevent their extensive usage in biomedical devices.[62] Research in ingestible and implantable systems is thus directed towards the development of biodegradable and biocompatible batteries that provide excellent performance while remaining compatible with the body's natural functions.[1,63] These also eliminate the need for surgeries to remove devices and open up new possibilities for ingestible systems. In this regard, Zn- and Mg-based batteries are becoming more popular. Both Zn and Mg have already shown their excellent performance in some implantable devices and other electronic appliances, as they are safe, cheaper, biocompatible, and sustainability-friendly.[64] Recently, several other groups have concentrated on edible and transient battery systems to enhance their *in vivo* applications.[65] Batteries are more popular than all other energy harvesting and storage systems and have been utilized in pacemakers, neurostimulators, drug delivery implants, capsule endoscopes, continuous glucose monitors, and wireless transmission devices.[66]

Although batteries are very popular in ingestible and implantable devices, they still possess several challenges, as follows:[6,59]

- **Size constraints and miniaturization**: conventional batteries come with larger sizes, so the current research focuses on miniaturized or micro-batteries. However, with miniaturization, the electrode surface area decreases, which further decreases the energy density.
- **Limited lifespan and replacement**: most batteries have a limited life cycle, so surgeries become an essential step for their replacement, which increases the patient's risk, healthcare costs, and discomfort.
- **Safety and biocompatibility concerns**: conventional batteries like Li-ion and silver-oxide batteries comprise toxic electrode materials and acidic electrolytes, making the battery highly toxic and also posing risks of explosions or leakage, thus limiting their applications in biomedical devices.
- **Low capacity for transient batteries**: as a replacement for conventional batteries, biodegradable or transient batteries provide very low energy output and thus restrict their applications to very low-powered electronics.

Considering the associated challenges, future batteries should be focused on autonomous, efficient, and safe power solutions that can operate inside the body for the long term. The major directions include:[6,59,67]

- **Advanced electrode materials:** the electrodes' physical surface area reduces as the batteries get smaller, which further lowers the energy output. The development of highly effective electrode materials that can store charges even with a small physical surface area and offer higher electrochemical activities should be the main focus of research in order to increase efficiency. The use of nanomaterials with higher electrocatalytic activities is getting much attention in this regard.
- **Biocompatible electrode materials and electrolytes:** as the primary focus is on *in vivo* applications, we need all components to be biocompatible and non-toxic to the human body. To overcome the limitations of biocompatibility, future research should focus on biocompatible electrode materials and non-toxic electrolytes. A detailed

study with the cells, as well as animal experiments, is required to establish their effectiveness.
- **Smart energy management:** to overcome the degradation and life cycle of the battery, future research should focus on smart energy management by integrating artificial intelligence and machine learning concepts into the devices for dynamic energy control and optimizing battery usage.

12.4.2 Supercapacitors

Similar to batteries, supercapacitors are another important energy storage system having a higher power density, longer cyclic stability, and faster charging–discharging rates.[68] They are specifically favored in applications requiring fast energy transfer. Their ability to charge quickly also makes them suitable for energy harvesting–storage hybrid systems. Although supercapacitors resemble batteries in many ways, starting with their electrode compositions or electrolyte, the charge storage mechanism is significantly different.[69] In batteries, diffusion-controlled redox reactions are the major charge-storing mechanism, whereas supercapacitors use an electrical double layer or faradaic reaction to store the charge on the surface of the electrodes, thus providing much faster reaction kinetics but with a lower energy density.[70,71]

Although supercapacitors are gaining popularity for their applications in ingestible and implantable devices, they still suffer from several limitations, such as:[63]

- **Low energy density:** the low energy density or storage capacity of supercapacitors makes them ineligible for several *in vivo* applications.
- **Self-discharge:** the fast discharge rate is considered another limitation, as supercapacitors are prone to self-discharge.
- **Miniaturization:** as for ingestible and implantable devices, supercapacitors need to be small, but reducing their size often leads to a significant decrease in energy.

Considering the challenges associated with the current supercapacitors, future directions can be focused on developing next-generation supercapacitors by:[63,72]

- **Increasing energy density:** the research may concentrate on a hybrid supercapacitor or supercapattery system, wherein one electrode functions as a battery and the other as a supercapacitor. This method can enhance the energy density of supercapacitors.
- **Overcoming self-discharge:** employing a gel electrolyte with a conductive channel or pseudocapacitive materials helps reduce voltage leakage or self-discharge in the supercapacitor.

- **Low-power electronics:** given the output of supercapacitors, they can be employed to power low-power electronics.
- **Coupling with energy harvesters:** integrating supercapacitors with energy harvesting technology can enhance efficiency, since the gathered energy can be utilized to charge the supercapacitor and subsequently applied in useful applications.
- **Efficient electrode materials:** as with miniaturization, the size of the electrodes decreases, which further decreases the electrochemical activity and power output. To enhance efficiency, research should focus on designing and developing highly efficient electrode materials that will provide higher electrochemical activities and can store charges even with a small physical surface area. Nanomaterials are becoming popular in this regard.

12.5 Power Management Systems Along with Their Challenges and Future Perspectives

A power management system (PMS) is considered an important component in implantable and ingestible biomedical devices, where its primary function is to ensure a controlled, effective, and stable power supply.[4] As already discussed, the output power of energy storage and harvesting devices is sometimes very low, where the PMS plays a crucial role in their long-term, autonomous, and safe functioning within the human body.[73] Similarly, in wireless power transfer (WPT) systems, the PMS is used to transform the incoming energy into a steady voltage that the device can use directly.[74,75] Using PMS in biomedical devices offers several advantages, including optimized energy usage, increased device lifespan, and reduced or eliminated need for surgical battery replacements, thereby minimizing both patient risks and expenses.[76] A PMS has different components such as an energy harvesting system (like mechanical, thermal, or biochemical), an energy storage system (like batteries or supercapacitors), and a power management unit (based on the applications, components will be different).[77]

Although PMSs are very useful in biomedical devices like implantable and ingestible devices, their applications are somehow hindered by their associated challenges, including:[78–80]

- **Complexity and size:** the miniaturization of PMS design is a very complex task. Most of the time, there is a size constraint in the devices, and thus, the integration of a PMS is somewhat difficult. Also, different components of the PMS contribute to their complexity.
- **Biological tissue interaction and biocompatibility:** when the PMS interacts with biological tissues, the performance is somehow impacted. Additionally, all the components of the PMS, from energy materials to electronics, must be biocompatible and free from any toxic effects.

- **Power consumption:** the power consumption in a PMS must be managed efficiently between all operations. Trade-offs between functionality and energy consumption must be thoroughly balanced, notably in systems without recharging ability.
- **Reliability in dynamic environments:** implantable and ingestible devices should be operational for a long time (such as several hours in the case of ingestible devices) to several years (in the case of implantable devices). Thus, the PMS should have the ability to regulate the voltage, protect against any leakage, and withstand the biological atmosphere in the long term.
- **Dynamic load management:** the PMS must be designed in such a way that it can manage the load (like sensing or frequency in transferring data) in terms of available power in the system.
- **Data security:** as the PMS may be applied in wireless data transmission from inside the body, it must integrate a security mechanism for operation and data handling without consuming extra energy.

Considering the above-mentioned challenges associated with PMSs, future research should focus on the following modifications:[78–80]

- **Advanced hybrid systems:** future PMSs should integrate multiple energy harvesting sources, like mechanical, thermal, and biochemical, along with different energy storage systems, thus designing a hybrid system. These hybrid system-based PMSs will have the option to optimize the amount of energy as per applications.
- **Low-power electronics and artificial intelligence/machine learning (AI/ML)-based control systems:** by utilizing ultra-low-power microcontrollers and integrated circuits in combination with AI/ML algorithms, energy consumption and usage patterns in PMSs can be optimized accordingly.
- **Biocompatible or biodegradable power units:** biocompatibility is an important phenomenon that one should consider while designing a PMS. Additionally, biodegradable batteries and circuits that degrade after completing their task, removing the need for recovery or any surgical retrieval, are desirable.
- **Wireless and battery-free designs:** the PMS can be integrated with a wireless power transfer system to receive useful data. There are continuous advancements in near-field and mid-field wireless power transfer, thus one can design a fully battery-free ingestible and implantable device, particularly for short-term ingestible sensors.

12.6 Conclusion

This chapter concludes the discussion from earlier chapters in this book, along with outlining the challenges and future perspectives on advancements in the field of implantable and ingestible energy harvesting technologies. The aim has been to explore possible energy harvesting and storage technologies and how energy materials fit into that scenario. We have summarized the challenges associated with each of the energy harvesting technologies (mechanical, thermal, and biochemical), as well as storage (battery and supercapacitor) and power management systems. We have identified three major challenges: (i) miniaturization of the devices, (ii) overall output in *in vivo* settings, and (iii) biocompatibility and long-term operation *in vivo*. The chapter has also summarized the scope of future research and how this could change the landscape in this proposed area, by working towards highly efficient novel materials, ultralow-power electronics, and innovative and hybrid systems. We are hopeful that, in this way, researchers can reveal the full potential of ingestible and implantable devices, enabling the development of autonomous, long-lasting solutions to provide real-time health monitoring and diagnostics (Table 12.1).

Acknowledgements

We acknowledge funding from the Novo Nordisk Foundation Challenge Programme in Energy materials with biological applications (EMGUT): grant ref. No. NNF22OC0072961 and European Research Council in Foldable, REconfigurable & Jagged devices for enhanced drug Absorption/seeding (FREJA), grant ref. No. 101054945.

References

1. S. Y. Yang, V. Sencadas, S. S. You, N. Z. X. Jia, S. S. Srinivasan and H. W. Huang, *et al.*, Powering Implantable and Ingestible Electronics, *Adv. Funct. Mater.*, 2021, **31**(44), 2009289.
2. S. Mukherjee, S. Suleman, R. Pilloton, J. Narang and K. Rani, State of the Art in Smart Portable, Wearable, Ingestible and Implantable Devices for Health Status Monitoring and Disease Management, *Sensors*, 2022, **22**(11), 4228.
3. J. Yun, S. Keerthana and S. R. Kwon, Miniaturized power-integrated and self-powered sensor systems for advanced biomedical applications, *Sens. Actuators Rep.*, 2025, **9**, 100260.
4. N. Golafshan, S. Patra, N. D. Huynh, N. Gadegaard and A. Boisen, Self-Powered Implantable and Ingestible Devices: Harvesting Energy Within the Body, *Mater. Today*, 2025, **87**, 403–421.
5. P. A. Thwaites, C. K. Yao, E. P. Halmos, J. G. Muir, R. E. Burgell and K. J. Berean, *et al.*, Review article: Current status and future directions of ingestible electronic devices in gastroenterology, *Aliment. Pharmacol. Ther.*, 2024, **59**(4), 459–474.
6. L. Zhai, J. Duan, T. Lin and H. Shao, Recent advances in implantable batteries: Development and challenge, *J. Alloys Compd.*, 2024, **979**, 173551.

7. T. Lu, S. Ji, W. Jin, Q. Yang, Q. Luo and T. L. Ren, Biocompatible and Long-Term Monitoring Strategies of Wearable, Ingestible and Implantable Biosensors: Reform the Next Generation Healthcare, *Sensors*, 2023, **23**(6), 2991.
8. M. Xu, Y. Liu, K. Yang, S. Li, M. Wang and J. Wang, *et al.*, Minimally invasive power sources for implantable electronics, *Exploration*, 2024, **4**(1), 20220106.
9. Z. Gao, Y. Zhou, J. Zhang, J. Foroughi, S. Peng and R. H. Baughman, *et al.*, Advanced Energy Harvesters and Energy Storage for Powering Wearable and Implantable Medical Devices, *Adv. Mater.*, 2024, **36**(42), 2404492.
10. G. Rebel, F. Estevez, P. Gloesekoetter and J. M. Castillo-Secilla, Energy Harvesting on Human Bodies, in *Smart Health: Open Problems and Future Challenges*, ed. A. Holzinger, C. Röcker and M. Ziefle, Springer International Publishing, Cham, 2015, pp. 125–159.
11. M. A. Hannan, S. Mutashar, S. A. Samad and A. Hussain, Energy harvesting for the implantable biomedical devices: issues and challenges, *Biomed. Eng. Online*, 2014, **13**(1), 79.
12. M. M. Mau, S. Sarker and B. S. Terry, Ingestible devices for long-term gastrointestinal residency: a review, *Prog. Biomed. Eng.*, 2021, **3**(4), 042001.
13. Y. Zou, L. Bo and Z. Li, Recent progress in human body energy harvesting for smart bioelectronic system, *Fundam. Res.*, 2021, **1**(3), 364–382.
14. S. Bairagi, Shahid-ul-Islam, M. Shahadat, D. M. Mulvihill and W. Ali, Mechanical energy harvesting and self-powered electronic applications of textile-based piezoelectric nanogenerators: A systematic review, *Nano Energy*, 2023, **111**, 108414.
15. R. Riemer and A. Shapiro, Biomechanical energy harvesting from human motion: theory, state of the art, design guidelines, and future directions, *J. Neuroeng. Rehabil.*, 2011, **8**(1), 22.
16. A. Khan, R. Joshi, M. K. Sharma, A. Ganguly, P. Parashar and T. W. Wang, *et al.*, Piezoelectric and triboelectric nanogenerators: Promising technologies for self-powered implantable biomedical devices, *Nano Energy*, 2024, **119**, 109051.
17. Y. Yu, H. Sun, H. Orbay, F. Chen, C. G. England and W. Cai, *et al.*, Biocompatibility and in vivo operation of implantable mesoporous PVDF-based nanogenerators, *Nano Energy*, 2016, **27**, 275–281.
18. Q. Zheng, B. Shi, F. Fan, X. Wang, L. Yan and W. Yuan, *et al.*, In Vivo Powering of Pacemaker by Breathing-Driven Implanted Triboelectric Nanogenerator, *Adv. Mater.*, 2014, **26**(33), 5851–5856.
19. Y. Yu, H. Li, D. Zhao, Q. Gao, X. Li and J. Wang, *et al.*, Material's selection rules for high performance triboelectric nanogenerators, *Mater. Today*, 2023, **64**, 61–71.
20. Z. Qian, R. Li, J. Guo, Z. Wang, X. Li and C. Li, *et al.*, Triboelectric nanogenerators made of polybenzazole aerogels as fire-resistant negative tribo-materials, *Nano Energy*, 2019, **64**, 103900.
21. Y. W. Cai, G. G. Wang, Y. C. Mei, D. Q. Zhao, J. J. Peng and N. Sun, *et al.*, Self-healable, super-stretchable and shape-adaptive triboelectric

nanogenerator based on double cross-linked PDMS for electronic skins, *Nano Energy*, 2022, **102**, 107683.
22. J. Sun, W. Li, G. Liu, W. Li and M. Chen, Triboelectric Nanogenerator Based on Biocompatible Polymer Materials, *J. Phys. Chem. C*, 2015, **119**(17), 9061–9068.
23. S. A. Graham, H. Patnam, P. Manchi, M. V. Paranjape, A. Kurakula and J. S. Yu, Biocompatible electrospun fibers-based triboelectric nanogenerators for energy harvesting and healthcare monitoring, *Nano Energy*, 2022, **100**, 107455.
24. B. Khan, R. T. Khalid, M. H. Masrur, M. Awais, N. Imdad and M. E. Hilal, *et al.*, Advances in polyvinylidene fluoride (PVDF) for self-powered wearable physiological monitoring and energy harvesting applications, *Nano Energy*, 2025, **143**, 111296.
25. K. Zeng, Y. Lin, S. Liu, Z. Wang and L. Guo, Applications of piezoelectric biomaterials in dental treatments: A review of recent advancements and future prospects, *Mater. Today Bio.*, 2024, **29**, 101288.
26. M. A. Parvez Mahmud, N. Huda, S. H. Farjana, M. Asadnia and C. Lang, Recent Advances in Nanogenerator-Driven Self-Powered Implantable Biomedical Devices, *Adv. Energy Mater.*, 2018, **8**(2), 1701210.
27. X. Xia, Q. Liu, Y. Zhu and Y. Zi, Recent advances of triboelectric nanogenerator based applications in biomedical systems, *EcoMat*, 2020, **2**(4), e12049.
28. Z. Che, S. O'Donovan, X. Xiao, X. Wan, G. Chen and X. Zhao, *et al.*, Implantable Triboelectric Nanogenerators for Self-Powered Cardiovascular Healthcare, *Small*, 2023, **19**(51), 2207600.
29. Y. Quan, E. Wang, H. Ouyang, L. Xu, L. Jiang and L. Teng, *et al.*, Biodegradable and Implantable Triboelectric Nanogenerator Improved by β-Lactoglobulin Fibrils-Assisted Flexible PVA Porous Film, *Adv. Sci.*, 2025, **12**(24), 2409914.
30. G. Jian, S. Zhu, X. Yuan, S. Fu, N. Yang and C. Yan, *et al.*, Biodegradable triboelectric nanogenerator as a implantable power source for embedded medicine devices, *NPG Asia Mater.*, 2024, **16**(1), 12.
31. Y. Tang, X. Liu, Y. Xiong, B. Xu and Y. Zhou, Emerging frontiers in triboelectric nanogenerator for biohealth apparatus, *Nano Energy*, 2025, **138**, 110844.
32. A. Nozariasbmarz, H. Collins, K. Dsouza, M. H. Polash, M. Hosseini and M. Hyland, *et al.*, Review of wearable thermoelectric energy harvesting: From body temperature to electronic systems, *Appl. Energy*, 2020, **258**, 114069.
33. T. Tomono, A new approach for body heat energy harvesting, *Int. J. Energy Res.*, 2019, **43**(9), 4967–4975.
34. W. Gao, Y. Wang and F. Lai, Thermoelectric energy harvesting for personalized healthcare, *Smart Med.*, 2022, **1**(1), e20220016.
35. Z. Tabaie and A. Omidvar, Human body heat-driven thermoelectric generators as a sustainable power supply for wearable electronic devices: Recent advances, challenges, and future perspectives, *Heliyon*, 2023, **9**(4), e14707.

36. S. Korkmaz and İA. Kariper, Pyroelectric nanogenerators (PyNGs) in converting thermal energy into electrical energy: Fundamentals and current status, *Nano Energy*, 2021, **84**, 105888.
37. P. Baskaran and M. Rajasekar, Recent progress in thermoelectric devices and applications, *Chem. Eng. J. Adv.*, 2025, **506**, 159929.
38. J. Si, X. Xiao, Y. Zhang, Y. Huang, S. Liang and Q. Xu, *et al.*, Pyroelectric Properties and Applications of Lithium Tantalate Crystals, *Crystals*, 2024, **14**(7), 579.
39. G. Rebel, F. Estevez, P. Gloesekoetter and J. M. Castillo-Secilla, Energy Harvesting on Human Bodies, in *Smart Health: Open Problems and Future Challenges*, ed. A. Holzinger, C. Röcker and M. Ziefle, Springer International Publishing, Cham, 2015, pp. 125–159.
40. G. Jian, S. Zhu, X. Yuan, S. Fu, N. Yang and C. Yan, *et al.*, Biodegradable triboelectric nanogenerator as a implantable power source for embedded medicine devices, *NPG Asia Mater.*, 2024, **16**(1), 12.
41. S. Verma, B. Padha, A. Ahmed, R. Singh, D. P. Dubal and S. Arya, Thermoelectric supercapacitors: materials, challenges and future outlook, *Prog. Energy*, 2024, **6**(4), 042002.
42. P. Baskaran and M. Rajasekar, Recent trends and future perspectives of thermoelectric materials and their applications, *RSC Adv.*, 2024, **14**(30), 21706–21744.
43. I. Chakraborty, R. T. Olsson, R. L. Andersson and A. Pandey, Glucose-based biofuel cells and their applications in medical implants: A review, *Heliyon*, 2024, **10**(13), e33615.
44. R. A. Bullen, T. C. Arnot, J. B. Lakeman and F. C. Walsh, Biofuel cells and their development, *Biosens. Bioelectron.*, 2006, **21**(11), 2015–2045.
45. R. Ibrahim, N. Shaari and A. H. Mohd Aman, Bio-fuel cell for medical device energy system: A review, *Int. J. Energy Res.*, 2021, **45**(10), 14245–14273.
46. Q. Zhu, J. Hu, B. Liu, S. Hu, S. Liang and K. Xiao, *et al.*, Recent Advances on the Development of Functional Materials in Microbial Fuel Cells: From Fundamentals to Challenges and Outlooks, *Energy Environ. Mater.*, 2022, **5**(2), 401–426.
47. P. Jalili, A. Ala, P. Nazari, B. Jalili and D. D. Ganji, A comprehensive review of microbial fuel cells considering materials, methods, structures, and microorganisms, *Heliyon*, 2024, **10**(3), e25439.
48. Y. Gao, M. Mohammadifar and S. Choi, From Microbial Fuel Cells to Biobatteries: Moving toward On-Demand Micropower Generation for Small-Scale Single-Use Applications, *Adv. Mater. Technol.*, 2019, **4**(7), 1900079.
49. P. Nadeau, D. El-Damak, D. Glettig, Y. L. Kong, S. Mo and C. Cleveland, *et al.*, Prolonged energy harvesting for ingestible devices, *Nat. Biomed. Eng.*, 2017, **1**(3), 0022.
50. A. Álvarez-Carulla, Y. Montes-Cebrián, J. Colomer-Farrarons and P. L. Miribel-Català, Self-Powered Point-of-Care Device for Galvanic Cell-Based Sample Concentration Measurement, *Sensors*, 2021, **21**(8), 2665.

51. P. Mostafalu and S. Sonkusale, Flexible and transparent gastric battery: Energy harvesting from gastric acid for endoscopy application, *Biosens. Bioelectron.*, 2014, **54**, 292–296.
52. A. Zebda, J. P. Alcaraz, P. Vadgama, S. Shleev, S. D. Minteer and F. Boucher, *et al.*, Challenges for successful implantation of biofuel cells, *Bioelectrochemistry*, 2018, **124**, 57–72.
53. A. Chauhan and P. Avti, Implantable Biofuel Cells for Biomedical Applications, in *Biofuel Cells*, 2021, pp. 69–95.
54. M. Sharifi, R. Pothu, R. Boddula and G. R. Bardajee, Trends of biofuel cells for smart biomedical devices, *Int. J. Hydrogen Energy*, 2021, **46**(4), 3220–3229.
55. A. K. Gupta and A. V. Krasnoslobodtsev, Fueling the Future: The Emergence of Self-Powered Enzymatic Biofuel Cell Biosensors, *Biosensors*, 2024, **14**(7), 316.
56. E. Katz and K. MacVittie, Implanted biofuel cells operating in vivo – methods, applications and perspectives – feature article, *Energy Environ. Sci.*, 2013, **6**(10), 2791–2803.
57. Z. Fredj, G. Rong and M. Sawan, Recent Advances in Enzymatic Biofuel Cells to Power Up Wearable and Implantable Biosensors, *Biosensors*, 2025, **15**(4), 218.
58. S. Wang, L. Wei, F. Wang, L. Wang and J. Mao, Advanced implantable energy storage for powering medical devices, *eScience*, 2025, 100409.
59. M. Yu, Y. Peng, X. Wang and F. Ran, Emerging Design Strategies Toward Developing Next-Generation Implantable Batteries and Supercapacitors, *Adv. Funct. Mater.*, 2023, **33**(37), 2301877.
60. X. Huang, L. Wang, H. Wang, B. Zhang, X. Wang and R. Y. Z. Stening, *et al.*, Materials Strategies and Device Architectures of Emerging Power Supply Devices for Implantable Bioelectronics, *Small*, 2020, **16**(15), 1902827.
61. P. Parvizi, M. Jalilian, A. M. Amidi, M. R. Zangeneh and J. R. Riba, From Present Innovations to Future Potential: The Promising Journey of Lithium-Ion Batteries, *Micromachines*, 2025, **16**(2), 194.
62. J. Ju, S. Dong, Y. Cui, Y. Zhang, B. Tang and F. Jiang, *et al.*, Leakage-Proof Electrolyte Chemistry for a High-Performance Lithium–Sulfur Battery, *Angew. Chem., Int. Ed.*, 2021, **60**(30), 16487–16491.
63. M. Yu, Y. Peng, X. Wang and F. Ran, Emerging Design Strategies Toward Developing Next-Generation Implantable Batteries and Supercapacitors, *Adv. Funct. Mater.*, 2023, **33**(37), 2301877.
64. L. Zhai, J. Duan, T. Lin and H. Shao, Recent advances in implantable batteries: Development and challenge, *J. Alloys Compd.*, 2024, **979**, 173551.
65. K. Rajaram, S. M. Yang and S. W. Hwang, Transient, Biodegradable Energy Systems as a Promising Power Solution for Ecofriendly and Implantable Electronics, *Adv. Energy Sustainability Res.*, 2022, **3**(9), 2100223.
66. A. B. Amar, A. B. Kouki and H. Cao, Power Approaches for Implantable Medical Devices, *Sensors*, 2015, **15**(11), 28889–28914.

67. U. Krishnamoorthy, P. Lakshmipathy, M. Ramya and H. H. Fayek, Navigating the future of healthcare with innovations and challenges in implantable battery technology for biomedical devices, *Discover Appl. Sci.*, 2024, **6**(11), 584.
68. A. Shuja, H. R. Khan, I. Murtaza, S. Ashraf, Y. Abid and F. Farid, *et al.*, Supercapacitors for energy storage applications: Materials, devices and future directions: A comprehensive review, *J. Alloys Compd.*, 2024, **1009**, 176924.
69. S. Sharma and P. Chand, Supercapacitor and electrochemical techniques: A brief review, *Results Chem.*, 2023, **5**, 100885.
70. D. Meena, R. Kumar, S. Gupta, O. Khan, D. Gupta and M. Singh, Energy storage in the 21st century: A comprehensive review on factors enhancing the next-generation supercapacitor mechanisms, *J. Energy Storage*, 2023, **72**, 109323.
71. M. Winter and R. J. Brodd, What Are Batteries, Fuel Cells, and Supercapacitors?, *Chem. Rev.*, 2004, **104**(10), 4245–4270.
72. H. Sheng, Y. Ma, H. Zhang, J. Yuan, F. Li and W. Li, *et al.*, Integration of Supercapacitors with Sensors and Energy-Harvesting Devices: A Review, *Adv. Mater. Technol*, 2024, **9**(21), 2301796.
73. Q. Ullah, T. Rauta, J. Keskikuru, J. Haverinen and P. Ruuskanen, Innovations in energy harvesting: A survey of low-power technologies and their potential, *Energy Rep.*, 2025, **14**, 671–692.
74. S. R. Khan, S. K. Pavuluri, G. Cummins and M. P. Y. Desmulliez, Wireless Power Transfer Techniques for Implantable Medical Devices: A Review, *Sensors*, 2020, **20**(12), 3487.
75. K. Obaideen, L. Albasha, U. Iqbal and H. Mir, Wireless power transfer: Applications, challenges, barriers, and the role of AI in achieving sustainable development goals - A bibliometric analysis, *Energy Strat. Rev.*, 2024, **53**, 101376.
76. Y. Liang, C. Zhang, R. Lin, J. Lin and J. Chen, Sustainable power solutions for next-generation medical devices, *Mater. Today Bio*, 2025, **33**, 102055.
77. C. Wang, Power Management Integrated Circuits for Implantable Devices, in *Wireless Power Technologies for Biomedical Devices*, ed. R. Das and H. Heidari, Springer Nature Switzerland, Cham, 2025, pp. 207–236.
78. V. Hueros, P. Risquez, P. Zumel, C. Fernandez and A. Barrado, Wireless Charger for Pacemakers Controlled from Primary Current Without Communication with Secondary Side, *Electronics*, 2025, **14**(8), 1563.
79. Y. Liang, C. Zhang, R. Lin, J. Lin and J. Chen, Sustainable power solutions for next-generation medical devices, *Mater. Today Bio.*, 2025, **33**, 102055.
80. S. R. Begum and A. Chandrasekhar, Opportunities and Challenges in Power Management Systems for Triboelectric Nanogenerators, *ACS Appl. Electron. Mater.*, 2023, **5**(3), 1347–1375.

Subject Index